Onshore Oilfield Surface Facilities: Process Safety Engineering Guide

Kun Fang

Onshore Oilfield Surface Facilities: Process Safety Engineering Guide

石油工业出版社有限公司
PETROLEUM INDUSTRY PRESS

Springer

Kun Fang
Beijing Branch Company
China Petroleum Engineering
and Construction Corporation
Beijing, China

ISBN 978-981-96-7829-7 ISBN 978-981-96-7830-3 (eBook)
https://doi.org/10.1007/978-981-96-7830-3

Jointly published with Petroleum Industry Press.
The print edition is not for sale in China (Mainland). Customers from China (Mainland) please order the print book from: Petroleum Industry Press.
ISBN of the Co-Publisher's edition: 978-7-5183-4905-0

The original submitted manuscript has been translated into English. The translation was done using artificial intelligence. A subsequent revision was performed by the author(s) to further refine the work and to ensure that the translation is appropriate concerning content and scientific correctness. It may, however, read stylistically different from a conventional translation.

This Springer imprint is published by the registered company Springer Nature Singapore Pte Ltd.
The registered company address is: 152 Beach Road, #21-01/04 Gateway East, Singapore 189721, Singapore

Preface

Petroleum and natural gas are flammable and explosive dangerous goods. The various treatment facilities involved in oilfield surface engineering are frequently operated, and the possibility of accidents is very high. Therefore, when designing oilfield surface engineering, we should follow the general design principles, adopt scientific and reasonable analysis methods, comprehensively analyze various possible hazards and the consequences they may cause, take sufficient and reasonable protective measures, and ensure the safety of operators, the normal operation of production equipment, and meet the requirements of environmental protection in actual production.

In order to achieve the above goals, the petroleum industry has formulated many oil and gas safety production operation manuals and design standards over the years. Through continuous summarization of experience and reference to foreign advanced practices, the process safety design of oilfield surface engineering has become more and more standardized and refined. This book systematically summarizes the above results, in order to provide guidance for the process safety design of onshore oilfield surface engineering. The process principles of offshore oilfield surface engineering are not significantly different from those of onshore oilfields, but the special environment at sea puts different requirements on offshore oilfield production equipment, automation level, and environmental protection. These requirements will affect the process safety design. Therefore, if the content of this book is applied to offshore oilfields, engineering technicians need to analyze specific conditions.

Chapter 1 of this book briefly introduces the process flow of oilfield surface engineering, some safety risks faced by oilfield surface engineering, and what process safety design should be carried out to avoid and respond to these safety risks; Chap. 2 details the safety principles of oilfield surface engineering process design; Chap. 3 introduces a process safety analysis method for oilfield surface engineering; Chaps. 4 and 5, respectively, introduce the safety protection schemes of oilfield surface engineering gathering and transportation systems and centralized treatment stations; Chaps. 6 and 7, respectively, introduce the design and calculation of safety valves and emergency vent valves; Chap. 8 introduces how to analyze and calculate the vent system; Chap. 9 introduces the torch system; Chap. 10 details the causes, hazards, and analysis methods of low-temperature effects.

To meet the practical needs of engineering applications, this book uses some non-statutory measurement units in some places. Please be aware of this when reading.

Due to the limited level of the compiler, there are inevitably many shortcomings and omissions, and readers are kindly asked to correct them.

Review Board at the date of publication:

Liu Zhongmin, General Manager, China Petroleum Engineering and Construction Company (CPECC);
Zhang Jiming, Execution Director, CPECC Beijing Branch Company;
Cheng Tianlin, CPECC Beijing Branch Company;
Zhu Lianwang, CPECC Beijing Branch Company;
Hui Xiaorong, CPECC Beijing Branch Company.

Beijing, China Kun Fang

About This Book

This book systematically summarizes the safety production operation manuals, design standards, and advanced practices formulated by the petroleum industry at home and abroad, in order to provide guidance for the process safety design of onshore oilfield surface engineering. If the content of this book is applied to the design of offshore oilfield surface engineering, engineering technicians need to analyze specific conditions.

This book can be used by engineers engaged in the design of onshore oilfield surface engineering, and can also be referred to by personnel in related professional fields.

Contents

About the Author

Kun Fang Currently the Chief Engineer of Beijing Branch Company of China Petroleum Engineering Construction Co., Ltd., has long been committed to the design and research of overseas oil and gas field surface engineering.

With a solid professional theoretical foundation and excellent R&D capabilities, Fang Kun has presided over and participated in the design of 9 billion-level mega overseas oil and gas field surface engineering projects, with a total design contract amount exceeding 2 billion yuan. As the person in charge of the detailed engineering project group of the three phases of the Hafaya oilfield in Iraq, the project's surface engineering investment reached 5.7 billion yuan, successfully building the world's largest heavy oil centralized treatment station of the same kind, which was praised by the Iraqi government as the "most successful benchmark" project, and received high recognition from the state, associations, and peers at home and abroad. In addition, he completed the total contract project of the first carbon capture project demonstration in the domestic printing and dyeing industry—the comprehensive utilization of carbon dioxide capture and resource from coal-fired boiler flue gas. In the field of scientific research, he served as the project leader of the major scientific and technological project of China National Petroleum Corporation "Research and Demonstration of Key Technologies for Large-Scale Carbon Dioxide Capture, Oil Driving and Burial Full Industry Chain", responsible for the development of million-ton absorption tower process package and supercritical carbon dioxide dehydration process research. He has published more than 10 papers in several internationally renowned journals such as Coatings (SCI), AMPP Conference (EI), International Conference on Optical Instruments and Technology (CA), International Journal of Hydrogen Energy (EI), etc. He has compiled nine standards, applied for 29 patents (including one PCT patent), and won more than 30 provincial and ministerial level and above awards, including one National Quality Engineering Gold Award, one Luban Award, nine provincial and ministerial level scientific and technological awards, and eight provincial and ministerial level first-class design awards.

Fang Kun serves as the deputy director of the New Energy Technology Professional Committee of the China Petroleum Engineering Construction Association, and has been awarded the title of Excellent Scientific and Technological Worker of China National Petroleum Corporation. He was rated as a scientific and technological expert by the China Construction Enterprise Management Association. He has won honors such as the Advanced Individual in Scientific and Technological Innovation of the China Petroleum Engineering Construction Association in 2023, and the Top Ten Excellent Project Managers of China Petroleum Engineering Construction Co., Ltd.

Chapter 1
Introduction

1.1 Brief Introduction to the Process Flow of Oilfield Surface Engineering

Surface engineering is a key component of the large system of oil and gas field development and production, referring to the total facilities for preliminary treatment and transportation of crude oil extracted from oil wells to the export station. Oilfield surface engineering can be divided into oil and gas gathering, transportation systems and crude oil treatment systems according to the process flow from the wellhead to the export station.

1.1.1 Oil and Gas Gathering System

Oil and gas gathering and transportation is the main and core part of oilfield surface engineering, mainly responsible for the collection, measurement, processing, storage, and export or transportation of oil and gas well products. The well fluid flows from the oil tree into the surface facilities, and then is transported to the central processing station for processing through the gathering and transportation system. The gathering and transportation system mainly includes well site facilities, single well pipelines, metering stations, gathering and transportation trunk lines, oil transfer stations, and transfer pipelines. The typical process flow diagram of the gathering and transportation system is shown in Fig. 1.1.

K. Fang, *Onshore Oilfield Surface Facilities: Process Safety Engineering Guide*,
https://doi.org/10.1007/978-981-96-7830-3_1

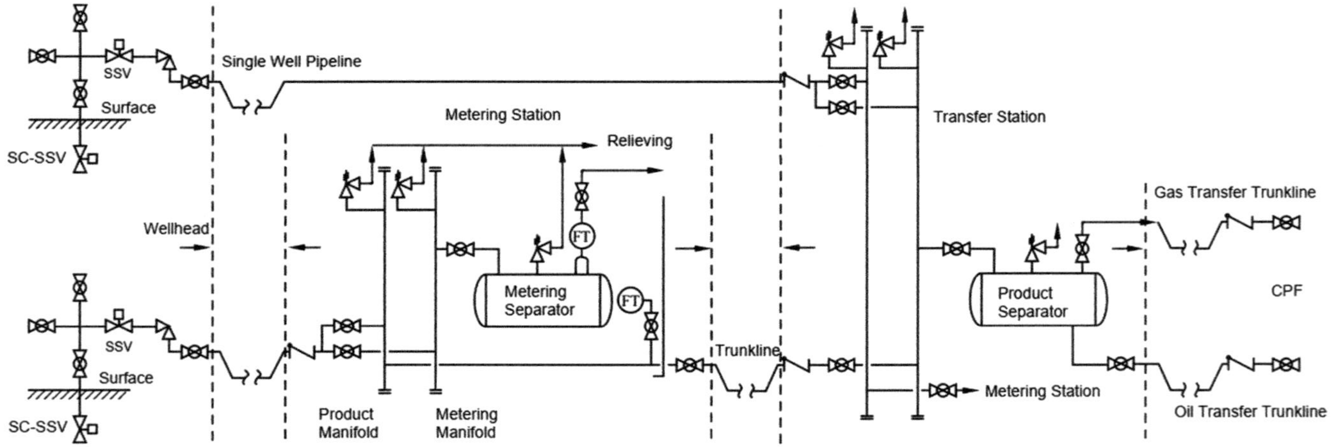

Fig. 1.1 Typical process flow diagram of the gathering and transportation system. *Note* The wellhead in the figure is a typical configuration and does not represent the configuration of the oil tree in every project.

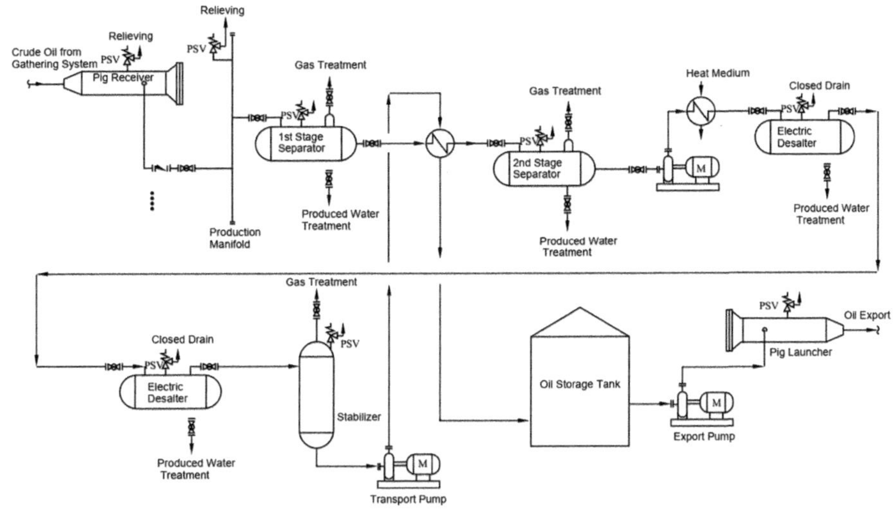

Fig. 1.2 Typical crude oil processing system in a central processing station

1.1.2 Crude Oil Processing System

The crude oil transported by the gathering and transportation system is processed in the central processing station, mainly including crude oil separation, crude oil heating, crude oil electro-dehydration, crude oil electro-desalting, crude oil storage, and qualified oil export systems. With the continuous improvement of environmental protection standards, the associated gas produced is generally pressurized and exported after treatment. A typical crude oil processing system in a central processing station is shown in Fig. 1.2.

1.2 Safety Risks and Accidents in Oilfield Surface Engineering

During the production process of oilfield surface engineering, there are mainly risks of overpressure, leakage, liquid overflow, gas breakthrough, negative pressure, and overtemperature [1].

1.2.1 Overpressure

Overpressure refers to the pressure in the process equipment exceeding the maximum allowable working pressure. The main causes of overpressure are:

(1) If the inlet flow exceeds the outlet flow, the pressure generated by the inlet source will exceed the maximum allowable working pressure of the process equipment. If the upstream flow control device fails, or there is a flow restriction device or blockage at the outlet of the process equipment, or overflow or gas channeling occurs in the upstream equipment, then the inlet flow will exceed the outlet flow.
(2) When heating with both the inlet and outlet shut off, overpressure can also be caused by the thermal expansion of the fluid. The consequence of overpressure is to cause the equipment to burst suddenly and cause hydrocarbon leakage. High pressure is a detectable abnormal state of overpressure, and once high pressure occurs, it indicates that overpressure may occur.

Overpressure protection is mainly divided into:

(1) Primary protection. The primary protection for overpressure of pressure equipment should be provided by PSH sensors, which shut off the inlet flow. If the vessel is a heated device, the PSH sensor should also shut off the fuel or heat source. The primary protection for atmospheric vessels should be provided by an appropriate venting system.
(2) Secondary protection. The secondary protection for overpressure of pressure equipment should be provided by PSV. The secondary protection for atmospheric vessels should be provided by a second-level venting system. The second-level venting system can be exactly the same as the first-level venting system, or it can be two orifice plates with self-acting PSVs or a separate PSV.

In process equipment containing gas and liquid, PSH sensors, PSVs or venting systems should be installed to detect or vent the pressure of the gas or vapor portion. The detection interface of the safety device should be opened as high as possible on the equipment to reduce the possibility of being blocked by impurities in the fluid. The installation of PSVs and venting systems on atmospheric tanks must comply with API Std 2000 Venting atmospheric and low-pressure storage tanks or other applicable standards.

1.2.2 Leakage

Leakage refers to the accidental escape of fluid in process equipment into the atmosphere. In this book, leakage means that the escaped fluid is a hydrocarbon.

1.2.2.1 Causes of leakage

The causes of leakage are corrosion, erosion, mechanical failure or overtemperature causing equipment damage, overpressure causing equipment rupture, or accidental damage caused by external forces.

1.2.2.2 Consequences and Detectable Abnormal States of Leakage

The consequence of leakage is: hydrocarbons are released into the atmosphere. Low pressure, backflow, and low liquid level are detectable abnormal states, indicating that leakage has occurred.

1.2.2.3 Leakage Protection

Leakage protection is mainly divided into:

(1) Primary protection. The primary protection for abnormal operating conditions caused by large leaks in pressure equipment should be provided by PSL sensors and PSVs, with PSL shutting off the inlet fluid and PSVs reducing backflow. The primary protection for leakage from the liquid part can also be provided by LSL, which shuts off the inlet fluid. In atmospheric equipment, the primary protection for liquid leakage is provided by LSL sensors, which shut off the inlet fluid. In process equipment, safety devices cannot detect small amounts of liquid leakage, and a leakage collection and discharge system should be used as primary protection at this time. Small amounts of gas leakage occurring in poorly ventilated areas cannot be detected by equipment sensing devices, and primary protection should be provided by a combustible gas detection system at this time.
(2) Secondary protection. The secondary protection for all detectable leaks and small amounts of gas leaks in poorly ventilated areas should be provided by the Emergency support system provided. Secondary protection for minor liquid leakage should be provided by the LSH sensor installed on the waste oil tank, which shuts off all equipment leaking to the waste oil tank.

1.2.2.4 Location of the Leak Safety Device

In process equipment containing liquid and gas parts, a PSL sensor should be installed to detect the pressure of the gas or vapor part. The PSL sensor should be installed as high as possible on the equipment to reduce the possibility of being blocked by impurities in the liquid. The PSV should be installed on the outlet pipeline of each device to prevent a large amount of backflow. The installation position of the LSL sensor should be a sufficient distance below the lowest operating liquid level to prevent accidental shutdown, but there should be enough volume between the LSL sensor and the liquid outlet to prevent gas breakthrough before the shutdown is completed.

1.2.3 *Liquid Overflow*

Liquid overflow refers to the discharge of liquid from the gas or vapor outlet of process equipment.

1.2.3.1 Causes of Liquid Overflow

The cause of liquid overflow is that the liquid inlet flow rate exceeds the outlet flow rate. This may be the result of upstream flow control device failure, liquid level control system failure, or liquid outlet blockage.

1.2.3.2 Consequences and Detectable Abnormal States of Liquid Overflow

The consequence of liquid overflow is overpressure or excessive liquid level of downstream equipment, or release of hydrocarbon substances into the atmosphere. High liquid level is a detectable abnormal state, indicating that overflow may occur.

1.2.3.3 Liquid Overflow Protection

Liquid overflow protection mainly includes:

(1) Primary protection. The primary protection for liquid overflow should be provided by the LSH sensor, which shuts off the inlet fluid of the equipment.
(2) Secondary protection. The secondary protection for liquid overflow to the atmosphere should be provided by the Emergency Support System (ESS). The secondary protection for liquid overflow to downstream equipment should be provided by the safety device of the downstream equipment.

1.2.3.4 Location of the Safety Device

The position of the LSH sensor should be a sufficient distance above the highest operating liquid level to prevent accidental shutdown, but there should be enough volume above the LSH sensor to prevent liquid overflow before the shutdown is completed.

1.2.4 *Gas Breakthrough*

Gas breakthrough refers to the discharge of gas from the liquid outlet in process equipment.

1.2.4.1 Causes of Gas Breakthrough

The cause of gas breakthrough is the failure of the liquid level control system or the inadvertent opening of the bypass valve of the liquid level control valve.

1.2.4.2 Consequences and Detectable Abnormal States of Gas Breakthrough

The consequence of gas breakthrough is overpressure of downstream equipment. Low liquid level is a detectable abnormal state, indicating that a gas breakthrough may have occurred.

1.2.4.3 Gas Breakthrough Protection

Gas breakthrough protection mainly includes:

(1) Primary protection. The primary protection for gas breakthrough should be provided by the LSL sensor, which shuts off the fluid inlet or liquid outlet.
(2) Secondary protection. The secondary protection for gas breakthrough to downstream equipment should be provided by the safety device of the downstream equipment.

1.2.4.4 Location of the Safety Device

The installation position of the LSL sensor should be a sufficient distance below the lowest operating liquid level to avoid accidental shutdown, but there should be enough volume between the LSL sensor and the liquid outlet to avoid gas breakthrough before the shutdown is completed.

1.2.5 Negative Pressure

Negative pressure refers to the pressure inside the process equipment being lower than the design crush pressure.

1.2.5.1 Causes of Negative Pressure

The cause of negative pressure is that the fluid discharge flow rate exceeds the inflow rate, which may be the result of the following factors: failure of the inlet or outlet control valve, blockage of the inlet pipeline when the fluid is discharged, or thermal shrinkage of the fluid when the inlet and outlet are shut down.

1.2.5.2 Consequences and Detectable Abnormal States of Negative Pressure

The consequence of negative pressure is equipment crush and leakage. Low pressure is a detectable abnormal state, indicating that negative pressure may have occurred.

1.2.5.3 Negative Pressure Protection

Negative pressure protection mainly includes:

(1) Primary protection. The primary protection for negative pressure of atmospheric equipment should be provided by an appropriate breathing system. The primary protection for pressure equipment subjected to negative pressure is provided by a gas supply system.
(2) Secondary protection. The secondary protection for atmospheric equipment should be provided by a second stage vent or a PSV. The secondary protection for pressure equipment subjected to negative pressure should be provided by a PSL sensor to shut off the inlet and outlet fluids.

1.2.5.4 Location of the Safety Device

The PSL sensor should be installed as high as possible on the equipment to reduce the possibility of fluid impurities blocking. The installation of the vent system and PSV should be in accordance with API Std 2000 Venting atmospheric and low-pressure storage tanks or other applicable standards.

1.2.6 Overtemperature

Overtemperature refers to the temperature exceeding the operating temperature designed for the process equipment. Affected by accidental events in the fire and flue gas heating equipment, overtemperature can be classified into overtemperature of the medium or process fluid and flue overtemperature. The overtemperature issues of non-fired heating equipment will be discussed in the analysis of individual equipment in later chapters.

1.2.6.1 Causes of Overtemperature

The reasons for the overtemperature of the medium or process fluid include:

(1) Due to device failure or accidental factors, the medium bypasses the fuel or flue gas control equipment, resulting in excessive input of fuel or heat.

(2) Extra fuel enters the combustion chamber through the air inlet, or combustible fluid leaks into the fire or flue gas heating chamber.
(3) In a closed heat transfer system (such as the heated medium circulating through the coil in the combustion chamber or flue gas heating chamber), the flow of the heated fluid is insufficient due to low flow.
(4) The liquid level is too low in the fired equipment with immersed fire tubes or flue gas tubes. In the fired equipment, flue overtemperature can be caused by any of the above reasons or by the accumulation of external substances (sand, scale, etc.) in the heat transfer part leading to insufficient heat transfer. Flue overtemperature in the flue gas heating equipment can be caused by the ignition of combustible medium leaking into the flue gas heating chamber.

1.2.6.2 Consequences of Overtemperature and Detectable Abnormal Conditions

The consequences of high temperature of the medium or process fluid may be to reduce the pressure-bearing capacity of the equipment, followed by leakage or rupture of the affected equipment; in a closed heat transfer system, if the medium is blocked in the coil, it will cause overpressure of the circulating coil; high temperature of the flue can ignite combustible materials on the flue surface. High temperature, low flow, and low liquid level are detectable abnormal conditions, indicating that overtemperature may have occurred.

1.2.6.3 Overtemperature Protection

Overtemperature protection mainly includes:

(1) First-level protection. The first-level protection for overtemperature of the medium or process fluid caused by excessive or extra fuel, heat, or medium leaking into the fired equipment or flue gas heating chamber should be provided by TSH sensors. If the overtemperature is caused by low liquid level, its first-level protection should be provided by LSL sensors. The TSH and LSL sensors on the fired equipment must shut off the fuel supply and the entry of combustible fluids. The TSH and LSL sensors on the flue gas heating equipment must shut off the fuel and heating source or change their direction. If the medium overtemperature is due to low flow in a closed heat transfer system containing combustible fluid, then the first-level protection should be provided by FSL sensors, which shut off the fuel supply entering the fired equipment or divert the flue gas flow from the flue gas heating equipment. The first-level protection for flue overtemperature should be provided by TSH sensors (installed on the flue), which shut off the fuel or flue gas source and the inflow of combustible fluids.
(2) Second-level protection. If the overtemperature of the medium or process fluid in the fired equipment is caused by excessive or extra fuel, then the second-level protection should be provided by TSH sensors (installed on the flue). If it

is caused by low flow, then the second-level protection should be provided by TSH sensors installed on the medium pipeline and on the flue. If it is caused by low liquid level, then the second-level protection should be provided by TSH sensors installed on the medium or process fluid pipeline and on the flue. If the overtemperature of the medium or process fluid in the flue gas heating equipment is caused by low liquid level or low flow, then the second-level protection should be provided by TSH sensors installed on the medium pipeline. These sensors should perform the same functions as the first-level protection. The second-level protection for flue overtemperature should be provided by the emergency support system and applicable FSV.

1.2.6.4 Location of Safety Devices

Temperature sensors, except for fusible or surface contact types, should be installed in thermocouple sheaths for easy removal and testing; in gas/liquid two-phase systems, TSH sensors should be installed in the liquid phase section; in tubular heaters, the heated medium flows through the coil in the fired equipment or flue gas heating chamber, in this case, TSH sensors should be installed as close as possible to the outlet of the pipeline near the heater; FSV should be installed at the outlet of the medium pipeline.

1.3 Process Safety Design Content

Process safety design should be based on the requirements of national safety production laws, regulations, and standard norms, actively absorb advanced foreign safety design concepts, eliminate or reduce hazards from the design source, and improve the quality of intrinsic safety design. The design should follow the principle of reasonably reducing risks, optimize design schemes, adopt suitable and reliable safety countermeasures, and reduce the risks in the life cycle of the construction project to the lowest feasible level.

Design flaws, equipment failures, incorrect operating conditions, unforeseen operating conditions, failure of hazard control, and human errors can all potentially cause process accidents. Process safety analysis is the systematic, comprehensive, and orderly method of identifying, assessing, and controlling hazards in processes and operations to prevent process safety accidents.

The scope of safety-related content in a project is broad, generally including: process safety, emergency shutdown/pressure relief systems, flare systems, pollution discharge, gas detection and alarm systems, fire systems, plant layout, passive fire protection, hazardous area division, personal safety protection, and public alarms and communication, etc.

The rest of the safety scope, such as safety valves (pressure relief valves) and flare systems, normal pressure venting and pollution discharge, gas detection and alarm, fire systems, plant layout, passive fire protection, hazardous area division, personal safety protection, public alarms and communication, etc., some have separate and complete reference standards, some have mature design practices, some are far from process safety, this book will not elaborate.

1.3.1 Selection of Safety Devices

Safety protection devices are divided into primary protection devices and secondary protection devices. Primary protection devices react faster, safer, or more reliable than secondary protection devices. Primary protection devices provide the highest level of protection, while secondary protection devices provide the next highest level of protection [2].

(1) Because the consequences of failure may vary in degree or order, a single safety device may not be able to provide complete primary or secondary protection, so several devices or systems may be needed. These devices or systems combined can provide the necessary level of protection. For example, a PSL sensor and an FSV can be used to prevent fluid from flowing to the leakage, these two devices can provide primary protection.

(2) The protection devices determined in the SAT can be connected with the necessary SDV or other final control devices to protect any process equipment in any process. It is important for users to understand the logic of the SAT and how the SAT is established.

(3) The positions of the SDV and other final control devices should be determined after studying the detailed process flow diagram and operating parameters. When an accident is detected in the process, the device can shut down the input source or divert the input source to other devices that can handle it safely, thus isolating the device from all input process fluids, heat, and fuel. If the process input is shut down, it is best to shut it down at the source.

(4) Unless there is a situation where a safety device is not needed, or its usual function can be replaced by other safety devices, all safety devices of each device should be considered and set up. The Safety Analysis Check Sheet (SAC) lists equivalent protection methods, so some devices can be omitted.

(5) If the device used is not mentioned in the SAC, then the SAT of the device can be established according to the above principles.

1.3.2 Protective Shutdown Actions

When abnormal operating conditions in process equipment are detected by safety devices or discovered by operators, all process fluid, heat, and fuel input sources should be shut down or diverted to other devices that can handle them safely. If a shutdown action is chosen, the process system input should be shut down at the energy source (wellhead, pump, compressor, etc.). The design should avoid the situation where shutting down the inlet of a process device leads to an abnormal state in the upstream device, which then causes its safety device to shut down the device. Because once this happens, each device in the process system will be shut down one by one from the back until the energy source is cut off. That is, in this case, the shutdown of a downstream device causes all other devices to be in an abnormal operating state, and they must all be protected by their safety devices, the interlocking effect depends on the operation of several additional safety devices, and it may put too much stress on the equipment.

(1) After the energy source is cut off, the inlet of the process equipment is usually shut down last to provide further protection or prevent the pressure or liquid level of the upstream equipment from being transmitted. If desired, the shutdown of the energy source should be simultaneous with or prior to the shutdown of the device inlet valve.
(2) There may be some special cases of running interlock shutdowns, for example:

① The inlet source of the separator changes frequently as the wellhead is periodically connected to the separator. When an abnormal state is detected, if the wellhead connected to the separator is directly shut down, the logic of the safety system must change each time a different wellhead is connected to the separator. This may cause negligence when changing the safety system logic. In this case, it is better to close the inlet of the separator and let the high pressure of the formed oil pipe close the oil well through the action of the PSH sensor. The rated pressure of the manifold and the oil pipe must be able to withstand the maximum pressure caused by this.
② The station receives production fluids from satellite wells through the oil pipe. Although the energy source of the system is the satellite wellhead, if an abnormal state is detected in the station, the SDV on the oil pipe entering the station must be shut down. If it is desired to shut down the SDV on the station's oil pipe after shutting down the satellite well, this can be done using the oil pipe PSH sensor installed at the satellite well location.
③ Compressors are equipped with automatic bypass valves during installation. When the compressor is shut down, it allows wellheads that can overcome pipeline pressure to continue production. If some wellheads cannot overcome pipeline pressure, these wellheads can be shut down through the PSH sensors on each oil outlet pipe to reduce potential safety system logic problems in ①.

1.3.3 Measures to Prevent Ignition

The main threat to surface engineering safety comes from the release of hydrocarbons. If the ignition of released hydrocarbons can be prevented, its impact can be mitigated. Therefore, preventing ignition is another protective measure that must be considered along with safety devices and emergency support systems. Ignition of hydrocarbons can be caused by electric arcs, flames, sparks, and incandescent surfaces. These ignition sources can be protected during design to reduce the possibility of hydrocarbons contacting ignition sources or to prevent gaseous hydrocarbons from reaching flammable concentrations.

This book classifies these methods as "Ignition Prevention Measures (IPM)". They include: ventilation, adherence to relevant electrical codes and recommended practices, rational arrangement of potential ignition sources, and protection of heat source surfaces. Each is introduced below.

1.3.3.1 Ventilation

The ignition of combustible gases requires their concentration mixed with air (oxygen) to reach the Lower Explosive Limit (L.E.L.). The designed safety system can shut off the source of hydrocarbons when an abnormal state is detected, reducing the amount of hydrocarbons released. Another method to prevent the formation of combustible mixtures is to ensure sufficient air volume, keeping the concentration of hydrocarbons below the Lower Explosive Limit (L.E.L.). To prevent the accumulation of combustible mixtures, the process area should be as open as possible to allow free air flow. Enclosed areas containing hydrocarbon processing equipment or combustion equipment should have sufficient ventilation to disperse gases or vapors before they reach the Lower Explosive Limit (L.E.L.). If the enclosed area is poorly ventilated, a combustible gas detector (ASH) should be installed to signal and shut off the source of hydrocarbons at a concentration set below the Lower Explosive Limit (L.E.L.).

1.3.3.2 Adherence to Relevant Electrical Codes and Recommended Practices

Protection against electrical source ignition is ensured by the following measures: designing and maintaining electrical equipment according to API RP 14F Recommended practice for design, installation, and maintenance of electrical systems for fixed and floating offshore petroleum facilities for unclassified and class I, division 1, and division 2 locations or other applicable standards, and classifying site areas according to API RP 500 Recommended practice for classification of locations for electrical installations at petroleum facilities classified as class I, division 1 and division 2.

(1) API RP 14F Recommended practice for design, installation, and maintenance of electrical systems for fixed and floating offshore petroleum facilities for unclassified and class I, division 1, and division 2 locations establishes guidelines for electrical equipment and wiring methods that can be safely used in hazardous and non-hazardous areas of oilfield surface engineering.

(2) API RP 500 Recommended practice for classification of locations for electrical installations at petroleum facilities classified as class I, division 1 and division 2 proposes methods for dividing areas around drilling rigs and production facilities for the safe installation of electrical equipment on fixed and mobile sites on land and at sea.

1.3.3.3 Rational Arrangement of Potential Ignition Sources

Potential ignition sources such as fire process equipment and certain rotating machinery usually have protective measures to reduce the possibility of igniting released hydrocarbons. Another protective measure is to place equipment in areas where the accidental release of hydrocarbons is unlikely. API RP 14J Recommended practice for design and hazards analysis for offshore production facilities provides guidelines for equipment layout. Other potential ignition sources are related to living, such as boilers, water heaters, stoves, dryers, etc. These devices should be placed in non-hazardous areas. If these devices use gas as fuel and are installed in poorly ventilated buildings, a combustible gas detector (ASH) should be installed to cut off the outdoor fuel shut-off valve.

1.3.3.4 Protection of Heat Source Surfaces

Any heat source surface with a temperature exceeding 204 °C (400°F) should be protected so that it is not exposed to places where liquid hydrocarbons spill or leak. Heat source surfaces with temperatures exceeding 385 °C (approximately 80% of the ignition temperature of natural gas) must also be protected so that they are not exposed to places where combustible gases and vapors accumulate. Protective measures can include insulation, shielding, water cooling, etc. However, if these heat source surfaces are on the surfaces of fixed or portable equipment within the areas defined by API Std 520 Sizing, selection and installation of pressure-relieving devices, insulation measures should be taken. Some mechanical equipment parts, such as turbochargers, exhaust manifolds, and those parts that cannot be insulated for fear of causing mechanical failures (including related pipelines), should adopt other protective measures.

1.3.3.5 Protection of Heating Equipment

Any equipment with a surface temperature exceeding 71 °C should take protective measures to prevent operators from accidentally touching it in normal work and walking areas. Protective measures can be taken: shields, barriers, or insulation. Some mechanical types of components such as turbochargers, exhaust manifolds, compressor manifolds, expansion tanks, and those parts that cannot be insulated for fear of causing mechanical failures (including related pipelines) are exceptions, in which case warning signs can be used.

1.3.4 Emergency Support Systems and Other Support Systems

The Emergency Support System (ESS) is a method of implementing specific safety functions that are common to the entire station. The ESS includes an Emergency Shutdown (ESD) system, fire detection system, gas detection system, ventilation system, leak collection and discharge system, oily tank, and SSSV system. These systems are the most basic systems for protecting facilities, and the system completes protection by initiating shutdown functions or reducing the consequences caused by escaping hydrocarbons. Other support systems include pneumatic supply systems, gas discharge systems, leak collection systems, and other service systems that can improve the safety performance of ground engineering. The pneumatic supply system provides control media for safety systems, and the gas discharge system can safely and controllably release gas into the atmosphere.

The ESS can reduce the consequences of hydrocarbon leaks in oilfield ground engineering, and the ESS includes:

(1) Combustible gas detection system. It detects the presence of leaking hydrocarbons and issues an alarm and shuts down the station before the gas concentration reaches the lower explosive limit (L.E.L.).
(2) Leak collection and discharge system. It collects leaking liquid hydrocarbons and triggers a station shutdown.
(3) Fusible plug circuit system. It detects the heat of a fire and triggers a station shutdown.
(4) Other fire system detection devices (flame, heat, smoke). They can be used to enhance fire detection capabilities.
(5) Emergency shutdown system. In the event of an abnormal state or accident observed by the operator, the emergency shutdown system can be manually triggered to shut down the station.
(6) Subsurface Safety Valve (SSSVS) system. It can be an automatically operated SSCSV, or an SCSSV triggered by the ESD system and/or fusible plug circuit.

1.3.4.1 Emergency Shutdown (ESD) System

The Emergency Shutdown (ESD) system is a manual control system configured in the station that acts globally. When this system is activated, it will shut down all production wells and process devices. This system can include a series of independent process shutdown systems that can be started separately. The action of the ESD system will lead to the cessation of all production activities in the station, including the shutdown of all pipeline SDVs. The designed ESD system should ensure that the power station and fire system can continue to operate when needed in an emergency [3].

The ESD system provides a means for workers to manually shut down the station when an abnormal situation is observed. The fusible elements in the fusible plug circuit system can be combined with the ESD control circuit.

The ESD station should be located in a convenient place, but also prevent unintentional startup. The ESD station should indicate its shutdown function and clearly mark the valve handle shutdown position. The manual ESD valve should be a quick-opening non-throttling type, so that the shutdown system can act quickly. Since the ESD system plays a key role in the safety protection system of ground engineering, high-quality anti-corrosion valves should be used. The ESD shutdown levels are shown in Fig. 1.3.

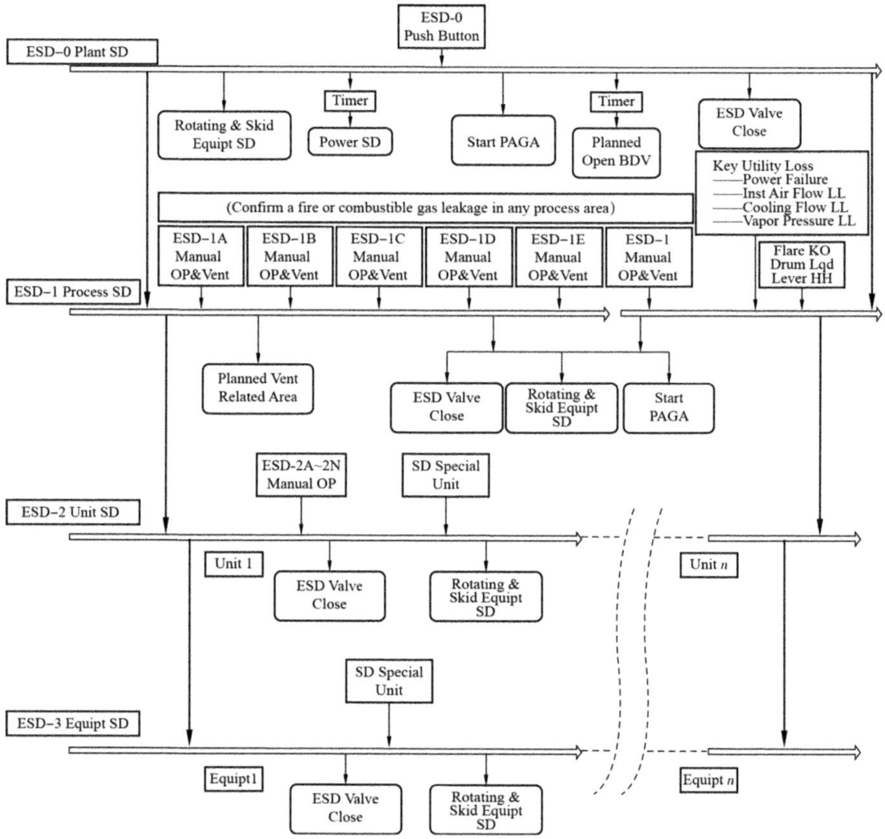

Fig. 1.3 ESD shutdown levels

1.3.4.2 Fire Detection System

1. Purpose of Using the Fire Detection System

Ground engineering should provide automatic fire detection so that workers can take measures as early as possible. The fire detection system includes a pneumatic fusible element (TSE) system and various electric fire detection devices [including flame detection devices (USH), heat detection devices (TSH) or smoke detection devices (YSH)]. The system should be installed in Class 1 and Class 2 areas divided according to API RP 500 Recommended practice for classification of locations for electrical installations at petroleum facilities classified as class I, division 1 and division 2, as well as buildings where people normally or occasionally sleep.

2. Pneumatic Fusible Plug Element Detection System

A reliable and widely used method for detecting fires is to use a pneumatic pipeline containing fusible elements in key positions. This system can send a signal to shut down other production facilities except for fire protection equipment. Fusible elements are usually metal plugs (the metal can be melted at a set temperature) or a section of fusible synthetic tubing.

3. Electric Fire Detection System

In addition to the pneumatic fusible plug element detection system, various electric fire detection devices (flame, heat, and smoke) are usually used in oilfield ground engineering. These detection devices are used to issue alarms, trigger shutdowns, or start fire extinguishing systems (such as media being CO_2 or water).

4. Installation and Operation

All electric fire detection devices should be tested by a nationally recognized laboratory and installed according to the manufacturer's recommended methods. In the absence of a main power supply at the station, the electric fire detection device should be able to work for at least 8 h.

The installation and operation of the fire detection system should be adapted to the area designed for protection. The fire detection system should be installed in process equipment, enclosed classified areas (enclosed hazardous areas), and enclosed non-classified areas (enclosed non-hazardous areas) as follows:

(1) Process equipment. Table 1.1 is a guide for the installation of fusible plugs. When fusible circuits or other devices (such as ultraviolet flame detectors, infrared sensors, etc.) replace fusible plugs, they should at least provide the same range of functions listed in Table 1.1. If the generated signal can initiate normal shutdown functions, fusible elements can be installed on various pneumatic pipelines of the safety system. Fusible elements should not be installed on flammable gas supply pipelines, otherwise gas will continue to leak after shutdown.

(2) Enclosed classified areas (enclosed hazardous areas). Fire (flame, heat, smoke) detection devices should be installed in enclosed classified areas (Class 1 or

Table 1.1 Fusible plug installation guide

Equipment		Fusible plug arrangement	Minimum number of fusible plugs
Wellhead		Install 1 per wellhead①	–
Manifold		Install 1 every 3 m (10 ft)①	2
Pressure vessel	Vertical vessel	Install one every 0.3 m (12 in) in diameter, up to 5	1
	Horizontal vessel	Diameter less than 1.2 m (48 in)—Install 1 every 1.5 m (5 ft)	2
		Diameter not less than 1.2 m (48 in)—Install 2 every 1.5 m (5 ft), in two rows	4
Atmospheric vessel		Install 1 each at the inlet, outlet, and manhole of each vessel	–
Fired vessels and flue gas heating equipment		Same as pressure vessels; in addition, install 1 on the outside of the fire barrier of the fired equipment	–
Heat exchanger (shell and tube)		Install 1 per section of the heat exchanger	2
Pump	Reciprocating	Install 1 above the rod seal box	–
	Centrifugal	Install 1 above each seal box	–
Compressor	Reciprocating	Install 1 per cylinder ②	–
	Centrifugal	Install 1 above the compressor	–
Engine	Spark	Ignition Install 1 above each carburetor or intake valve②	–
	Diesel engine	Install 1 on each fuel injection nozzle②	–
	Gas turbine	Install 1 each on the fuel solenoid valve, regulating valve, and PT pump②	–

① Not applicable to underwater wellheads or manifolds
② Or equivalent content. *Note* If fusible tubes or other devices (such as ultraviolet flame detectors, etc.) are used as fusible plugs, they should at least provide the same content listed in the table

Class 2 areas) divided according to API RP 500 Recommended practice for classification of locations for electrical installations at petroleum facilities classified as class I, division 1 and division 2. These devices can automatically shut down the hydrocarbon release source in the enclosed classified area. Fusible plug systems of flammable or non-flammable gases can also be used in these areas.

(3) Enclosed non-classified areas (enclosed non-hazardous areas). In all rooms where people normally or occasionally sleep, smoke detection devices that can initiate audible alarms should be installed. In houses with heat sources (such as water heaters, clothes dryers, kitchen stoves, fireplaces, space heaters, etc.), smoke and/or rate-of-rise detectors that can initiate audible alarms should be installed. In buildings where people normally or occasionally sleep, fusible plug systems using flammable gases for fire detection cannot be installed (even if these buildings are in classified areas, because it is similar to a hydrocarbon release source).

1.3.4.3 Combustible Gas Detection System

1. Purpose of Use

In offshore stations, the accumulation of combustible gas in the atmosphere poses a threat to safety. Accumulation of combustible gas is likely to occur in enclosed areas. There are two ways to enhance safety, one is good ventilation, and the other is to install a combustible gas detector (ASH) system. The combustible gas detector (ASH) system should warn workers by sound and/or light alarm when low concentration flammable gas or vapor appears, and when the concentration approaches the lower explosive limit (L.E.L), it should shut off the gas source or remove all ignition sources.

2. Installation

Combustible gas detection sensors should be installed in the following locations:

(1) In all enclosed classified areas (Class 1 and Class 2) as defined by API RP 500 Recommended practice for classification of locations for electrical installations at petroleum facilities classified as class I, division 1 and division 2.
(2) In enclosed areas with poor ventilation that house natural gas engines.
(3) In buildings where people regularly or occasionally sleep and that contain sources of flammable gases. Gas detectors should be tested by a nationally recognized testing laboratory (NRPL) and meet ANSI/ISAS 12.13.

Part I "Performance Requirements for Combustible Gas Detectors". The installation, operation, and maintenance of the gas detection system should follow ANSI/ISAS 12.13 Part II "Installation, Operation, and Maintenance of Combustible Gas Detector Instruments". In enclosed areas containing combustible gas compressors, the minimum number of sensors required is: one for each compressor set, plus one for every three sets or part of the set.

Note: This requires at least two sensors to be installed in an enclosed compressor room. In enclosed buildings containing other gas engines (such as motor generators or pumps), at least one sensor should be installed for each engine.

In addition to buildings where people regularly or occasionally sleep or enclosed areas where flammable gases are transported by compressors, either sufficient ventilation or a gas detection system can be provided. Similarly, in enclosed areas in the station where continuous DC power is not provided, if the enclosed area does not contain an ignition source, does not contain a compressor transporting flammable gas, and does not have a building where people regularly or occasionally sleep, the gas detector can be canceled.

3. Operation and Maintenance

The gas detector should be able to detect the concentration of at least two gases. When the gas concentration is not higher than 25% of the L.E.L, an audible alarm signal should be issued (a light alarm signal should be installed in high-noise areas, because in these areas, the audible alarm is ineffective) to warn the operator. When the gas concentration is not higher than 60% of the L.E.L, the automatic correction action should be initiated, such as closing the gas source's SDV or cutting off the power supply of the equipment, this method is not applicable to hazardous areas (classified areas). As an alternative to the total process shutdown, the alarmed area can be isolated, such as closing the fuel inlet valve of the generator room. To ensure that the situation does not become more dangerous, the form of automatic correction action should be carefully considered. If the shutdown of machinery or equipment may generate an ignition source, consider initiating a fire extinguishing system, such as a Halon system, before shutdown. For gas detection systems that monitor more than one area, the location of the detected combustible gas or vapor should be indicated.

Note: In special cases, when detecting lower gas concentrations, the provisions of Chapter 48 of API RP 500 Recommended practice for classification of locations for electrical installations at petroleum facilities classified as class I, division 1 and division 2 should be followed to reduce area classification.

1.3.4.4 Adequate Ventilation

Adequate ventilation refers to ventilation conditions (natural or artificial ventilation) sufficient to prevent the accumulation of a large amount of vapor-air mixtures with concentrations higher than their flammable (explosive) lower limit (L.E.L) 25%. See the additional explanation of Article 4.6 of API RP 500 Recommended practice for classification of locations for electrical installations at petroleum facilities classified as class I, division 1 and division 2, including recommended successful methods.

1.3.4.5 Leak Collection and Discharge System

The leak collection and discharge system can collect and discharge leaked liquid hydrocarbons to a safe place. All equipment that may produce leaks or overflows should be equipped with fences, sewage trenches, or drip pans to drain the waste oil into the waste oil tank. On structures without process vessels or other equipment that may produce leaks or spills, the installation of a leak collection and discharge system is optional (such as structures with only wellheads, pipe manifolds, pipelines, cranes, and/or instrument air washers).

All gravity discharge pipelines should be designed to prevent gas from escaping from the waste oil tank through the discharge port. The typical installation method is to install a water seal on each discharge port or each discharge pipe manifold, or to install a total pipe network water seal on the waste oil tank inlet pipeline. The installation of check valves is not applicable in this case, and it cannot be used as an option to protect the water seal. Pressure discharge and gravity discharge cannot be mixed before the waste oil tank inlet.

1.3.4.6 Waste Oil Tank

The waste oil tank can be a tank, a closed-top waste oil drum, or an open-top waste oil drum. All waste oil tanks should be equipped with an automatic discharge function to handle the maximum inlet flow. The purpose of installing a venting system on the atmospheric waste oil tank is to safely disperse hydrocarbon vapors. Depending on the design and location of the waste oil drum, the vent pipe of the waste oil drum can achieve this purpose without installing a flame arrester. Due to the possibility of blockage caused by corrosion, low flow/low pressure (no static electricity) and distance from potential ignition sources/flashback sources, the flame arrester near the water seal position on the waste oil drum can be canceled.

A properly designed open-top oily waste drum can be used temporarily to collect deck rainwater or droplets and discharge treated production water and treated sand. Unless in an emergency, containers (such as torch scrubbers, condensate oil aggregators, and various fuel filters) should not directly discharge liquid hydrocarbons into the open oily waste drum. The open-top oily waste drum should prevent the outflow of hydrocarbons (overflow and/or bottom outflow), and preventive measures should be decided after analyzing each situation individually. Some factors to consider include: the length of the oily waste drum, the properties of the liquid, the maximum flow rate, the effect of fluctuations, and the range of tidal changes.

1.3.4.7 Other Support Systems

The integrity of the safety system of the station processing facilities depends on the normal operation of several other support systems. These ancillary support systems are as important as other parts of the ground engineering safety protection system and should be well maintained. This chapter discusses the pneumatic supply system and the gas venting system.

The pneumatic supply system provides power and supplies gas for instruments. The gas venting system guides the gas released from the process equipment to a safe place and eventually releases it into the atmosphere.

1. Pneumatic Supply System

(1) Purpose of use. The pneumatic supply system provides control media for the safety system of the station processing facilities. Safety and shutdown systems generally require a power supply suitable for valve actuator operation, as well as a low-pressure instrument gas supply.
(2) Gas supply quality. The normal operation of the safety system depends on the gas supply system, so a reliable high-quality gas source is very important. Good gas supply quality should be:
① Does not contain liquid hydrocarbons.
② Does not contain water and water vapor.
③ Does not contain solid matter.
④ Non-corrosive.

(3) Gas source. The usual start-up control medium is air, natural gas, or nitrogen. When air is used as the gas source, the system should be designed to prevent air from mixing with hydrocarbon gases from the process system or utility system under normal and abnormal conditions. If another gas source is used, the replacement medium must not form a flammable mixture when combined with the original pneumatic source.
(4) Supply and response. The scale of the gas supply system should ensure that sufficient pressure and volume of gas are supplied to all equipment. The gas supply should be calculated to adapt to the maximum demand that may occur at any time, and to ensure that any safety device (such as PSH, BSL, ESD stations, etc.) does not exceed 45 s when shutting down equipment or the station. To achieve this response, the size of the gas supply pipeline, the size of the safety device exhaust port, and the application of auxiliary rapid discharge facilities should be considered. The size of the gas supply and discharge pipelines should be selected to achieve the best discharge conditions. Due to its volume and flow characteristics, too long or too thin pipelines require too much discharge time. When designing instruments and control systems, API RP 500 Recommended practice for classification of locations for electrical installations at petroleum facilities classified as class I, division 1 and division 2 should be used as a guide.

2. Gas Venting System

The gas venting system provides a means to guide the gas discharged from the process equipment to a safe place under normal conditions (torch, venting) and abnormal conditions (venting), and eventually discharge it into the atmosphere. The so-called safe place should be a place where the gas concentration can be diluted to below L.E.L., so as not to pose a hazard to the equipment, or a place where the gas can be safely burned.

These systems start at the normal gas outlet or pressure relief device of the process equipment and end at the selected safe place. These systems can be various, they can be an outlet short pipe on a control valve or an independent PSV, or a pipe network connecting several valve outlets. If the gas is discharged from a pressure vessel under normal operating conditions (torch, venting), an oil separator is needed to remove the liquid hydrocarbons in the gas.

The discharge point of the gas venting system can be through a vertical, cantilevered, or underwater pipe to discharge the gas into the atmosphere. In some cases, the discharge point can be far from the station. When choosing a safe discharge point, the following issues should be considered:

(1) Personnel safety.
(2) Discharge volume.
(3) The relative position with other equipment, especially fire containers or other ignition sources, housing, fresh air intake systems, and helicopter and berthing facilities.
(4) Dominant wind direction, if underwater discharge, consider the dominant flow direction.
(5) Design considerations. The design of the atmospheric gas venting system should follow API Std 520 Sizing, selection, and installation of pressure-relieving devices, API Std 521 Pressure-relieving and depressuring systems, and API Std 2000 Venting atmospheric and low-pressure storage tanks and ASME BPVC Section VIII [4]. The system should be designed so that the back pressure generated at the maximum instantaneous flow rate, including inertial force, does not exceed the working pressure of the unit with the lowest rated pressure in the device. Fire arresters can be used in venting systems to reduce the risk of internal equipment ignition due to external fire sources. The flare gas purifier should be a pressure vessel, and its design should enable it to handle the maximum predicted flow rate.

References

1. API Standard 521 Pressure-relieving and depressuring systems, Seventh Edition, June 2020.
2. API RP 14E Recommended practice for design and installation of offshore production platform piping systems, Fifth Edition, March 2007.
3. API RP 14C Recommended practice for analysis, design, installation, and testing of basic surface safety systems for onshore production platforms.
4. API Std 520, Part I Sizing, Selection, and Installation of Pressure-Relieving Devices—Part I—Sizing and Selection.

Chapter 2
Safety Principles of Oilfield Surface Engineering Process Design

Process safety exists throughout the lifecycle of the device, including but not limited to design, commissioning, normal production, and shutdown maintenance stages. This chapter focuses on the principles that process professionals should follow during the design stage. By analyzing and studying the device, installing safety protection devices, and formulating related operating procedures, the safety of operators and the normal operation of production devices can be ensured. For other safety and health issues, such as ignition source control, device layout, gas detection systems, fire detection systems, sound and light alarm systems, forced ventilation systems, fire systems, noise control, and escape evacuation channels, etc., refer to relevant safety professional literature, which is not repeated here.

2.1 Equipment Design Temperature and Design Pressure

Reasonable equipment design temperature and design pressure are important means to ensure equipment safety and reduce the occurrence of leaks, which can prevent or reduce the possibility of dangerous events [1].

2.1.1 Design Pressure

The design pressure of the equipment should at least be the combination of the most unfavorable operating temperature and pressure anticipated during operation. It is used in design to determine the minimum allowable wall thickness of the equipment or to determine the physical characteristics of different parts of the equipment, i.e., the design pressure is the basis of equipment mechanical design. The design pressure should not be less than the maximum pressure encountered under normal

K. Fang, *Onshore Oilfield Surface Facilities: Process Safety Engineering Guide*,
https://doi.org/10.1007/978-981-96-7830-3_2

operation, maintenance, and accident conditions (excluding fire). In the case where the maximum allowable working pressure is not determined, the design pressure can be used to replace the maximum allowable working pressure [2].

The maximum allowable working pressure refers to the maximum allowable gauge pressure at the top of the vessel under operating conditions and design temperature. This pressure is determined by the calculation results of each part of the vessel when the nominal wall thickness is applied. It does not include the corrosion allowance and the additional metal wall thickness required for loads other than pressure. The maximum allowable working pressure is the pressure setting basis for the pressure relief device that protects the vessel.

Usually, the maximum allowable working pressure is the same as the design pressure. If the steel thickness exceeds the required thickness during manufacturing, the maximum allowable working pressure can be higher than the design pressure.

The operating pressure is the pressure during normal operation in the equipment and piping system. The maximum operating pressure (Maximum operating pressure, MOP) is generally taken as 1.05 times the operating pressure or the operating pressure plus 1 bar, aiming to ensure the flexibility of the control system. The maximum operating pressure should also consider conditions such as startup conditions and compressor shutdown stable pressure. It is particularly noteworthy that for LPG storage facilities, the maximum operating pressure is the saturated vapor pressure corresponding to the specific LPG component at the maximum operating temperature.

2.1.1.1 Pressure Vessel

The design of pressure vessels mainly follows the relevant ASME specifications. If the pressure vessel is not included in the pump or compressor system, its design pressure can be considered according to the principles in Table 2.1.

The above design pressure margin is mainly used to ensure the normal operation of the safety valve. The maximum operating pressure must consider the set points and accuracy of related pressure instruments. Before the safety valve operates, pressure alarms and pressure shutdowns must be normal and reliable. If the expected operating pressure will fluctuate within a reasonable range, the margin between the operating pressure, pressure alarm set point, pressure shutdown alarm point, and design pressure should be appropriately increased.

Table 2.1 Pressure vessel design pressure

Maximum operating pressure (gauge), bar	Design pressure (gauge), bar
$0 < MOP < 10$	MOP + 1.0 (Minimum 3.5)
$MOP > 10$	MOP × 110% (if considering the use of pilot-operated safety valves, 1.05 times MOP can be accepted)

In the process of vessel design, the design pressure at the bottom of the vessel must take into account the hydrostatic head and pressure drop.

2.1.1.2 Pump

For process pipelines and equipment connected to the pump, if they operate under full liquid conditions, their design pressure should consider the maximum outlet pressure of the pump. The design pressure of the pump suction line up to the first isolation valve is the same as the pump outlet design pressure. However, if only a single pump is installed, the overpressure problem of the inlet pipeline does not need to be considered.

1. Centrifugal Pump

Under normal circumstances, safety valves are not set at the outlet of the centrifugal pump. If the downstream system of the centrifugal pump operates with full liquid, its design pressure is taken from the larger value calculated by formulas (2.1) and (2.2):

$$p_d = p_{s,Op} + 0.098 F_{Shutoff} \cdot \Delta H \cdot \rho_{max} \cdot F_{API} \tag{2.1}$$

$$p_d = p_{s,Max} + 0.098 \Delta H \cdot \rho_{max} \cdot F_{API} \tag{2.2}$$

where

P_d—Pump outlet design pressure (gauge pressure), bar;

$p_{s,Op}$—Maximum operating pressure of the upstream tank of the pump + maximum static head (from high liquid level to pump centerline), bar;

$p_{s,Max}$—Relief pressure of the upstream tank of the pump + maximum static head (from high liquid level to pump centerline), bar;

ΔH—Head under rated flow, m;

ρ_{max}—Maximum liquid density, kg/m;

$F_{Shutoff}$—Shutoff factor, 1.25 for constant speed centrifugal pumps, 1.30 for variable speed centrifugal pumps;

F_{API}—Correction factor (API 610), when rated head ≤ 75 m, $F_{API} = 1.10$; when 75 m < rated head < 300 m, $F_{API} = 1.08$; when rated head ≥ 300 m, $FAPI = 1.05$.

2. Positive Displacement Pumps

Since reciprocating pumps and other positive displacement pumps continuously increase pressure, the outlet pipeline of the positive displacement pump needs to set a safety valve according to full discharge. If the downstream system of the positive displacement pump operates with full liquid, its design pressure is calculated according to formulas (2.3) and (2.4):

$$p_d = p_{Rated} + 1\text{bar}\,[\text{rated outlet pressure (gauge pressure)} < 10\text{ bar}] \tag{2.3}$$

$$p_d = p_{Rated} \times 1.1 \left[\text{rated outlet pressure (gauge pressure)} \geq 10\text{ bar}\right] \quad (2.4)$$

where

p_{Rated}—Rated outlet pressure, refers to the outlet pressure of the pump under rated flow.

Note the following two points:

(1) If there is an existing safety valve downstream of the positive displacement pump, the pump design pressure cannot be less than the set point of the existing safety valve.
(2) If multiple different types of positive displacement pumps are connected to the same outlet manifold, the design pressure takes the maximum design pressure of each pump.

2.1.1.3 Compressors

The compressor system is similar to the pump system. All compressor inlet separators must consider full vacuum conditions.

1. Centrifugal/Axial Compressors

The design pressure of the compressor inlet separator and cooler needs to consider the settling out pressure. For multi-stage compressors, it also needs to analyze whether the whole machine settling out pressure or single-stage settling out pressure should be used. The inlet section design pressure is 1.05 times the settling out pressure. The compressor outlet design pressure can consider 1.1 times the maximum operating pressure, and at least 1.5 bar higher than the maximum operating pressure. Without significantly increasing the cost, the compressor can be designed for full pressure to make the system simpler and safer.

2. Positive Displacement Compressors

The design pressure of the outlet section of the positive displacement compressor should be high enough to prevent the safety valve from frequently opening due to fluctuations in operating pressure. The setting of the outlet section safety valve set point can refer to Table 2.2.

Table 2.2 Recommended safety valve set point overview

Rated outlet pressure (gauge pressure) p_{Rated}, bar	Minimum margin of outlet section safety valve set point
$p_{Rated} \leq 10$	1 bar
$10 < p_{Rated} \leq 170$	10%
$170 < p_{Rated} \leq 240$	8%
$240 < p_{Rated} \leq 345$	6%

2.1.1.4 Shell and Tube Heat Exchangers

The design pressure of each part of the heat exchanger should consider the most extreme temperature and pressure combination. For the condition of heat exchange tube rupture, according to the 10/13 principle of API standard, if the design pressure of the low-pressure side is not less than 10/13 of the high-pressure side design pressure, the low-pressure side does not need to set a pressure relief device. The design pressure of the shell and tube heat exchanger is set according to the pressure vessel.

2.1.1.5 Atmospheric and Low-Pressure Storage Tanks

Storage tanks with a design pressure (gauge pressure) exceeding 1.03 bar are treated as pressure vessels, and the design standard follows ASME specifications [3, 4].

Atmospheric and low-pressure storage tanks usually refer to tanks designed and manufactured according to API standards.

1. API 650 Tanks

Under normal circumstances, the design of atmospheric tanks and low-pressure tanks should comply with API Std 650 Welded tanks for oil Storage. Atmospheric and low-pressure storage tanks are usually vertical cylinders with flat or conical tops, including factory-manufactured tanks and large on-site welded tanks, with a maximum internal pressure (gauge pressure) of 0.172 bar. Depending on the type of fluid stored, the tank can be left open or purged with nitrogen or fuel gas. The minimum design positive pressure (gauge pressure) for a purged tank should be 5 mbar to meet emergency venting requirements.

2. API 620 Tanks

API 620 tanks are commonly used in situations where the internal pressure of the tank is high or where the liquid must be kept under a certain pressure due to hazards, personnel safety, or product evaporation losses. They are usually vertical cylinders with conical or domed tops. The maximum internal pressure (gauge pressure) of an API 620 tank does not exceed 1.03 bar [5]. The recommended design pressures for API 650 tanks and API 620 tanks are shown in Table 2.3.

2.1.1.6 Flare System

API RP 14E Recommended practice for design and installation of offshore production platform piping systems recommends a flare system that includes a flare drum, flare manifold, and flare knockout drum, with a minimum design pressure (gauge pressure) of 3.5 bar. Total reports that the design pressure (gauge pressure) of the flare system is usually between 3.5 and 15 bar. The design pressure of the flare system should

Table 2.3 Atmospheric and low-pressure tank design pressure

Tank design standard	Operating pressure (gauge pressure), mbar	Design pressure (gauge pressure), mbar
API Std 650	≤4.2	7/−2.5
	≤12.3	20/−6.0
	≤30	50/−6.0
	≤104	172/−6.0
API Std 620	≤303	500/−6.0
	≤625	1030/−46

consider the maximum working pressure or maximum back pressure that may exist in the system.

2.1.1.7 Vacuum System

If there are no other protective measures (such as seal gas and breather valves), equipment that may face vacuum conditions must be designed for full vacuum. Equipment designed for full vacuum should be able to withstand a pressure of at least 3.5 bar (gauge pressure).

Vacuum conditions need to be analyzed specifically for each problem, mainly the following operations can lead to vacuum:

(1) Start-up, shutdown, and regeneration blowdown, etc.
(2) Equipment filled with liquid is isolated and cooled.
(3) Equipment filled with condensable gas is isolated and cooled, such as steam blowdown. The steam blowdown condition can consider the maximum operating temperature of steam as 150 °C. It should be noted that full vacuum design may require the setting of reinforcement rings and vacuum testing, leading to increased investment.

2.1.2 Design Temperature

The design temperature is the temperature used for equipment mechanical design, including the maximum design temperature and the minimum design temperature. The setting of the maximum design temperature is shown in Table 2.4.

Table 2.4 Maximum design temperature

Operating temperature TO, °C	Design temperature, °C
$-39 \le TO \le -30$	85/−45
$-30 < TO \le 60$	85/Lowest operating temperature
$60 < TO \le 343$	Maximum operating temperature +25
$TO > 343$	Specific analysis based on the selected material and process requirements

The setting of the maximum design temperature also needs to consider the following points:

(1) Maximum ambient temperature. The design temperature cannot be lower than the blackbody temperature produced by solar radiation.
(2) Failure of the air cooler. The maximum operating temperature downstream of the air cooler is the maximum operating temperature at the inlet of the air cooler minus 20% of the temperature difference under normal conditions.
(3) Failure of the refrigerant. In shell and tube heat exchangers, the design temperature of the hot fluid outlet should be at least equal to the hot fluid inlet temperature, and the design temperature on the cold fluid side should not be less than the maximum operating temperature of the hot fluid.
(4) The design temperature does not consider the temperature caused by fire conditions.
(5) The design temperature of the underground sewage tank should be below 100 °C to meet the stress requirements.
(6) The maximum design temperature of the flare knockout drum needs to consider the hottest flow.

The minimum design temperature should usually be taken as the minimum of the following three temperatures:

(1) Minimum operating temperature −5 °C.
(2) Minimum ambient temperature −5 °C.
(3) The lowest temperature caused by depressurization.

Normally, the low temperature produced by the depressurization of liquefied gas or small relative molecular mass compressed gas is much lower than the operating temperature.

2.2 ESD Emergency Shutdown Principles

The ESD system is used for the safe and reliable emergency shutdown of process systems and their related equipment, to minimize the risks caused by process abnormalities or external events, prevent equipment damage or personal injury. If the consequences of an accident are minor, the ESD system can also prevent equipment shutdown to protect the operation of the device as much as possible. Emergency shutdowns can be activated automatically or by operators.

2.2.1 ESD System Objectives

The ESD system aims to achieve the following objectives:

(1) Personnel protection.
(2) Equipment protection.
(3) Environmental protection.
(4) Minimize production losses of the facility.
(5) Continuity of production (by minimizing unnecessary downtime).

The ESD system design mainly includes the following:

(1) Set up manual ESD-0, ESD-1, and ESD-2 buttons to initiate a full-field shutdown, process column/device area shutdown, unit/single equipment/sled shutdown.
(2) Automatically detect abnormal external conditions, operations, or equipment status.
(3) Isolate the factory/system/equipment/pipeline/pipe, to minimize ignition sources.
(4) Set up manual pressure relief facilities to reduce the dangerous consequences caused by abnormal conditions.
(5) Interact with the process control system to alert operators through audio and visual alarms.
(6) Interact with the public broadcast system to remind operators to intervene.

2.2.2 ESD System Independence

All ESD system equipment, such as field sensors and actuators and their related circuits, should be exclusive to the ESD system and operate independently of other monitoring control and alarm systems. The ESD system should be completely independent and not rely on any other system's communication connections or interfaces. The ESD system will independently perform the following operations:

(1) Perceive abnormal operations or equipment conditions.
(2) Shut down and/or isolate equipment as needed, and vent as appropriate.
(3) Provide manual isolation and/or venting facilities.
(4) Interact with other systems and initiate related shutdowns, such as signals from the fire detection system will cause the ESD system to execute a shutdown.

2.2.3 ESD System Startup

The ESD system can be started for shutdown by manual buttons or automatic systems. Operators should be able to conveniently access the ESD button at a suitable location in the control room or on the human–machine interface.

The ESD button should be equipped with a protective cover, shield, or similar parts to reduce the risk of accidental startup.

2.2.4 ESD System Operation

Even if the moving equipment is in local operation mode, the ESD system should be able to stop it or trip it.

For sled-mounted equipment with their own control panels such as gas compressors, torch systems, and thermal media systems, the ESD remote stop/trip signal will be sent to its local control panel (LCP) or unit control panel (UCP), and the LCP or UCP will implement the stop/trip operation. The sled-mounted equipment control panel should upload any shutdown actions to the DCS for recording.

2.2.5 ESD System Reset

Once the ESD system puts the process in a safe state, it will remain in a safe state until the ESD system is reset. The ESD system cannot be reset before the cause of the ESD system startup is clear.

The ESD system has two reset methods, namely group soft start and field manual reset. For key valves in the production system such as boundary valves, inlet/outlet pipeline valves, and emergency relief valves, they need to be manually reset one by one on site. Once the field manual reset is completed and the ESD system is running well, the affected valves can be reopened in a controlled sequence through the human–machine interface. Field reset must be hard-wired to the ESD system. For non-critical emergency shutdown valves belonging to the same process column or unit, a group soft start can be used within the process column or unit for convenience. Lower-level ESD logic must be reset before higher-level reset.

2.2.6 Emergency Shutdown Valve (ESDV)

The ESDV has the following characteristics:

(1) On/off type.
(2) Bubble level seal.

(3) Valves located in the fire area have fault protection functions.
(4) Actuator fireproof (only applicable to valves located in the fire area).
(5) Equipped with on/off limit switches.
(6) Equipped with detection equipment.

Shutdown valves are generally arranged at the boundaries of different systems and areas with different pressure levels, and factors such as fire area layout and equipment pressure relief requirements also need to be considered.

2.2.7 Startup/Maintenance Override

The design of startup override for process and utility systems should minimize the impact on safety systems. Process override switches help Devices or equipment that are in an abnormal state (such as low fluid level) before starting the car. The process override switch cannot affect the alarm function of the process control system, and the process system will be in an alarm state before the process parameters reach a reasonable position. The maintenance override switch will bypass the trip starter to achieve sensor maintenance or online function testing. The maintenance override switch comes with an alarm function and is only applicable to input signals and must not bypass the startup permission.

2.2.8 Cascade Effect

The cascade effect refers to the shutdown of a device leading to process abnormalities and then causing other devices/systems to stop immediately. The design should try to avoid process cascade emergency shutdowns to prevent the escalation of abnormal situations due to the intervention of the ESD system. If it is judged that this situation cannot be avoided, then the ESD system should actively shut down all affected devices.

2.2.9 ESD Levels and Hierarchies

When abnormal operations or equipment conditions occur, protective measures should be suitable for the severity of the situation. In the emergency shutdown system, events that need to be urgently shut down are divided into several levels according to their causes, consequences, and impacts. The highest level is to evacuate after a full-field shutdown and depressurization, and the lowest level is to automatically

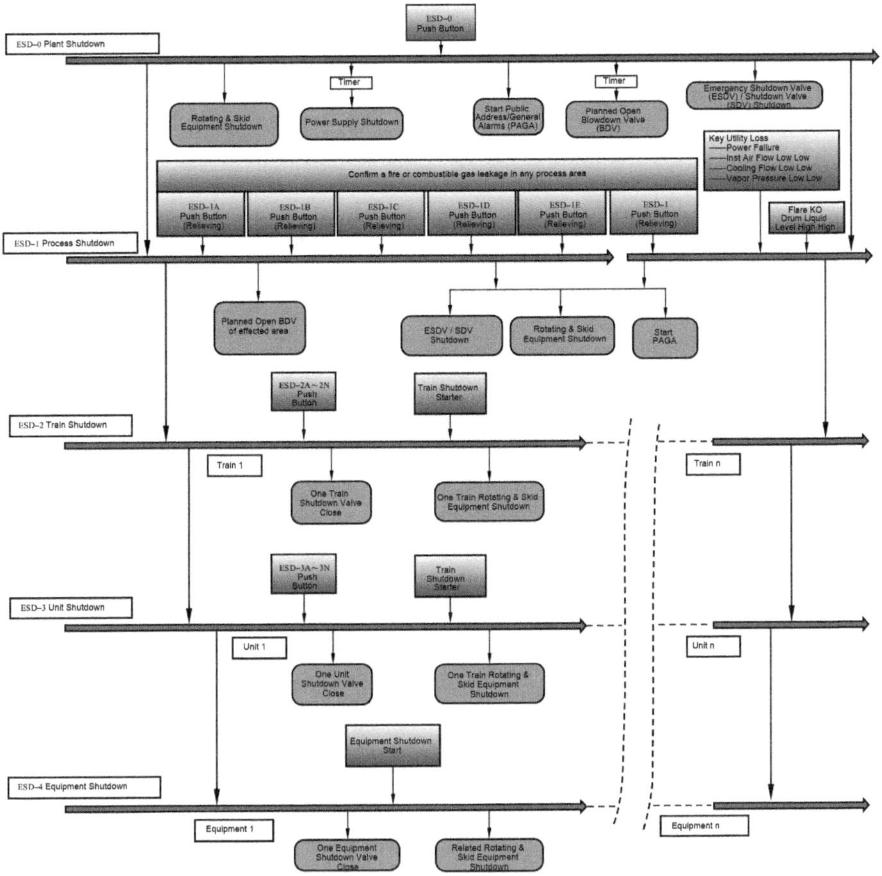

Fig. 2.1 ESD hierarchy diagram

shut down a single device when it exceeds the operating limit. Generally, to protect the health and safety of personnel, the environment, and assets, while ensuring the normal operation of the device as much as possible, five ESD safety protection levels will be set: Level 0, Level 1, Level 2, Level 3, and Level 4 (Fig. 2.1).

Detailed information about ESD levels is generally reflected in the cause and effect diagram. The main causes, consequences, and brief descriptions of ESD actions are as follows:

2.2.9.1 ESD-0

ESD-0 represents the highest level of device protection, which can also be defined as a global emergency shutdown. The entire factory will be shut down, depressurized, and abandoned:

(1) Except for tank seal gas and torch purge gas, other facilities are shut down. The seal gas system is kept running normally to prevent air from entering the tank.
(2) Start all safety shut-off valves.
(3) Shut down all moving equipment except the instrument air system and torch system.
(4) Shut down all fire equipment.
(5) Depressurize the gas system. If any of the situations such as war, earthquake, air raid, flood, and terrorist attack occur, it can only be manually activated

ESD-0 level shutdown. The ESD-0 system will initiate corresponding sound and light alarms and public broadcast alarms.

2.2.9.2 ESD-1

ESD-1 is the second highest level of ESD, which can also be defined as process emergency shutdown. The process emergency shutdown will shut down the entire factory, and the public systems (instrument air system, nitrogen system, fire system, emergency power supply, and diesel system) and living systems (fresh water system) will operate normally. ESD-1 shutdown can also be a cascade shutdown of ESD-0.

The following situations will trigger ESD-1 shutdown:

(1) Automatic start: Cascade shutdown of ESD-0, low instrument air pressure, high liquid level of torch separator tank, power station shutdown, etc.
(2) Manual start: Confirm fire, fire gas detection alarm, confirm process control system failure, etc.

The ESD-1 system will initiate corresponding sound and light alarms and public broadcast alarms.

2.2.9.3 ESD-2

ESD-2 shutdown corresponds to the processing column level and may be a cascade shutdown of ESD-1.

Automatic start: Cascade shutdown of ESD-1, high processing column inlet pressure, high processing column inlet liquid level, etc.

The ESD-2 system will initiate corresponding sound and light alarms and public broadcast alarms, and trigger ESD-3 shutdown.

2.2.9.4 ESD-3

ESD-3 shutdown corresponds to the processing unit level and may be a cascade shutdown of ESD-2.

Automatic start: Cascade shutdown of ESD-2, high processing unit inlet pressure, high processing unit inlet liquid level, etc.

The ESD-3 system will initiate corresponding sound and light alarms and public broadcast alarms, and trigger ESD-4 shutdown.

2.2.9.5 ESD-4

ESD-4 level is the lowest ESD level, only shutting down a certain device. It can be automatically triggered by the device protection system, or manually started by the operator through a button.

Factors that can trigger ESD-4 shutdown include: low pump suction pressure, vibration protection, overload protection, etc.

The ESD-4 system will initiate corresponding audio and visual alarms and public broadcast alarms.

2.3 Principles of Venting and Draining

During the stages of project startup, shutdown maintenance, detection, and modification, it is necessary to vent and drain the equipment and pipelines in order to safely carry out the next operation. The design of the venting and draining collection system should consider the following:

(1) Venting and draining must meet the project's HSE requirements.
(2) Before venting and draining, depending on the characteristics of the material, the material inside the equipment, pipelines, and instruments should be transferred as much as possible to other process systems, vented to the torch or other suitable locations, reducing the material inventory inside the equipment, pipelines, and instruments to the lowest safe operation level, and then carry out the draining operation and recover the process material.

2.3.1 Pressure Vessel Venting and Draining

All pressure vessels must have at least one vent and one drain to discharge the process material to a safe location for maintenance and testing.

The vent should be set at the highest point of the container or its connected pipeline; the drain should be set at the lowest point of the container or its connected pipeline and be able to drain completely.

For recommendations on the size of the venting and draining interfaces of the container, see Table 2.5.

If the container is horizontally arranged and the seam-to-seam length is greater than 6 m, additional drain interfaces should be considered.

Table 2.5 Recommended sizes for equipment venting and draining interfaces

Equipment volume V m^3	Vent interface in	Drain interfacing
$V \leq 1.5$	1	1
$1.5 < V \leq 6$	1	1 1/2
$6 < V \leq 17$	1	2
$17 < V \leq 71$	1 1/2	3
$V > 71$	2	>3

Table 2.6 Recommended sizes for pipeline venting and draining interfaces

Pipeline size D in	Vent interface in	Drain interface in
$D \leq 4$	3/4	3/4
$6 < D \leq 16$	3/4	1
$D > 18$	1	2

2.3.2 *Pipeline Venting and Draining*

The pipeline system needs to have a sufficient number of venting and draining points to ensure effective venting and draining of the pipeline system. When piping, try to avoid liquid bags at the high points of liquid pipelines and gas bags at the low points of gas pipelines. On the process PID, the positions, sizes, and types of venting and draining points required for process operation should be marked. Venting and draining points dedicated to hydrostatic testing are not required on the PID. For recommendations on the size of the pipeline venting and draining interfaces, see Table 2.6.

When selecting the size of the drain interface, consider the viscosity and density of the material. For high viscosity and high pour point fluids, the size of the drain interface can be appropriately increased. To achieve a reasonable flow rate, the size of the open and closed drain pipe is generally at least 6 in. The slope of the drain pipeline is determined by the project. Generally, a 2 in blowout port is set at the end of the open and closed drain pipe and its sub-drain pipe.

2.3.3 *Drainage System*

To collect and treat liquids during normal operation, abnormal operation, maintenance stages, and emergency conditions, appropriate drainage facilities need to be set up according to the location of the process device and the nature of the material. The setting of the drainage system can refer to Table 2.7.

Table 2.7 Drainage classification and piping network

Drainage classification	Typical source	Drainage piping network
Non-degassed hydrocarbons	Process facilities under certain operating pressure: separators, towers, compressors, pigging barrels, and station manifolds, etc.	Closed drain system
Oil-containing wastewater	(1) Water dripping from pumps, compressors, containers, and other process equipment during daily operation (2) Wastewater from non-sealed sampling points	Open drain system
Occasional oil-containing wastewater	(1) Wastewater produced by non-hydrocarbon treatment devices: water contaminated with lubricating oil from power stations, water injection pumps, and other equipment (2) Oil-containing wastewater from the containment of crude oil or condensate storage	
Non-oil wastewater	(1) Uncontaminated rainwater from the containment of crude oil or condensate storage facilities (2) Public system wastewater: wastewater from water supply tanks, fire water tanks, or other water treatment facility	Clean water drainage system
Non-oil water	Green belts, roads, clean areas, etc.	
Domestic wastewater	Sanitary facilities, kitchens, residential areas, etc.	Domestic wastewater network
Chemical reagents	Laboratories, dosing pumps, chemical storage tanks, etc.	Specific piping or open drainage systems

Closed drainage systems and open drainage systems need to be separated to prevent pressurized gases from converging to process equipment through the open drainage system. In the open and closed drainage systems, the sewage pipeline should maintain a certain slope to allow the fluid to flow into the sewage tank by gravity. When considering the number of sewage tanks, the overall plan should be taken into account to ensure a reasonable depth of the sewage tanks.

2.3.4 *Venting System*

If the following conditions are met, the gas can be directly discharged into the atmosphere:

(1) Direct discharge into the atmosphere does not violate local pollution and noise-related laws and regulations.
(2) The gas is not toxic or corrosive.
(3) When flammable gas is directly vented, it should be ensured that the vented gas has been diluted to below the lower explosion limit before it comes into contact with the ignition source. When direct venting is not allowed or not feasible, the gas should be discharged into the flare system, for related design details, see Chap. 8.

2.4 Isolation Principles

The isolation principles discussed in this book only apply to the mechanical isolation of fluid systems, not to electrical equipment. The following functions can be fachieved through mechanical isolation:

(1) Ensure safe isolation before entering the container or performing maintenance work.
(2) Isolate specific process equipment from other operating equipment.
(3) Isolate the process equipment from the sewage system.
(4) Isolate the process equipment from the venting system.
(5) Isolate the process equipment from the public system.

2.4.1 *General Requirements*

All systems containing hazardous media or potential hazards need to take isolation measures. Hazardous media often include the following categories:

(1) Fluids above the autoignition point.
(2) Flammable liquids that may flash to form a vapor cloud when leaked.
(3) Flammable gases that may form a vapor cloud when leaked.
(4) Fluids that are prone to forming hydrates or solid deposits, causing blockages, and thus leading to hazards.
(5) Liquids with a flash point below 60 °C (including methanol).
(6) Toxic substances that may cause serious irreversible damage unless rapid repair measures are taken, such as high-concentration hydrogen sulfide gas, etc.
(7) Hot medium fluids above 100 °C or cold medium fluids below −10 °C.
(8) Fluids with an operating pressure (gauge pressure) greater than 20 bar.

Table 2.8 Minimum isolation requirements for equipment maintenance

Isolation method	Long-term isolation	Entering container	Hot work	Non-hot heat work	Cold processing
Duration	>24 h			<24 h	<24 h
Active isolation	√	√	√	√	√
Baffle	√	√	√	√	√
Double isolation bleed valve group	×	×	×	√	√
Single isolation bleed valve group	×	×	×	×	√
Single isolation valve	×	×	×	×	√①

① Only applicable to non-toxic and harmless systems with a pressure rating not greater than 150# and a temperature less than 150 °C

The following principles are generally followed when isolating:

(1) Before entering the container, active isolation or the installation of a baffle should be carried out at all nozzle flanges as close to the container as possible. Before personnel enter, the equipment should be depressurized, drained, and purged according to the prescribed procedures.
(2) Hot work or flange disassembly should not be carried out before the active isolation equipment (baffle or figure-eight blind plate) is moved to the closed position.
(3) Before any work is carried out on high-risk hydrocarbon systems (high pressure, high temperature, large inventory), it is recommended to carry out active isolation, install baffles, or double isolation bleed valve groups (Double block and bleed, DBB) valves.

The minimum isolation requirements during equipment maintenance can refer to Table 2.8. Due to different isolation application scenarios, the content to be considered also varies.

2.4.1.1 Shutdown Maintenance Isolation

Shutdown maintenance means that the entire plant, unit, or system has been isolated and has been vented, drained, and purged. The system is at atmospheric pressure and does not contain hydrocarbons or other substances. When shutting down a unit, all feed, product, and utility pipelines should be safely isolated at the boundary.

2.4.1.2 Skid-Mounted Equipment Isolation

All inlet and outlet pipelines of skid-mounted equipment need to consider isolation, and the isolation method needs to be confirmed during the detailed design stage.

2.4.1.3 Personnel Protection Isolation

Personnel are only allowed to enter confined spaces after active isolation measures such as installing blind flanges or blind plates at all interfaces (except for atmospheric ventilation interfaces).

When entering individual equipment, note:

(1) All containers, tanks, and confined spaces should be actively isolated on all nozzles by means of blind plates or removable pipe sections. The isolation location should be as close as possible to the container, usually on the nozzle.
(2) Isolate the main feed pipelines and product pipelines of the equipment.
(3) There should be a ventilation hole on the isolated side of the equipment.
(4) If liquid is drawn from the container, isolation can be performed at the suction port of the pump.

2.4.1.4 Special Operation Isolation

An example of a special operation requiring active isolation is the loading and unloading of tower packing. This operation may be performed periodically, and permanent active isolation facilities are usually installed.

2.4.1.5 Pressure Test Isolation

Pressure test blind plates are not usually installed. However, if equipment such as heat exchangers needs to be repaired and tested online, isolation blind plates should be installed for leak testing and pressure testing. Valves sometimes also need to be tested for airtightness, but usually only temporary blind plates are installed instead of permanent blind plates or valves.

2.4.1.6 Isolation of Bypass Pipelines during Operation

During normal operation, the isolation of bypass pipelines during operation should consider the following factors:

(1) During normal operation or in emergencies, any pipeline is usually isolated by closing the valve. After isolation, the residual liquid in the pipeline should be left in place.

(2) If the liquid has a high pour point, considerations should be made for pipeline flushing, slope, and gravity drainage during design.
(3) If the pipeline is only used for circulation and flushing before startup, the pipeline should be depressurized and isolated by blind cutting during normal operation.
(4) When leaks from the startup pipeline or bypass pipeline can cause serious product contamination, a blind cutting device should be installed.
(5) Temporary pipelines that are no longer in use should be removed.

2.4.1.7 Single Column Isolation in Multi-column Units

When there are no less than two columns in a unit, consider shutting down and isolating one column, while the remaining column units continue to operate.

2.4.1.8 Isolation of Multi-stage Commissioning Units

If the project is commissioned in multiple stages, consider shutting down and isolating some early commissioning equipment, while keeping the remaining stage facilities running normally. Isolation measures that can be taken include blind cutting isolation devices or removable pipe sections.

2.4.1.9 Online Maintenance Isolation

Under normal circumstances, devices that can be repaired without stopping production if there are backups mainly include electro-desalting, pumps, compressors, heat exchangers, and control valves, etc.

2.4.2 Detailed Isolation Principles

Device isolation methods are roughly divided into active isolation and valve isolation. The process safety requirements for isolation and the inherent integrity and isolation speed of the isolation measures determine which isolation measure to take.

(1) Active Isolation: Ensure 100% physical isolation of personnel or products from hazards or pollution sources, and no leakage is allowed. It is suitable for situations where personnel enter containers or to prevent minor leaks from causing pollution.
(2) Valve Isolation: Valve isolation is suitable for situations where the isolation requirements are not as strict as active isolation, such as devices that need to be isolated for a short period.

2.4.2.1 Active Isolation

Active isolation is suitable for the following situations:

(1) The main equipment to be isolated can be tested to ensure that it does not contain gas and the workplace is safe. Several devices in the interconnected pipeline that do not contain stop valves can be considered as one device.
(2) Isolate part of the device for major repairs.
(3) Isolate the common system source of a device, such as the gas and fuel of a fire device, etc.
(4) To prevent pollution, such as preventing inert gas, fuel gas, or instrument air from entering the process system and causing pollution during normal operation.

The following are several ways of active isolation:

(1) Remove the pipe section—disassemble part of the pipeline, install blind flanges at both ends of the removed pipeline, and the flange design pressure is the same as the pipeline design pressure.
(2) Insert plate isolation—according to the pipeline system specifications, insert a rated pressure insert plate or rotate the figure-eight blind plate. The selection of the isolation device depends on the pipeline size and flange level, see Table 2.9. Figure 2.2 shows several typical active isolation combinations.

Table 2.9 Isolation device selection overview table

Pipeline Size	Pressure Level ①, ②					
	150#	300#	600#	900#	1500#	2500#
25						
50			slip blind flange			
80						
100						
150						
200						
250						
300						
350				Ring joint, spade or removable pipe section		
>350						

① The selection of isolation equipment mainly considers the use of ring joints or spades when the weight of the figure-eight blind plate exceeds 25 kg
② The process specialty should refer to this table to display the figure-eight blind plate, ring joint or spade on the P&ID drawing, to indicate the need for active isolation. The piping specialty will consider various constraints (such as weight and the operational flexibility of using the figure-eight blind plate, etc.) and determine whether the removable pipe section is more suitable

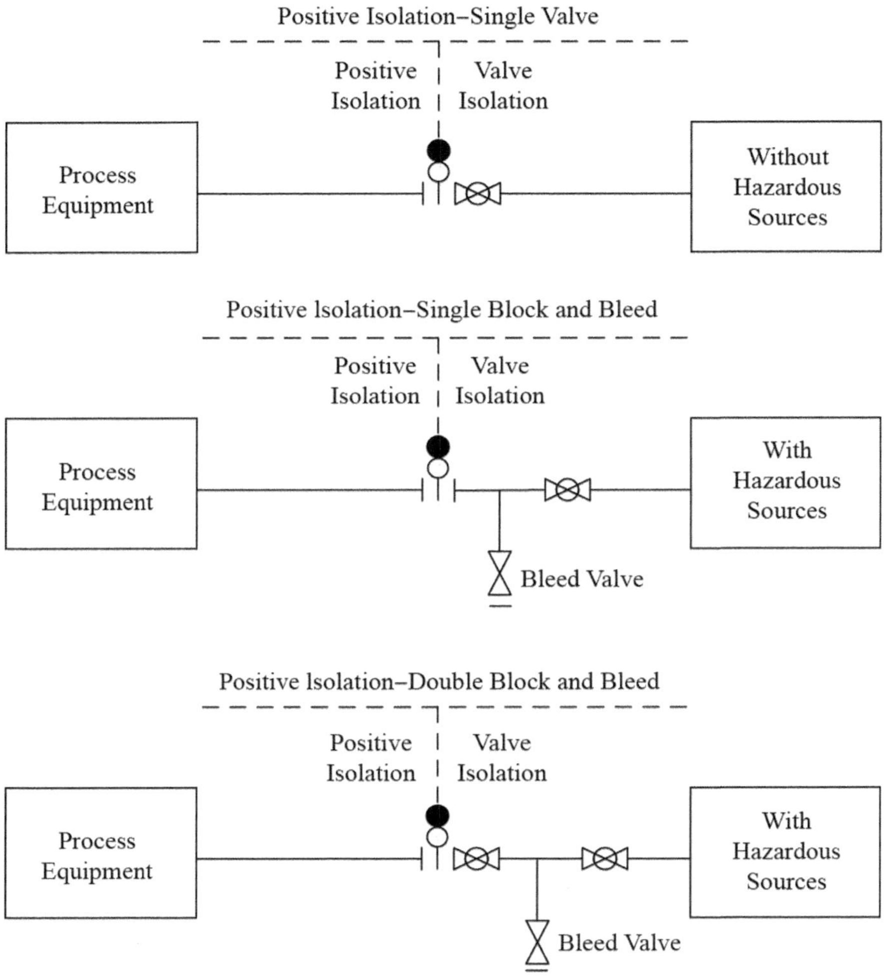

Fig. 2.2 Typical active isolation device combination

2.4.2.2 Valve Isolation

1. Valve Isolation Methods

Valve isolation has the following methods:

(1) Double block and bleed (DBB).
(2) Single block and bleed (SBB).
(3) Single valve isolation (SVI).

Before performing active isolation, valve isolation is usually used to maintain the seal. If used for maintenance or inspection work, a valve integrity check should be performed to ensure that no gas or fluid passes through after the valve is closed. Valve isolation should be achieved using a reliable manually operated valve, which should have airtight closure function. Pressure safety valves and check valves should not be used for isolation. Valves driven by actuators cannot be used as the only isolation valve, but can be used as the downstream isolation valve for the DBB valve group.

(1) Double Block and Bleed Valve Group

Several typical double block and bleed valve groups are shown in Figs. 2.3, 2.4 and 2.5. The vent valve ensures that there is no high pressure difference on the second isolation valve, thus ensuring the isolation integrity of the valve group. The vent valve discharges on the spot, and its operation can be monitored to confirm whether the isolation valve is working.

The DBB valve group shown in Fig. 2.3 consists of two independent isolation valves connected by a pipe section. The vent valve is installed on the pipe section

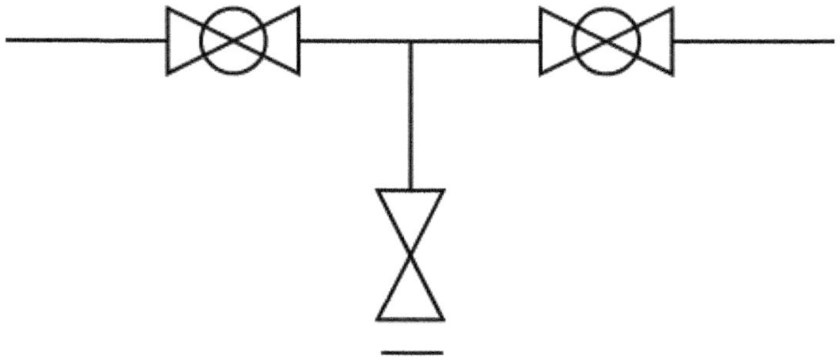

Fig. 2.3 DBB type 1: two independent isolation valves + vent valve

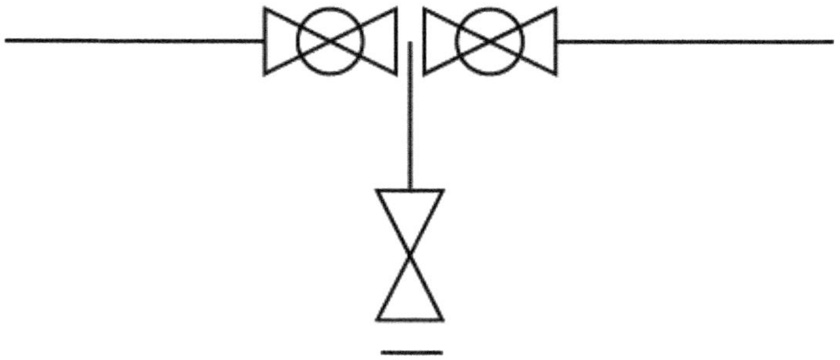

Fig. 2.4 DBB type 2: two integrated isolation valves + vent valve

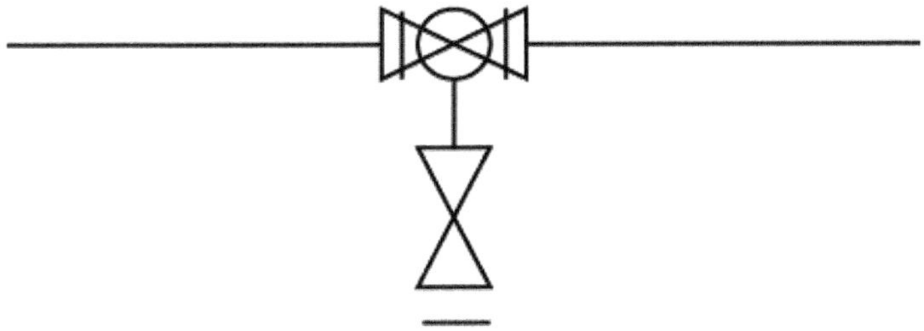

Fig. 2.5 DBB type 3: integrated DBB valve

between the two isolation valves. The pipe section between the isolation valves should be as short as possible to reduce emissions. The operating space for operating the vent valve should be considered in the design.

The second DBB valve group consists of two directly connected isolation valves and a vent valve (Fig. 2.4).

The third DBB valve is a double-seat valve, each valve forms a seal against the fluid or pressure, and the fluid between the two valve faces is discharged through the discharge port of the valve body (Fig. 2.5).

This integrated double-seat valve with a vent port is not a true double block and bleed valve group, a single failure may cause both isolations to fail at the same time, so this type of valve cannot be used for the isolation of process pipelines and equipment.

The vent valve of the DBB valve group is generally a 3/4 in or 1 in valve with a blind flange. If the fluid is acidic or the cut-off valve exceeds 20 in, the size of the vent valve should not be less than 1 in.

(2) Single Block and Bleed Valve Group

The single block and bleed valve group is shown in Fig. 2.6. The vent valve is placed on the equipment side of the isolation valve for leak testing before disconnecting the flange. If a leak occurs, additional shutdown measures will be needed to isolate the system. The vent valve should be discharged on the spot, which is conducive to monitoring it to judge whether the isolation valve is effective. The single block and bleed valve group can be used to isolate systems with a rated pressure pound level below 600#.

(3) Single Valve Isolation

Single valve isolation uses a single isolation valve for isolation, and the isolation integrity is lower. Single valve isolation is a standard practice for isolating systems, suitable for situations with no special sealing requirements, low risk, and non-entry operations (Fig. 2.7).

The above-mentioned valve isolation methods can be summarized in Table 2.10.

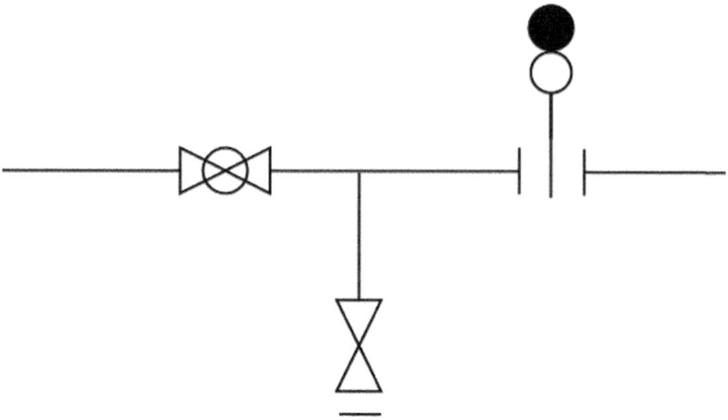

Fig. 2.6 Single block and bleed valve group

Fig. 2.7 Single valve isolation

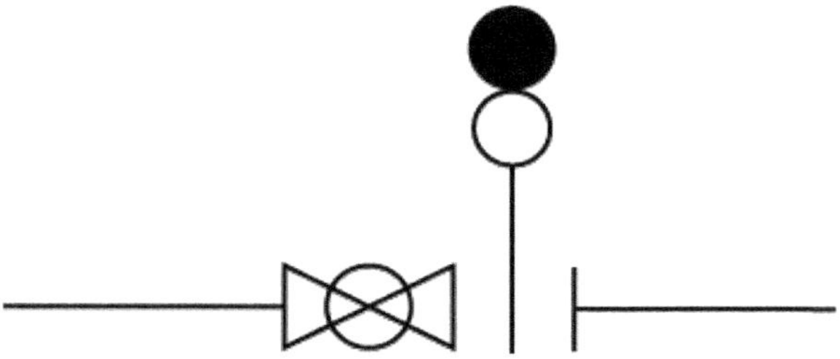

2. Applications of Valve Isolation

(1) In petroleum and natural gas installations, the following situations require the use of Double Block and Bleed (DBB) valve assemblies to achieve active isolation:

① All pipelines of ANSI Class 600# and above.

② All manifolds in tank areas where product contamination may occur.

③ Isolation of high-temperature fluids (temperatures not lower than the ignition point) systems.

④ Isolation between metering systems and verification systems.

⑤ If the equipment may be discontinued during the operation of the unit, or if it is necessary to isolate the equipment but still make it available for normal operation of the unit at any time.

(2) For other situations not mentioned above, Single Block and Bleed (SBB) and Single Valve Isolation (SVI) can be used to achieve the required isolation.

Table 2.10 Common valve isolation overview

Type	Description	Diagram	Remarks
1	Single valve isolation	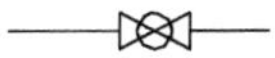	Non-active isolation
1B	Single block and bleed valve	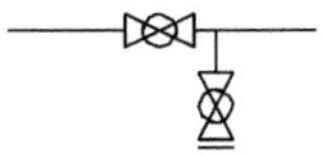	
2	Double block and bleed valve	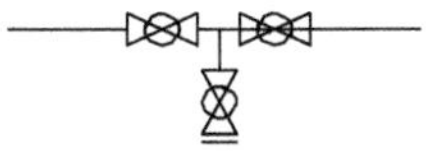	
3	Double block and bleed valve (connect to the vent and drain system)	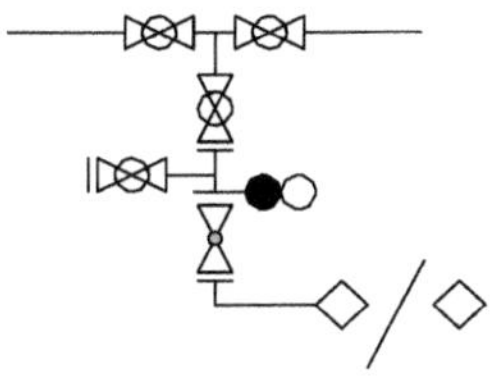	
4	Single block and bleed valve group + eight character blind flange/insert plate	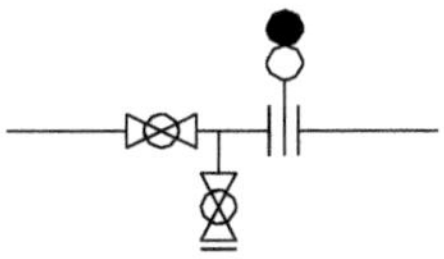	Proactive isolation
4B	Double block and bleed valve group + removable short section	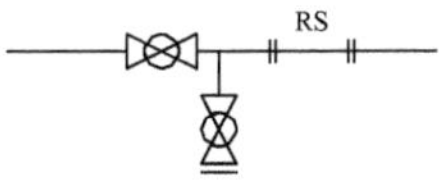	
5	Double block and bleed valve group double block and bleed valve group	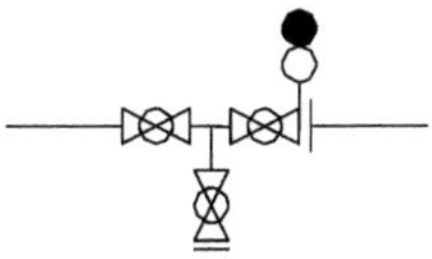	
5B	Double block and bleed valve group + removable short section	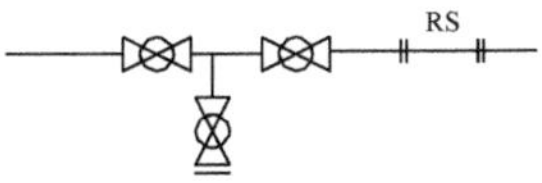	

Table 2.11 defines the selection principles for SVI/SBB and DBB.

3. Locking and Unlocking of Valves

If a valve is operated incorrectly, causing equipment or device damage or unsafe conditions, the isolation valve should be locked in the appropriate position "Locked Open" (LO) or "Locked Closed" (LC). If the valves must be operated in a specific order, a mechanical interlock system must be used to ensure the correct operation sequence.

Table 2.11 SVI/SBB and DBB selection principles

Fluid type	Valve purpose①	Operating pressure (gauge), bar		
		≤150#	150#–300#	≥600#
Process fluid; hazardous public system medium; hazardous chemicals; toxic medium②	V	SVI/SBB	SVI/SBB	DBB
	I	SVI/SBB	DBB	DBB
Harmless public system medium; harmless chemicals③	V	SVI/SBB	SVI/SBB	SVI/SBB
	I	SVI/SBB	SVI/SBB	SVI/SBB

① V refers to valves that need to achieve active isolation, I refers to valves that do not need to achieve active isolation
② The isolation requirements for hazardous chemicals (such as concentrated acids/highly corrosive chemicals) should be evaluated based on specific situations (such as using DBB to protect personnel from physical contact with alkali solutions)
③ To protect personnel from any toxic substances, maintenance/operating personnel should always wear a respirator when installing active isolation devices or when approaching equipment and instrument pipelines

Generally, the use of locks on valves should be minimized. Only when incorrect operation of the isolation valve could cause equipment damage or unsafe conditions, should the isolation valve be locked in the open or closed position. This is to keep the use of locks at the lowest reasonable level to avoid these devices not achieving the locking purpose due to frequent operation.

The following lists the applications where valves need to be locked or sealed, and the types of locking devices. Locking devices will be used for:

(1) PSV isolation valves (for multiple PSV key interlock systems).
(2) Manual valves of flare vent lines.
(3) Main isolation valves in fire water headers and foam systems.
(4) Valves that must be opened to ensure a pressure relief path (for example, valves that must be opened to allow for thermal expansion).
(5) Isolation valves downstream of Blowdown valves (BDV).

4. Types of Valves Used for Isolation

The use of ball valves for isolation is recommended in the following situations:

(1) Safety valve isolation.
(2) Blowdown valve (BDV) isolation.
(3) Pipelines that need to be cleaned.

Isolation valves for other situations can refer to the project pipeline material regulations. For operational or maintenance purposes, isolation may sometimes be required between different units. Only active isolation measures can be used in this case, and a local pressure gauge should also be installed to display the pressure near the blind or spacer plate.

2.4.2.3 Equipment Isolation Methods

1. Pumps

For pumps installed in parallel, isolation devices can be installed at the inlet and outlet valves, as shown in Fig. 2.8. The cut-off valve on the pump suction and return pipeline connected to the container within 10 m in the horizontal direction from the container branch does not need a cut-off valve, unless there is a special requirement to isolate the container inventory. For the suction or return pipeline from the container to the pump, if there is no special requirement for container isolation and there is

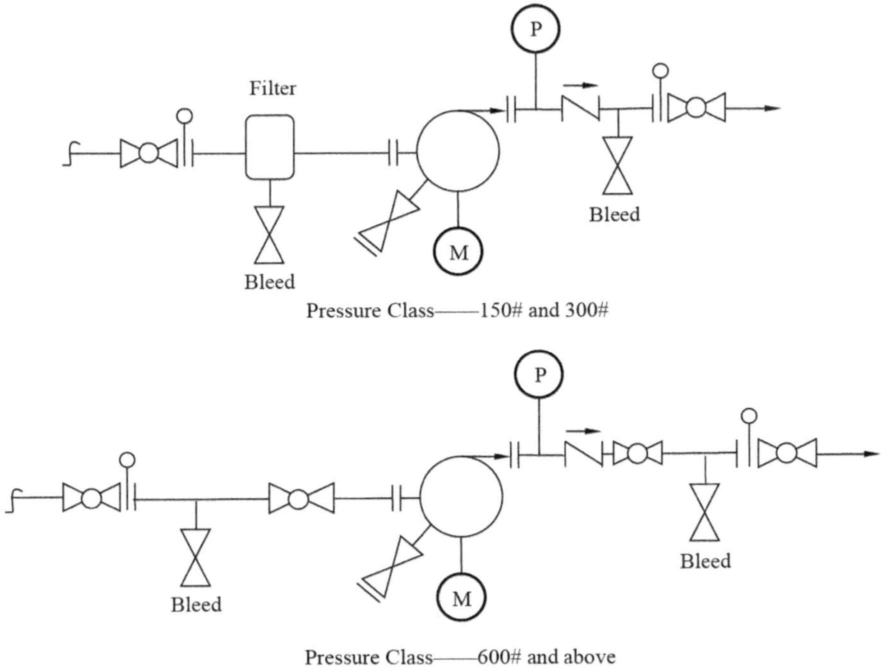

Fig. 2.8 Pump isolation setting

an isolation valve within 10 m in the horizontal direction, there is no need to install another isolation valve.

2. Heat Exchanger

All heat exchangers should be equipped with isolation devices. Isolation valves are usually not installed on air coolers, as they may lead to isolation of the pressure relief valve. If the fluid in the air cooler is corrosive and prone to scaling, and online maintenance of individual tube bundles is considered, valves can be used. In most cases, a permanent figure-eight blind plate or ring can be installed at the pipe mouth.

Plate and frame heat exchangers usually have the same isolation requirements as shell and tube heat exchangers. Venting points, drainage points, and blowdown points should be provided in the isolation valve of each plate and frame heat exchanger for safe depressurization, drainage, and blowdown before maintenance.

For shell and tube heat exchangers, removable pipe sections should be installed on the tube side for easy tube bundle extraction. If the tower itself already has sufficient isolation devices, there is no need to install isolation devices on the tower kettle reboiler.

If the heat exchanger needs online maintenance, the isolation measures shown in Fig. 2.9 should be adopted.

3. Pig Receiver and Launcher

The isolation of the pig receiver and launcher is very critical for safe pigging operations. The principle of active isolation applies to the pig receiver and launcher. After the pig receiver and launcher are depressurized and isolated, they need to be flushed and blown down, and then the quick-opening blind plate is opened. To ensure safe operation and prevent the pig receiver and launcher from accidentally opening the quick-opening blind plate under pressure, a system interlock device should be installed.

4. Compressor

Ring spacers should be installed on the suction and discharge valves of the compressor, but they should not be set on the suction port of the air compressor that sucks air from the atmosphere.

5. Control Valve

The type of isolation applied to the control valve depends on the following factors (Fig. 2.10):

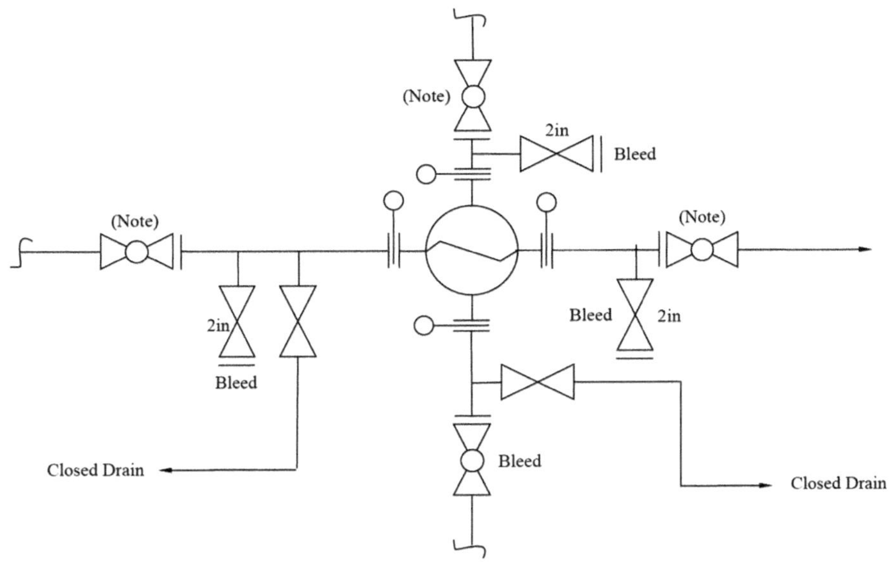

Fig. 2.9 Heat exchanger maintenance isolation setting. *Note* The decision to use DBB isolation can be made by referring to the detailed rules of isolation principles

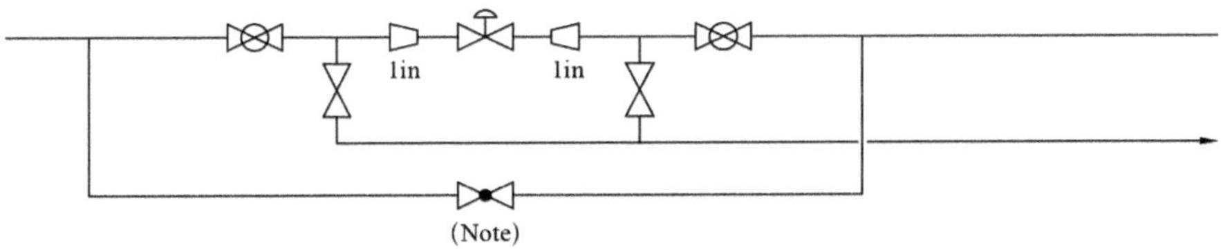

Fig. 2.10 Control valve isolation. *Note* If the pressure class is not less than 600# and there are toxic gases such as H_2S in the system, the isolation valve should be in DBB form.

(1) Availability of the spare column: If the control valve is part of the entire spare column, bypass or isolation is not required. Similarly, if omitting the bypass of the control valve will not harm plant safety or cause any production loss, the bypass and isolation can be omitted. Otherwise, the control valve will need a bypass and the necessary isolation devices installed.
(2) If the control valve bypass is difficult to manually control or may cause dangerous situations (such as split-range control), a bypass should not be installed.

6. Safety Valve

The inlet and outlet isolation valves of the safety valve are all full bore valves. If multiple safety valves vent to the atmosphere, a tail pipe should be set for each safety valve separately. The isolation setting of the safety valve is shown in Figs. 2.11 and 2.12.

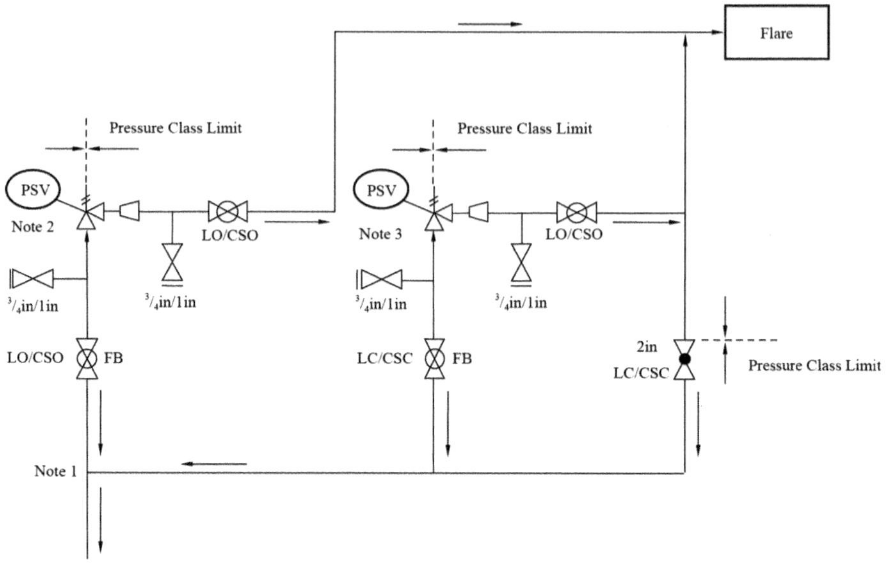

Fig. 2.11 Sealed vent system. *Note 1* From the vessel or pipeline connected to the vessel (no intermediate valve). *Note 2* If the pressure class is not less than 600#, install an additional isolation valve. *Note 3* The spare safety valve needs to be installed on site, not stored in the warehouse

7. Blowdown Valve (BDV)

During shutdown or maintenance, to isolate the emergency blowdown valve and the flare system, a lock-open isolation valve and a 3/4 in vent valve should be installed downstream of the emergency blowdown valve. No isolation valve can be installed upstream of the emergency blowdown valve (Fig. 2.13).

2.4.3 Isolation of Public/Process Systems

Temporary pipelines are usually used for blowdown. In the following cases, isolation valves should be installed on all public systems:

(1) Each sub-header on the main header of the process unit.
(2) The entrance of each user.
(3) The exit of each cooling water user.

The isolation of the dosing system is shown in Fig. 2.14.
Temporary nitrogen purge isolation is shown in Fig. 2.15.

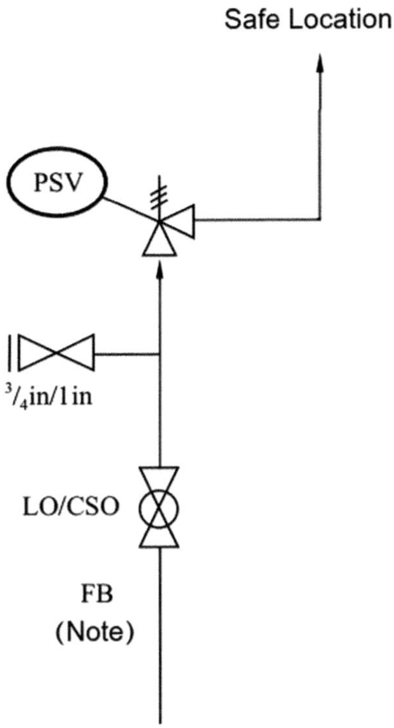

Fig. 2.12 Open vent system. *Note* From the vessel or pipeline connected to the vessel (no intermediate valve)

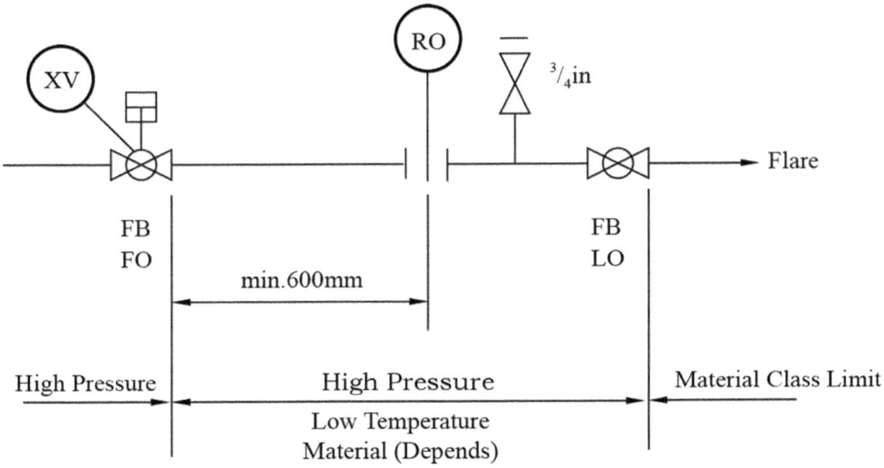

Fig. 2.13 Emergency blowdown valve isolation setting

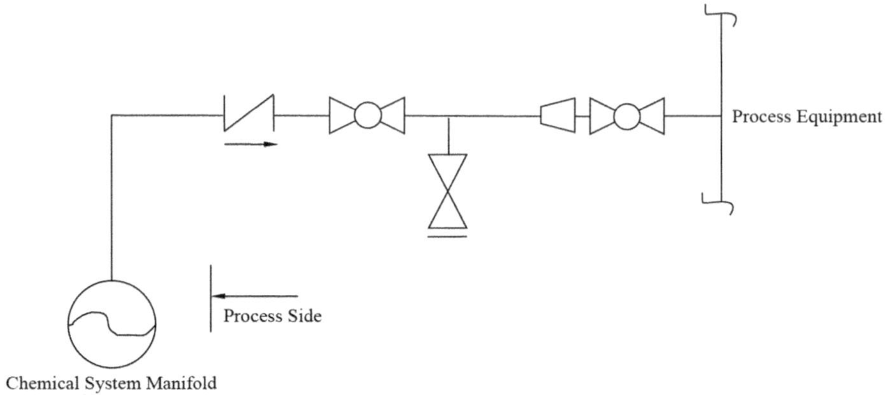

Fig. 2.14 Isolation of the chemical dosing system

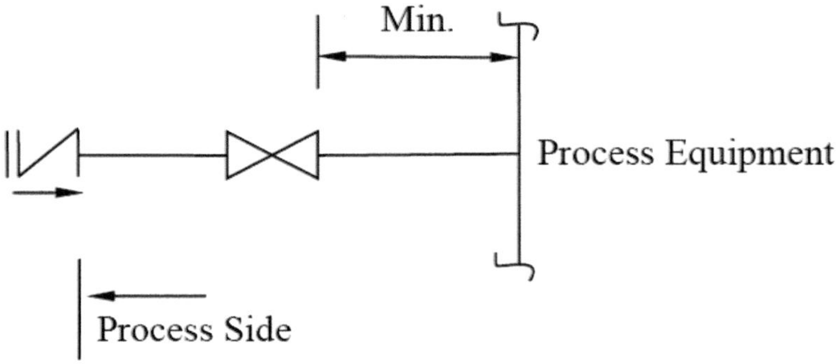

Fig. 2.15 Temporary nitrogen purge isolation

Permanent nitrogen purge isolation is shown in Fig. 2.16.

During equipment shutdown maintenance, the equipment venting pipeline and torch system should be isolated to ensure reliable isolation between the equipment and the torch. The isolation setting is shown in Fig. 2.17.

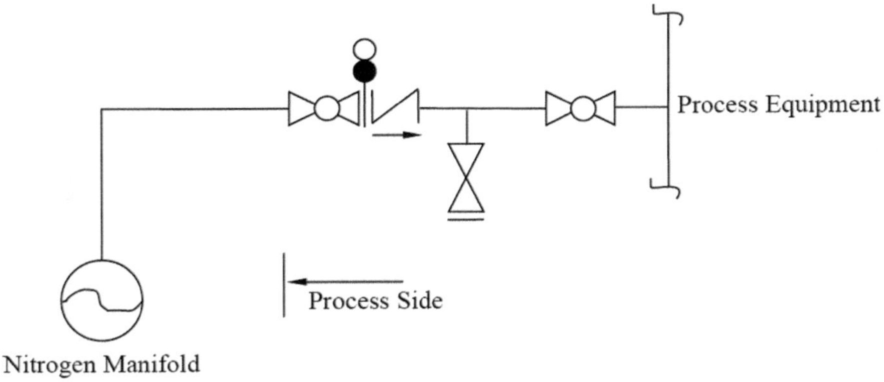

Fig. 2.16 Permanent nitrogen purge isolation

Fig. 2.17 Equipment venting/torch system isolation setting

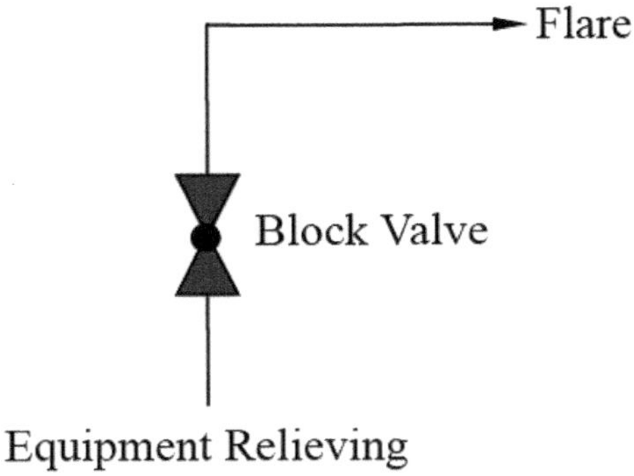

References

1. API Standard 521 Pressure-relieving and depressuring systems, Seventh Edition, June 2020.
2. SY/T 10043 Depressurization and depressurization system guide.
3. ASME BPVC SECTION VIII DIVISION 1 Rules for construction of pressure vessels, July 2013.
4. ASME BPVC SECTION VIII DIVISION 2 Alternative rules for construction of pressure vessels, July 2013.
5. API Std 620 Design and construction of large, welded, low-pressure storage tanks, Eleventh Edition, February 2008.

Chapter 3
Process Safety Analysis and HSE Analysis

With the advancement of science and technology, the processes and equipment of oil and gas production facilities are becoming more and more complex, which also makes the probability of accidents in oil and gas production facilities increasingly higher, and the degree of harm greatly increased. Once an accident occurs in an oil and gas production facility, it will not only cause personal injury and property loss, but may also cause significant social problems. An effective process safety management system is the basis for preventing major process safety accidents.

Process safety analysis is the most technical and challenging key element in implementing process safety management, and it is the core content of the process safety management system. There are many methods of process safety analysis. The Occupational Safety and Health Administration (OSHA) in the United States issued 29 CFR 1910.119 "Process Safety Management", and AQ/T 3034—2010 "Implementation Guidelines for Process Safety Management in Chemical Enterprises" recommended in China have recommended some safety analysis methods [1]. Mainly include:

(1) What if method. By asking a series of "What if…?" questions, hazards related to the process are identified. It is more suitable for relatively simple process systems. The usual practice is to ask and answer questions for each process step in the natural order of the process, from raw materials to products, and to specifically analyze the situations of equipment failure and operation errors.
(2) Safety checklist method. The safety checklist method is a typical qualitative safety analysis method. It uses accumulated experience and lessons from past accidents to improve the safety of process systems. According to the pre-prepared safety checklist, the process design or operating process system is checked item by item according to the items listed in the list to ensure that the items listed in the table meet the relevant requirements and are not omitted or ignored. It is the most traditional method in safety work. At present, the

K. Fang, *Onshore Oilfield Surface Facilities: Process Safety Engineering Guide*,
https://doi.org/10.1007/978-981-96-7830-3_3

safety checklist method is the most basic, simplest, and most widely used analysis method in safety system engineering, which can be used at any stage of engineering activities or process cycles.

(3) What if method + Safety checklist method. This is a combination of the two methods mentioned above, which makes up for the shortcomings of using the two methods separately.

(4) Hazard and Operability Study (HAZOP). The purpose of this method is to identify the sources of danger in the process, propose corresponding control measures for the sources of danger, prevent deviations and deviations during the process production stage, and thus control the safety of the process.

HAZOP can help find events to prevent accidents, including: design defects, situation diagnosis errors, operation errors, alarm failures, shutdown system failures, management system failures. Using the HAZOP method is not only beneficial to safety, but also beneficial to improving the reliability and operability of the process system. Overseas oilfield surface engineering covers central processing stations, natural gas processing plants, power stations and produced water treatment stations, etc. Due to the involvement of various production processing devices in dangerous materials such as crude oil and associated gas, it is necessary to apply process safety management methods and techniques to prevent production accidents in overseas oilfield surface engineering.

(5) Failure Mode and Effects Analysis (FMEA). According to the characteristics of the system (subsystems, equipment, components), the system is divided according to actual needs, and the possible types of failures and their impacts are analyzed.

(6) Fault Tree Analysis (FTA). FTA uses a logical method to vividly carry out hazard analysis. This method is intuitive, clear, with a clear thought process, strong logic, and can be used for qualitative analysis as well as quantitative analysis.

(7) Event Tree Analysis (ETA). This is a method of identifying hazards by inferring possible consequences from the initial time in the order of accident development. It starts with an initial time, divides into stages according to the order of accident development, and analyzes step by step. Each possible subsequent event of an event can only take one of the two completely opposite states (success or failure, normal or fault, safe or dangerous, etc.), and gradually develops towards the result until the system failure or accident stops.

(8) Other equivalent methods.

Regardless of the method chosen, process safety analysis should cover the following content:

(1) Hazards of the process system.
(2) Review of past events that may have serious consequences.
(3) Engineering and management measures to control hazards and the consequences of failure.

(4) On-site facilities.
(5) Human factors.
(6) The range of possible impacts on personnel safety and health after loss of control.

Appropriate safety analysis methods should be selected at different stages of the system lifecycle. This book mainly targets the design stage of onshore oilfield surface engineering, so the safety checklist method is used as the process safety analysis method.

3.1 Overview of Process Safety Analysis

This section mainly targets the oil and gas processing facilities of onshore oilfield surface engineering, proposes recommended practices for the design, installation, and testing of its safety protection system, discusses the basic concepts of the station safety protection system, and briefly explains the protection methods and system requirements [2].

3.1.1 Symbols, Abbreviations, and Symbols of Safety Devices

When describing or involving safety systems, in order to maintain uniformity, it is necessary to use uniform symbols, abbreviations, and symbols for each safety device. This book summarizes the commonly used safety device symbols to Table 3.1 based on the American Instrument Society standard ISA-S5.1 commonly used internationally.

3.1.2 Hazard Identification of Process Safety Analysis

The dangers in the process usually come from two aspects: the danger of the materials involved themselves and the danger brought by the process treatment process (process technology and process equipment). When identifying hazards, it is not enough to only consider the nature of the material, but also the production process and conditions. For example, water itself has no explosion hazard, but if the temperature and pressure of the production process exceed the boiling point of water, the presence of water has the danger of steam explosion. Analyzing specific production processes and conditions can exempt some hazardous substances from further analysis and evaluation. For example, if the flash point of a substance is higher than 400 °C, and production is carried out at room temperature and normal pressure, the possibility of causing a major fire can be ruled out. The process of identifying hazards is the process of process safety analysis.

Table 3.1 Safety device symbols

Sensing and self-acting safetydevices				
Item	Item safety device name		Symbol	
	General	American instrument society	Single device	Combined device
Backflow	Check valve	Flow safety valve	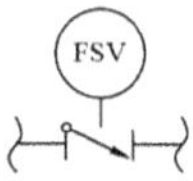	
Burner flame	Burner flame detector	Burner safety low limit		
Liquid level	High liquid level sensor	Liquid level high		
	Low liquid level sensor	Liquid level low		
Pressure	High pressure sensor	Pressure safety high		
	Low pressure sensor	Pressure safety low		
	Relief valve or safety valve	Pressure safety valve	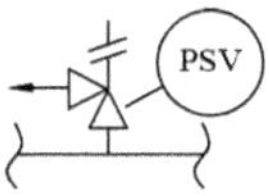	
	Emergency pressure relief	Pressure safety valve	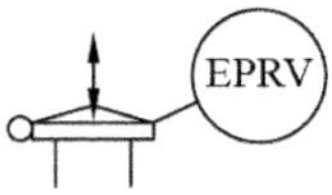	
Pressure or vacuum	Pressure-vacuum relief valve	Pressure safety valve	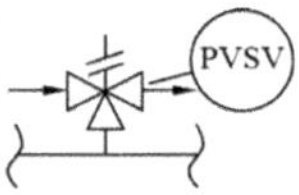	

(continued)

Table 3.1 (continued)

Sensing and self-acting safetydevices				
Item	Item safety device name		Symbol	
	General	American instrument society	Single device	Combined device
	Venting	None		
Temperature	High temperature sensor	Temperature safety high	TSH	TSHL
	Low temperature sensor	Temperature safety low	TSL	
Flame	Flame arrestor or flue arrestor	None	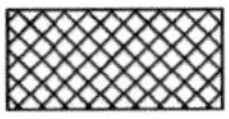	
Fire	Flame detector (ultraviolet or infrared)		USH	
	Heat detector	Temperature safety high	TSH	
	Smoke detector (ionization type)		TSE	
	Fusible material	Temperature safety element	ASH	

(continued)

Table 3.1 (continued)

Sensing and self-acting safetydevices				
Item	Item safety device name		Symbol	
	General	American instrument society	Single device	Combined device
Combustible gas concentration	Combustible gas detector	Analyzer safety high	ASH	
Toxic gas concentration	Toxic gas detector		OSH	
Sand	Erosion detector			NSL
Actuated valve				
Purpose		General symbol		
Wellhead surface safety valve		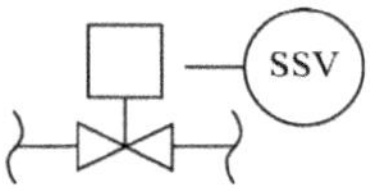		
Blowdown valve		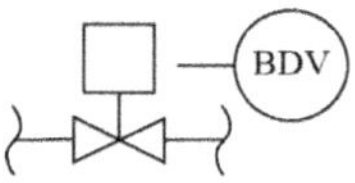		
Shutdown valve		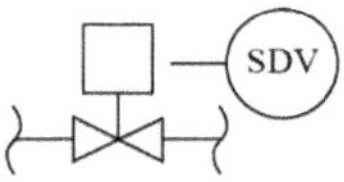		

3.1.3 Two Levels of Process Safety Protection

If the operating conditions in the process deviate and accumulate excessively, such as over-high, over-low, over-pressure, negative pressure, over-temperature, etc., it may lead to dangers such as overflow, gas breakthrough, leakage, fire or explosion, causing casualties, equipment damage and serious environmental pollution. Therefore, safety design should be carried out first in the process, considering the setting of safety protection devices. The process control system can be considered as the basis of safety protection. It stabilizes the process within the process conditions required for normal operation, generally setting control valves for process parameters including pressure, temperature, liquid level and flow rate, controlling them within the high and low alarm limits. When the process control system cannot control the deviation, it is handled by process safety protection.

Process protection is generally considered to have two levels:

(1) The first level of safety protection can be considered as the Emergency Shutdown (ESD) system. When the process control fails or cannot handle the process deviation, the process parameters deviate beyond the high and low alarm lines, and further deviate, reaching the high high or low low alarm limit, at this time it will automatically trigger an emergency shutdown, playing a shutdown, isolation role, preventing further accumulation of deviation, and uncontrollable situations. The ESD system performs different levels of shutdowns based on the deviation situation. In high-level ESD shutdowns triggered by fires and earthquakes, in addition to triggering SDV for shutdown, BDV will also be triggered for pressure relief.
(2) The second level is facilities such as PSV, which is the last protection to prevent equipment from bursting due to overpressure. In the case of previous process and safety protection failures, it handles hydrocarbons and other combustible or toxic fluids entering the torch to achieve the final protection of process equipment under overpressure accident conditions. The following is an example of separator accident condition analysis (Table 3.2).

   ① For overpressure conditions, the high-pressure sensor (PSH) provides primary protection by cutting off the inlet. If this device fails, the safety valve (PSV) provides secondary protection.
   ② For large gas leak conditions, the low-pressure sensor (PSL) cuts off the inlet, and the check valve (FSV) prevents downstream gas from flowing back to the leak point, providing primary protection.
   ③ Large oil leaks are similar to ②, with the low liquid level sensor (LSL) cutting off the inlet, and the check valve (FSV) preventing downstream liquid from flowing back to the leak point, providing primary protection. The waste oil tank and its LSH provide secondary protection, i.e., before the oil leak forms pollution, it will trigger the secondary protection sensor.

Table 3.2 Separator accident conditions

Accident condition	Primary protection	Secondary protection
Overpressure	PSH	PSV
Large gas leak	PSL, FSV	ASH, Minimize ignition source
Large oil leak	LSL, FSV	Waste oil tank (LSH)
Small gas leak	ASH, minimize ignition source	Fire detection
Small oil leak	Waste oil tank (LSH)	Manual monitoring
Inlet flow exceeds outlet flow	LSH	Vent scrubber
High temperature	TSH	Leak detection device

④ For gas leaks, before a fire occurs, it will trigger fire detection and protection equipment, serving as the primary protection for small gas leaks and secondary protection for large gas leaks.

⑤ Small oil leaks are primarily protected by the waste oil tank (LSH), with no automatic secondary protection. Manual monitoring is required before pollution occurs.

⑥ For high-temperature conditions, excessive temperature will cause WAMP to drop below the PSV set point. The high-temperature sensor (TSH) cuts off the inlet or heat source, providing primary protection, and the leak detection device provides secondary protection.

⑦ For conditions where the inlet flow exceeds the outlet flow, the high liquid level sensor (LSH) provides primary protection; if the gas phase outlet enters the atmosphere, then the PSH (to prevent triggering the safety valve) or the LSH in the downstream scrubber provides secondary protection, i.e., a scrubber must be installed downstream of the container discharging into the atmosphere.

It is necessary to analyze the system to determine the necessity of each device. For example, if the process conditions cannot cause overpressure in the container, then some devices are not necessary; if heating the container does not affect its maximum working pressure, then the TSH can be removed.

3.1.4 Premises for Safety Protection System Analysis and Design

The recommended method for safety protection system analysis and design is based on the following premises:

(1) For safe operation, the design of process facilities should be based on feasible engineering practices.

(2) Safety systems should have two levels of protection to prevent or reduce the impact of equipment failures in the process system. These two levels of protection should be independent of the control devices for normal operation of the process system. Usually, to have a wider range of action, these two levels of safety device types should be different in function. Two identical devices have the same performance and may have the same inherent weaknesses.

(3) The two levels of protection should be the highest level (first level) and the next highest level (second level). For specific situations, it is necessary to judge to determine these two highest level protection systems. For example, equipment rupture caused by overpressure can be provided with two levels of protection by PSH and PSL. Before the pressure is too high, to prevent equipment rupture, PSH shuts down the affected equipment; after the rupture occurs, PSL shuts down the affected equipment. However, sometimes PSV is chosen instead of

PSL because it can release excess volume to a safe place to prevent equipment rupture. Moreover, PSV reacts quickly to prevent equipment rupture, while PSH cannot effectively and quickly perform actions in some situations.
(4) Using proven and accepted system analysis methods for production processes can determine the minimum safety requirements for process equipment. If this analysis is applied to an independent device, assuming the worst input and output conditions, then this analysis will be effective for that device in any process system.
(5) All process equipment in surface engineering refers to the process equipment included in the entire process system from the wellhead to the most downstream discharge point. Therefore, all process equipment and functions are incorporated into the safety system.
(6) When sufficiently protected process equipment is combined into a facility, no additional safety threats will be generated. Therefore, if all process equipment safety devices are reasonably combined into a safety system, the entire facility is protected.

All safety analyses are conservatively assumed to occur without operator intervention in the event of a failure or during the process of successive occurrences, in order to thoroughly identify all the root causes of accidents or the final consequences they cause.

3.1.5 Safety Procedure Flowchart

Figure 3.1 is a safety procedure flowchart, which shows how accidental events can cause personal injury, pollution, or facility damage. It also indicates where safety devices and methods should be used to prevent the expansion of events. In fact, the release of hydrocarbons is a significant factor threatening safety, so the safety system mainly prevents hydrocarbons from being released from the process system, and if there is, it reduces its adverse effects. Its main purposes are:

(1) Prevent accidental events that can lead to the release of hydrocarbons.
(2) If it happens, shut down the process system or the affected part of the process system to prevent hydrocarbons from flowing to the leak point or overflowing.
(3) Collect and recover liquid hydrocarbons flowing out of the process system and disperse gases.
(4) Prevent the ignition of released hydrocarbon substances.
(5) Shut down the process system in case of fire.
(6) In addition to the equipment that has already had accidents, prevent accidental events that may cause the release of hydrocarbons from other equipment.

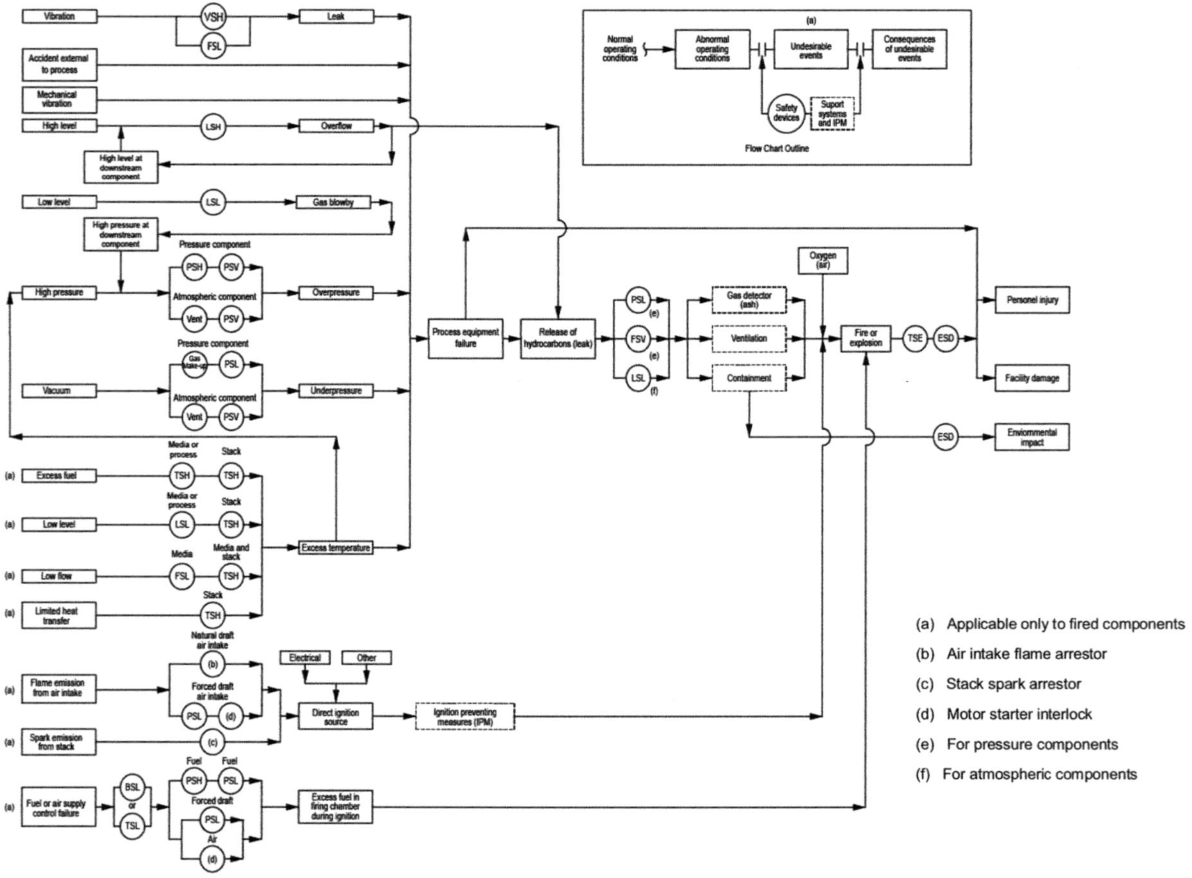

Fig. 3.1 Safety procedure block diagram of oilfield surface engineering production facilities

3.2 Process Safety Analysis Process

3.2.1 Design Steps

Process safety starts from process design and ends when handed over to the instrument specialty for the preparation of cause and effect diagrams. It can generally be divided into the following steps:

Prepare principle-type documents → carry out system description and design → carry out unit description and design → carry out unit safety analysis → improve safety description → complete instrument control description cause and effect diagram [3].

The flowchart is shown in Fig. 3.2.

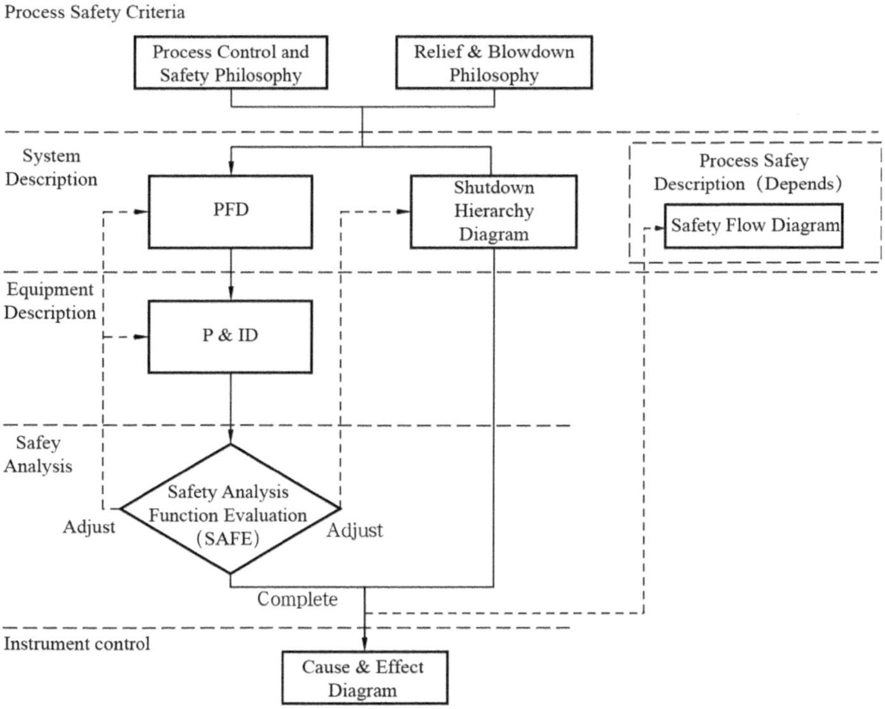

Fig. 3.2 Process safety design flowchart

3.2.2 *Process Description*

(1) The first step is to prepare principle-type documents. This mainly includes the Safeguarding Philosophy, Process Control Philosophy, and Relief and Blowdown Philosophy.

Some projects now combine the Safeguarding Philosophy and Process Control Philosophy into one document, called the "Process Control and Safeguarding Philosophy".

(2) The second step is to carry out system description and design. This mainly completes the overall system description and design of the project, describing both the process and safety protection aspects.

The process aspect includes the Block Flow Diagram (BFD) and Process Flow Diagram (PFD); the safety protection aspect includes the Safety Flow Diagram (SFD) and Shutdown Hierarchy Diagram. Some projects call the Shutdown Hierarchy Diagram the Shutdown Logic Diagram.

(3) The third step is to carry out unit description and design. Complete the process flow, control, and safety protection description of each equipment unit. The reports to be completed include: Process and Utility Description, Process Control Description; the drawings to be completed include: Piping & Instrumentation Diagram (P&ID).
(4) The fourth step is to carry out unit safety analysis. Carry out detailed safety analysis for each equipment unit and complete the Safety Analysis Function Evaluation (SAFE).
(5) The fifth step is to improve safety facilities. After completing all unit safety analyses, there may be additions, deletions, or adjustments to the safety protection facilities of the unit. At this time, the P&ID and Shutdown Hierarchy Diagram should be upgraded and improved according to the results.
(6) The sixth step is to complete the Cause and Effect Diagram.

3.2.3 *Brief Introduction to Documents*

3.2.3.1 Report-Type Documents

1. Process Control and Safety Protection Principles

This document should describe the major principles of process control and safety protection separately.

(1) The process control aspect should introduce the setting of the system's self-control level, and for different level stations involved in the project, such as well-heads, metering stations, transfer stations, central processing stations, power stations, natural gas processing plants, etc., specify the corresponding control level, signal transmission method, etc., to provide guidance for future design.

(2) For safety protection, the ESD system and Blowdown system should be introduced. For the ESD system, the ESD shutdown level should be clearly introduced, and each level's triggering content should be explained through the shutdown hierarchy. For example, define that a level 0 shutdown is triggered by an earthquake or war situation to abandon the station operation; a level 1 shutdown is triggered by a fire or power outage to shut down the process and emergency process venting, etc.; a level 2 shutdown is a process column shutdown; a level 3 shutdown is a unit shutdown; a level 4 shutdown is an independent equipment shutdown.

2. Process and Utility Description

Mainly describe the process and utility system flow of the project, and also introduce the design parameters of individual equipment in detail, such as equipment processing capacity, operating and design pressure, temperature, and other parameters.

3. Process Control Description

Describe how the parameters (temperature, pressure, flow rate, etc.) of the process system are maintained in a normal state. This document is generally completed by the process and instrument specialties together.

3.2.3.2 Drawing-Type Documents

1. PFD and P&ID

For the PFD, it should be noted that the instrument signals of the interlocked SDV do not need to be reflected in the diagram, only the position of the SDV needs to be indicated. For the P&ID, it should be noted that there should be a preliminary P&ID drawing before the unit safety analysis, but the consideration of safety protection facilities is not comprehensive. After the safety analysis step, the P&ID drawing should be further improved.

2. Safety Flow Diagram

The English name of this document is "Safeguarding flow diagram", abbreviated as SFD; some use "Process safeguarding flow schemes". This drawing is based on PFD, removes the description of process control, and emphasizes the description of safety protection, such as high and low alarms for liquid level and pressure, and the shutdown information triggered, focusing on the representation of safety shutdown and isolation SDV, pressure relief BDV and safety protection PSV/TSV, etc.

Unlike the PFD, which only shows the position of the SDV, the SFD should also show the associated signals that trigger the SDV. For example, the high liquid level of the first-stage separator, and the shutdown signals of the 0th, 1st, and 2nd stages all cause the shutdown of the inlet SDV.

This document reflects the overall arrangement of safety facilities from a macro perspective. Because it reflects a comprehensive range of safety protection facilities, in actual operation, this document can be completed after the SAFE file is completed.

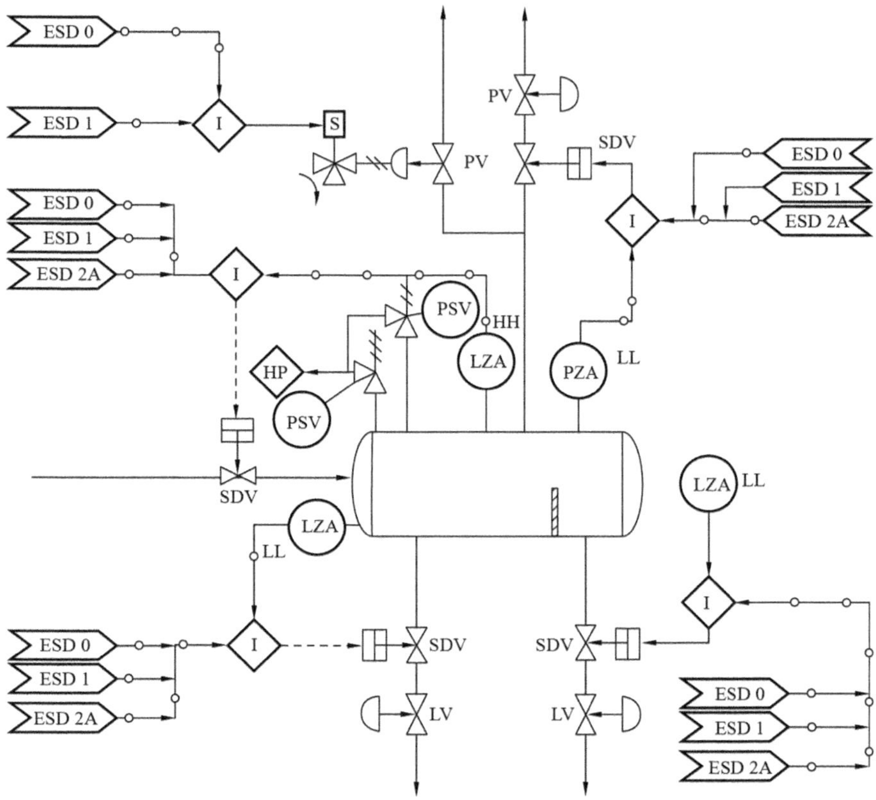

Fig. 3.3 Typical separator safety flow chart

The safety process of a typical separator is shown in Fig. 3.3.

3. Shutdown Hierarchy Diagram

The "Shutdown Hierarchy Diagram", some projects call it the "Shutdown Logic Diagram". It mainly shows the specific action valves triggered by the shutdown levels defined in the safety protection principle document, represented by a hierarchical block diagram.

4. Safety Analysis Table

The English name of the Safety Analysis Table file is "Safety Analysis Function Evaluation", abbreviated as SAFE file, which can also be directly translated as Safety Analysis Function Evaluation Table (Table 3.3).

API RP 14C Analysis, design, installation, and testing of safety systems for offshore production facilities and SY/T 10,033 "Recommended practices for the analysis, design, installation, and testing of safety systems for offshore production platform topsides" provide a set of experience methods for safety analysis, analyzing

Table 3.3 Safety Analysis Evaluation Table (SAFE) example

Safety Analysis Table									
				Operating function					
				Switch off or control olevice sign					
Process equipment		Fixtures signs	Substitute protection						
Signs	Object of protection		SAC reference	Substitute device					

all possible accidents (pressure, temperature, liquid level abnormalities) and their consequences, providing protective measures that can be used to prevent the occurrence of consequences and mitigate damage, they can analyze each system and unit of the project one by one, and record the analysis results in the corresponding format table, completing the safety analysis evaluation table.

This document and the cause and effect table look very similar, but there are differences, mainly:

(1) Different purposes. The compilation of SAFE is to analyze the implementation of each protective measure of all process system facilities item by item, with the purpose focusing on internal safety analysis and inspection during the design process; the cause and effect diagram focuses on the interlock shutdown or venting results after the process safety design is completed, with the purpose of programming for the execution of the safety system.

(2) Different coverage. SAFE focuses on the process deviations (over-temperature, over-pressure, etc.) and protective measures of individual equipment, analyzing each piece of equipment one by one, and completing the summary of each system and equipment. The cause and effect diagram includes not only the protection of each system and each piece of equipment, but also covers the interlock signals of higher-level shutdowns (such as 0-level abandonment, 1-level fire shutdown).

3.3 Process Safety Analysis Example

The preparation of the SAFE file is divided into three steps: the first step is to select the SAT table (Safety Analysis Table) for the corresponding equipment; the second step is to select the SAC table (Safety Analysis Checklist) for the corresponding equipment; the third step is to analyze item by item according to the requirements of SAC, and fill in the results on SAFE. The core process is: Safety Analysis Table (SAT) → Safety Analysis Checklist (SAC) → Safety Analysis Function Evaluation Table (SAFE).

The Safety Analysis Table (SAT) and Safety Analysis Checklist (SAC) correspond to individual equipment (analysis objects). SAFE is a document that records and summarizes the analysis of each piece of equipment in the system after completion. API RP 14C Analysis, design, installation, and testing of safety systems for offshore production facilities appendix provides ten different types of safety analysis tables and safety analysis checklists, which are: single well pipeline, wellhead injection pipeline, manifold, pressure vessel, atmospheric vessel, fired equipment, pump, compressor set, pipeline (platform—onshore facilities), heat exchanger (tube and shell). It includes the basic process facilities of offshore platforms (oilfield stations). In specific systems, directly apply these ten different types of safety analysis checklists, and fill in SAFE according to the safety analysis checklist. The following uses the separator as an example to introduce step by step.

3.3.1 SAT

3.3.1.1 Analyze P&ID, Find the Corresponding Category

After receiving a separator safety analysis task, first get the draft P&ID, analyze which type of equipment it belongs to, and find the corresponding SAT table. The separator belongs to the pressure vessel, and its corresponding SAT table should be used. The typical separator P&ID is shown in Fig. 3.4. Typical safety devices for pressure vessels include: LSHH, LSLL, PSHH, PSV, oil, gas and water outlet check valves FSV (Flow safety valve), etc.

3.3.1.2 Pressure Vessel SAT Table

Find its corresponding SAT for the pressure vessel, see Table 3.4. As can be seen from Table 3.4, the accidental events affecting the pressure vessel are overpressure, negative pressure, overflow, gas breakthrough, leakage, and overtemperature (when the vessel is heated), and it also gives the general reasons that may cause various accidental events and detectable abnormal states.

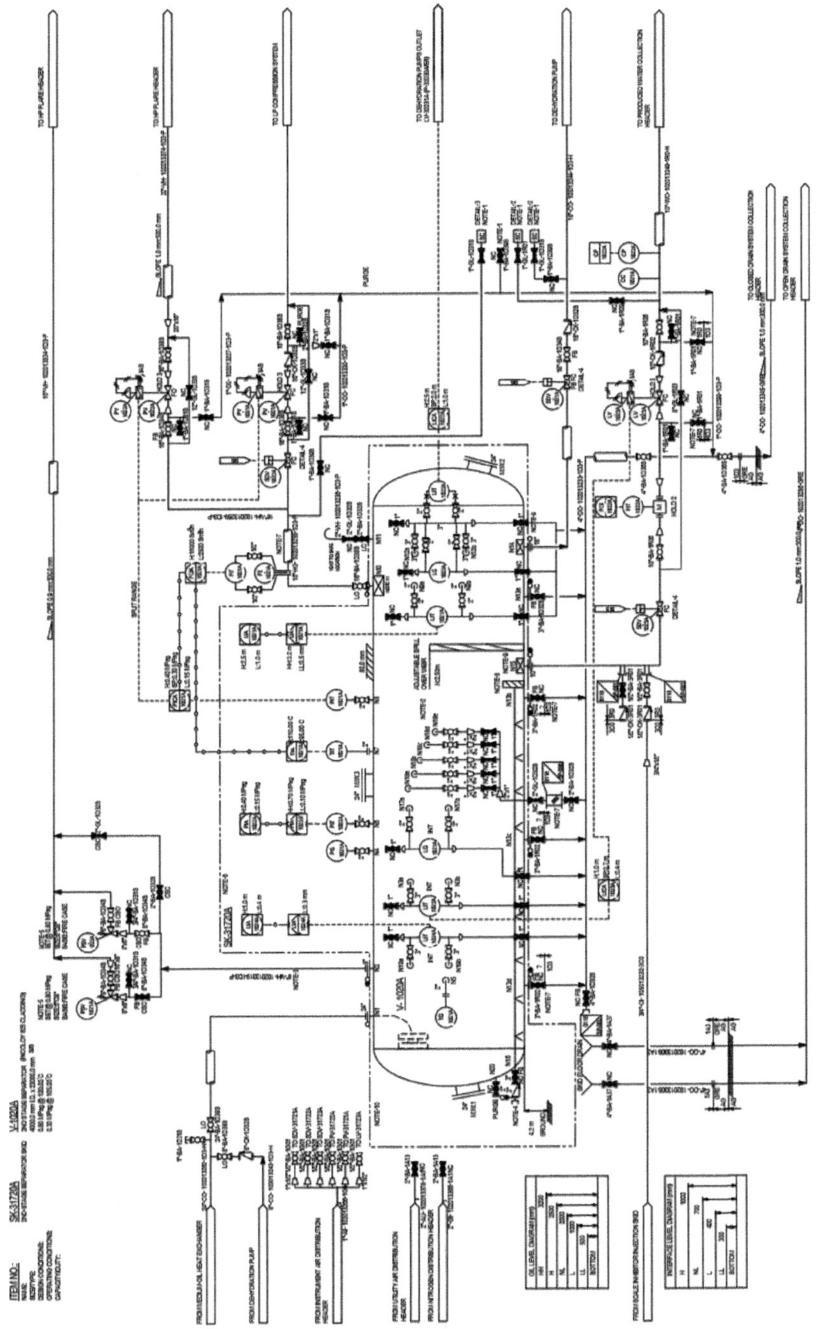

Fig. 3.4 The typical separator P&ID

Table 3.4 SAT: pressure vessel

Accidental event	Cause	Detectable abnormal conditions by equipment
Overpressure	Outlet pipe blockage or throttling Inlet flow exceeds outlet flow Gas breakthrough (upstream equipment) Pressure control system failure Thermal expansion Excessive heat input	High pressure
Negative pressure (vacuum)	Outlet flow exceeds inlet flow Cold shrinkage Outlet opening Pressure control system failure	Low pressure
Overflow	Inflow of liquid exceeds outflow of liquid; Segment blockage Liquid outlet blockage or throttling Liquid level control system failure	High liquid level
Gas breakthrough	Liquid outlet flow exceeds inlet flow Liquid outlet opening Liquid level control system failure	Low liquid level
Leakage	Pipe material deterioration Erosion Corrosion Impact damage Vibration	Low pressure Low liquid level
Over temperature	Temperature control system failure Inlet temperature too high	High temperature

The main causes of accidents are equipment malfunctions, process abnormalities, and accidental incidents, but all major causes within the same category will lead to the same accidents. For example, excessive pressure could be due to a blocked or throttled outlet pipe, or it could be due to an excessive upstream feed or a malfunctioning pressure control system.

Safety monitoring systems can detect these abnormal conditions and trigger actions through automatic control mechanisms, which can prevent accidents or mitigate the consequences of these incidents.

3.3.2 SAC

3.3.2.1 Find the SAC Table

Find the SAC table corresponding to the pressure vessel: Safety Analysis Checklist "SAC—Pressure Vessel" (Table 3.5).

3.3.2.2 Use the SAC Table for Item-by-Item Analysis

Table 3.5 is a tool for discussing the safety analysis of individual equipment. Assuming that a piece of equipment is considered as a single unit, under the worst possible input and output conditions, SAC lists the safety devices needed to protect it.

When this equipment is connected with other process equipment, specific conditions under which the recommended device may not be set are listed below it. This is because the safety devices on other equipment can provide equivalent protection, or because in specific processes, not setting it will not jeopardize safety.

For example, in Table 3.5, A.4.a.1 of the pressure vessel requires the installation of PSH, but if A.4.a.2 is met, i.e., if the input source comes from a pump or compressor, and the pressure they generate cannot exceed the MAWP of the vessel, PSH is not required. After completing PSH, analyze each safety device such as PSL, PSV, LSH, LSL and check valve individually.

3.3.3 Safe

3.3.3.1 Complete the SAFE Table for a Single Device

Find the SAFE table for the pressure vessel and fill in the analysis results (Table 3.6).

In Table 3.6, if a PSH is set for the separator, fill in its position number after PSH in the Device, and fill in its corresponding SAC reference number, A.4.a.1, and mark " × " in the cell corresponding to the SDV number that is interlocked and shut down. If the upstream input source comes from a pump and the pressure it generates cannot exceed the design pressure of the container, it means that the requirement of A.4.a.2 is met, and a PSH does not need to be set, and the position number does not need to be written. Write A.4.a.2 under the SAC reference number for alternative protection.

If the alternative protection meets A.4.a.4, that is, the gas outlet is connected to the downstream equipment with a suitable pipeline without a cut-off valve or regulating valve, and this equipment is protected by a PSH that also protects the upstream equipment. In addition to writing A.4.a.4, also list the instrument position number and equipment position number.

Table 3.5 SAC—pressure vessel

Serial number	Facility	Number	Optional item	Corresponding protection in PID	SAFE noted as
A.4 Pressure vessel					
a	PSH	1	Install PSH	√	A.4.a.1
		2	The input source comes from a pump or compressor, and the pressure they generate cannot exceed the MAWP of the vessel		
		3	The input source is not a wellhead oil pipeline, production manifold or pipeline, and each input source has a PSH protection, which also protects the vessel		
		4	The gas outlet is connected to the downstream equipment with a suitable pipeline without a cutoff valve or regulating valve, which is protected by a PSH that also protects the upstream equipment		
		5	The vessel is the last stage gas scrubber of the flare, vent or blowdown system, and its design pressure can withstand the maximum allowable gathering pressure		
		6	The vessel operates at atmospheric pressure and has an appropriate venting system		

(continued)

Table 3.5 (continued)

Serial number	Facility	Number	Optional item	Corresponding protection in PID	SAFE noted as
b	PSL	1	Install PSL	√	A.4.b.1
		2	The minimum pressure during operation is atmospheric pressure		
		3	Each input source is protected by PSL, and there is no control device or throttling device between PSL and the vessel		
		4	The vessel is a gas scrubber or small separator rather than a process equipment and is provided with sufficient protection by the downstream PSL or perfect design function (such as the vessel is a gas scrubber of the pneumatic safety system or the last stage gas scrubber of the flare, vent or blowdown)		
		5	The gas outlet is connected to the downstream equipment with a suitable pipeline without a cutoff valve or regulating valve, which is protected by a PSL that also protects the upstream equipment		
c	PSV	1	Install PSV	√	A.4.c.1
		2	Each input source is protected by a PSV, which is set to a pressure no higher than the MAWP of the vessel, and a PSV is installed on the vessel to protect against fire heat and thermal expansion		
		3	Each input source is protected by a PSV, which is set to a pressure no higher than the MAWP of the vessel, and at least one of them will not be isolated from the vessel		
		4	The PSV on the downstream equipment can meet the venting requirements of the vessel and will not be cut off from the vessel		

(continued)

Table 3.5 (continued)

Serial number	Facility		Number	Optional item	Corresponding protection in PID	SAFE noted as
			5	The vessel is the last stage gas scrubber of the flare, vent or blowdown system, and its design pressure can withstand the maximum allowable accumulation pressure, and there are no internal or external flow obstructions, such as fogging gas, back pressure valve or fire arrestor		
			6	The vessel is the last stage gas scrubber of the flare, vent or blowdown system, and its design pressure can withstand the maximum allowable accumulation pressure, and a rupture disc or safety cap (PSE) is installed to bypass internal or external flow obstructions, such as fogging gas, back pressure valve or fire arrestor		
d	LSH	Oil	1	Install LSH	√	A.4.d.1
			2	The downstream equipment of the gas outlet is not a flare or blowdown system, and it can safely handle the maximum liquid carryover		
			3	The characteristics of the vessel do not require the treatment of the separated liquid phase		
			4	The vessel is a small catcher, and its internal liquid can be manually discharged		
		Water	1	Install LSH	√	A.4.d.1
			2	The downstream equipment of the gas outlet is not a flare or blowdown system, and it can safely handle the maximum liquid carryover		
			3	The characteristics of the vessel do not require the treatment of the separated liquid phase		

(continued)

Table 3.5 (continued)

Serial number	Facility		Number	Optional item	Corresponding protection in PID	SAFE noted as
			4	The vessel is a small catcher, and its internal liquid can be manually discharged		
e	LSL	Oil	1	Install LSL	√	A.4.e.1
			2	The liquid level in the vessel is not automatically maintained, and the vessel does not have an immersion heater that can withstand overheating		
			3	The downstream equipment of the liquid outlet can safely handle the maximum gas flow that can be discharged through the liquid outlet, and there is no immersion heater in the vessel that can withstand overheating; the throttling device on the discharge pipeline can be used to limit the gas flow		
		Water	1	Install LSL	√	A.4.f.1
			2	The liquid level in the vessel is not automatically maintained, and the vessel does not have an immersion heater that can withstand overheating		
			3	The downstream equipment of the liquid outlet can safely handle the maximum gas flow that can be discharged through the liquid outlet, and there is no immersion heater in the vessel that can withstand overheating; the throttling device on the discharge pipeline can be used to limit the gas flow		
f	FSV		1	Install FSV at each outlet	√	A.4.f.1
			2	The maximum amount of hydrocarbon material refluxed from downstream equipment is negligible		
			3	The control device on the pipeline can effectively reduce reflux		

Table 3.6 Pressure vessel SAFE

Safety analysis table					Operational function		Primary separator oil outlet SDV	Secondary separator oil outlet SDV	Secondary separator gas outlet SDV	Secondary separator water outlet SDV	Oil outlet dehydration pump
					Shutdown or Control	Device mark		SDV-1021A	SDV-1022A	SDV-1023A	Tr ip
Process equipment		Device mark		Substitute protection							
Mark	Protection object			SAC reference	Substitute device		1	2	3	4	5
V-1020A	Secondary separator	PSH	PIAHH-1022A	A.4.a.1			×				
		PSL	PIALL-1022A	A.4.b.1				×	×	×	
		PSV	PSV-1021A	A.4.c.1							
			PSV-1022A	A.4.c.1							
		LSH (oil)	LIAHH-1021A	A.4.d.1			×				
		LSH (water)	N/A	A.4.d.3							
		LSL (oil)	LIALL-1021A	A.4.e.1							×
		LSL (water)	LIALL-1024A	A.4.e.1						×	
		FSV (oil)		A.4.f.1							

(continued)

Table 3.6 (continued)

Safety analysis table					Operational function		Primary separator oil outlet SDV	Secondary separator oil outlet SDV	Secondary separator gas outlet SDV	Secondary separator water outlet SDV	Oil outlet dehydration pump
					Shutdown or Control	Device mark		SDV-1021A	SDV-1022A	SDV-1023A	Tr ip
Process equipment		Device mark		Substitute protection							
Mark	Protection object			SAC reference	Substitute device		1	2	3	4	5
		FSV (gas)		A.4.f.1							
		FSV (water)		A.4.f.1							

After completing the PSH, fill in the PSL, PSV, LSH, LSL, and check valves one by one.

3.3.3.2 SAFE of the Entire System

After completing the SAFE of each device, summarize them to complete the SAFE file of the entire system.

3.4 Process Equipment Safety Analysis Checklist

3.4.1 Single Well Pipeline

The single well pipeline (Flowline) refers to the pipe section between the oil nozzle outlet of the wellhead device and the next processing station, responsible for transporting well fluids (hydrocarbons). After the well fluid is lifted to the surface by the downhole device, it enters the surface device after pressure reduction by the oil nozzle, and then is transported to the downstream device through the single well pipeline.

If self-spraying extraction is used in the early stage of overseas oil field development, once downstream problems such as valve mis-closure causing overpressure and pipeline corrosion causing leaks occur, the single well pipeline will experience overpressure and low pressure conditions. If overpressure and low pressure protection measures are not considered during design, it will cause safety accidents. The boundary point between downhole and surface engineering is generally set after the oil nozzle, and its safety protection scheme is a unified system. Many times, surface engineering and downhole engineering need to be considered together. For the sake of discussion, we only consider it separately from the perspective of surface engineering, and will also supplement where it is necessary to consider it together. In order to determine a safe and reasonable safety device, a safety analysis of the entire pipeline must be conducted. The accidental events affecting the single well pipeline are overpressure and leakage. The typical safety devices of the wellhead and pipeline are shown in Figs. 3.5, 3.6 and 3.7, and the safety analysis of the single well pipeline is referred to Tables 3.7 and 3.8.

3.4.2 Injection Pipeline

The injection pipeline sends fluids used for gas lift or reservoir injection into the wellbore.

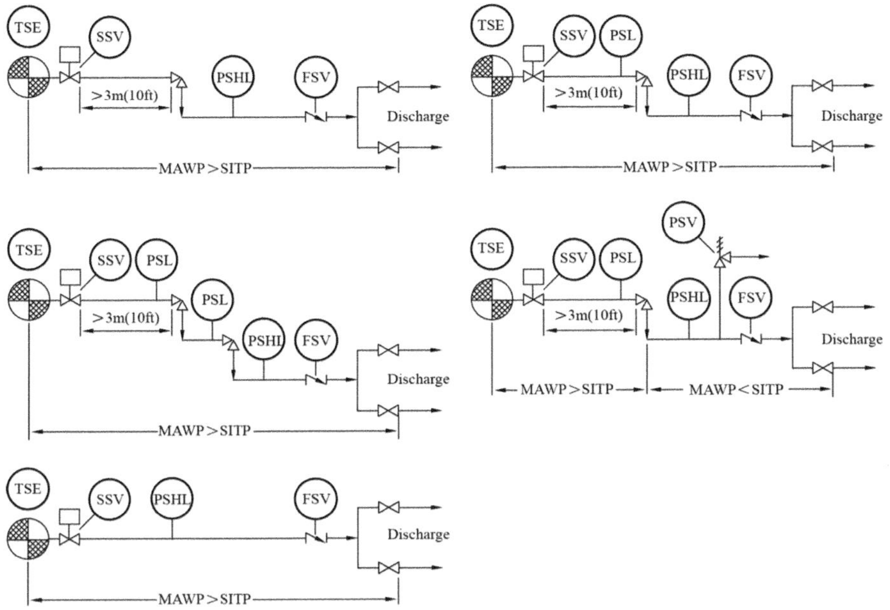

Fig. 3.5 Safety device: tree network pipeline

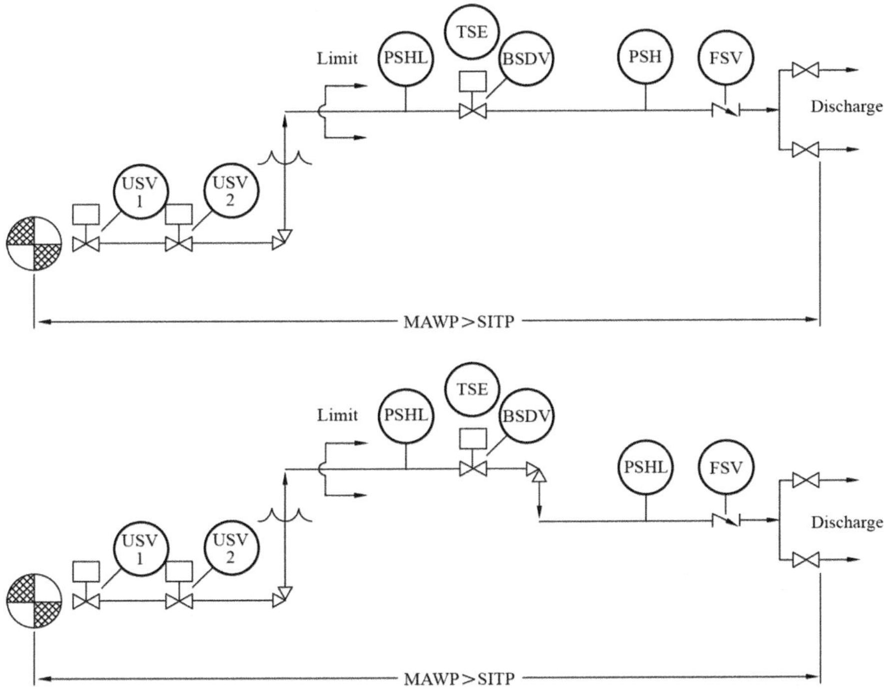

Fig. 3.6 Safety device: subsea network pipeline

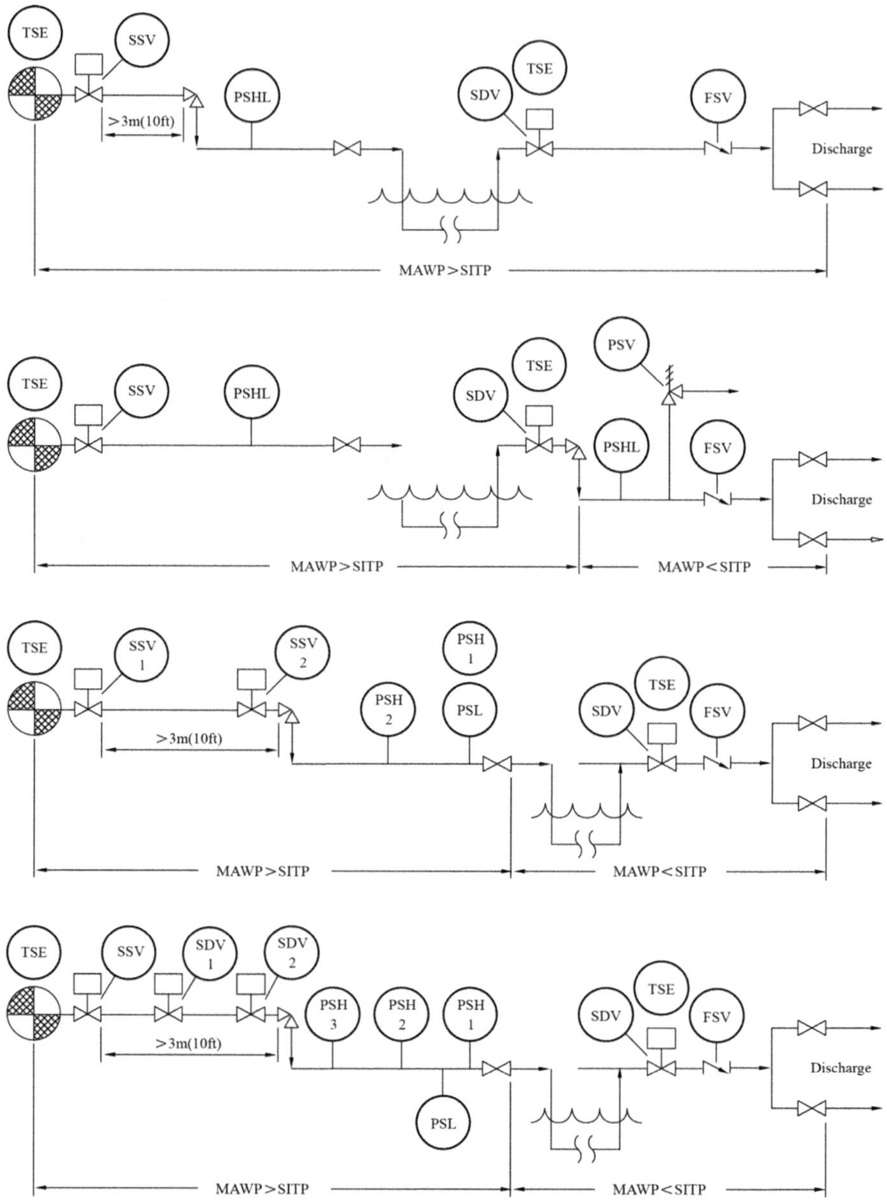

Fig. 3.7 Safety device: satellite network

Table 3.7 Safety analysis table: single well pipeline

Accidental event	Reason	Detectable abnormal state of equipment
Overpressure	Pipeline blockage or throttling Blockage downstream of the oil nozzle Hydrate blockage Failure of upstream flow control device Change in oil well status Closure of downstream valve of single well pipeline	High pressure
Leakage	Material deterioration Erosion Corrosion; Impact Vibration	Low pressure

Pressure protection is usually provided by the PSH and PSL of the injection source (such as a compressor or pump), which shut off the fluid inflow. If the PSH and PSL also protect the injection pipeline, wellhead, and other equipment, then these devices are not needed on the injection pipeline. If the designed injection pipeline can withstand the maximum pressure that the injection source can apply, then a PSV is not needed. Usually, the PSV installed on the injection source can also protect the injection pipeline, wellhead, and other equipment.

The recommended safety devices for the injection pipeline are shown in Fig. 3.8, and the safety analysis table and safety analysis checklist are shown in Tables 3.9 and 3.10.

3.4.3 Manifold

The manifold receives fluids produced from two or more wellheads and distributes these produced fluids to the required process systems, such as low pressure, medium pressure or high pressure production and testing separation facilities. As shown in Fig. 3.9, the safety analysis table and safety checklist are shown in Tables 3.11 and 3.12.

3.4.4 Pressure Vessels

Pressure vessels process hydrocarbons under pressure, performing gas–liquid separation, dehydration, storage, and buffering. Some pressure vessels require heating when in use. This chapter only involves the role of heat input on the processing

Table 3.8 Safety analysis checklist: wellhead pipeline

Serial number	Facility	Number	Optional item	Corresponding protection in PID	SAFE noted as
A.1 Wellhead pipeline					
a	PSH	1	Install PSH		
		2	The maximum allowable working pressure of the oil outlet pipe section is greater than the maximum shut-in pressure and there is downstream PSH protection		
b	PSL	1	Install PSL		
		2	The oil pipe section is located between the wellhead and the first throttle device and its length is less than 3 m (10ft), or it is installed underwater (reasonably close to this distance)		
c	PSV	1	Install PSV		
		2	The MAWP of the oil pipe section is greater than the shut-in pressure		
		3	The volume of the oil pipe upstream of the cut-off valve can give the SDV enough time to shut off before exceeding the MAWP, and two SDVs are installed (one of which can be an SSV), as well as independent PSH, relays, and monitoring points		
		4	The oil pipe section is protected by the upstream PSV		

(continued)

Table 3.8 (continued)

Serial number	Facility	Number	Optional item	Corresponding protection in PID	SAFE noted as
		5	The oil pipe section is protected by the downstream PSV, the PSV cannot be separated from the oil pipe section, and there are no throttling and flow limiting devices between the PSV and the pipe section		
		6	The oil pipe section is protected by HIPPS (High Integrity Pressure Protection System)		
d	FSV	1	Install FSV		
		2	Protected by the FSV on the downstream pipeline		

section of the heated vessel. Pressure vessels related to compressors (not compressor cylinders) should be protected according to the requirements of this chapter.

The recommended safety devices for pressure vessels are shown in Fig. 3.10, and the safety analysis table and safety checklist are shown in Tables 3.13 and 3.14.

3.4.5 Atmospheric Vessels

Atmospheric vessels are used to process and temporarily store liquid hydrocarbons. Some applications require heat input into the vessel, this chapter only discusses the impact of heat input on the processing section of the atmospheric vessel. Containers used to store diesel fuel and chemicals are not included, as they are auxiliary equipment of the production process system, not part of the process system.

3.4.5.1 Pressure Safety Devices (Vent and PSV)

Atmospheric vessels should be protected by a properly sized venting system to prevent overpressure and underpressure. API Std 2000 Venting atmospheric and low-pressure storage tanks can be used as a guide for calculating the size of the

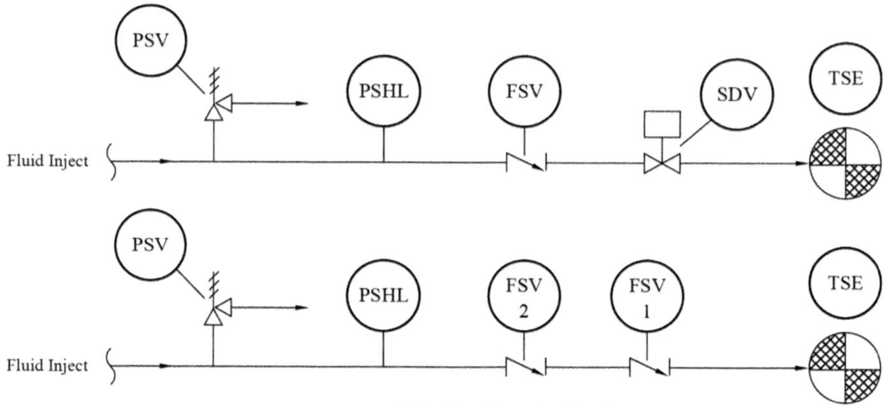

(a) Wellhead Injection Flowline

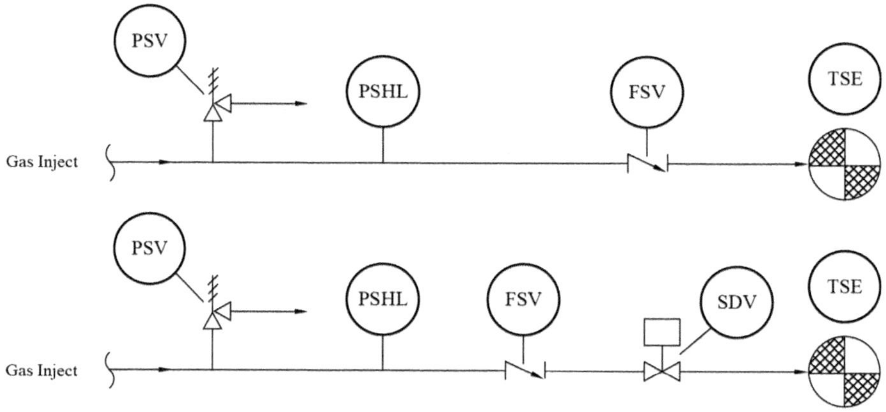

(b) Wellhead gas lift pipeline

Fig. 3.8 Safety devices: tree wellhead injection pipeline

Table 3.9 Safety Analysis Table (SAT): wellhead injection pipeline

Accidental event	Causes	Detectable abnormal conditions of equipment
Overpressure	Outlet pipeline blockage or throttling Hydrate blockage Upstream control system failure Formation blockage Formation backflow	High pressure
Leakage	Pipe material deterioration Erosion Corrosion Impact damage Vibration	Low pressure

Table 3.10 Safety Analysis Checklist (SAC): wellhead injection pipeline

Serial number	Facility	Number	Optional item	Corresponding protection in PID	SAFE noted as
A2 Wellhead injection pipeline					
a	PSH	1	Install PSH		
		2	Pipeline and equipment are protected by the upstream PSH		
b	PSL	1	Install PSL		
		2	Pipeline and equipment are protected by the upstream PSL		
c	PSV	1	Install PSV		
		2	The MAWP of the pipeline and equipment is greater than the maximum pressure that the injection source can apply		
		3	Pipeline and equipment are protected by the upstream PSV		
		4	Pipeline and equipment are protected by HIPPS (High Integrity Pressure Protection System)		
d	FSV	1	Install one or several FSV		

venting system. A flame arrestor should be installed in the venting system to prevent backfire. A vacuum pressure relief device (PSV) or a second-stage venting system should be installed on the atmospheric vessel to protect it, in case the control device of the first-stage venting system becomes dirty and blocks fluid flow.

3.4.5.2 Liquid Level Safety Devices (LSHH and LSLL)

Atmospheric vessels should be protected by LSHH sensors to cut off the inlet fluid and prevent liquid overflow. Except for filling operations continuously monitored by operators or vessels where overflow can be input to other process equipment. If the

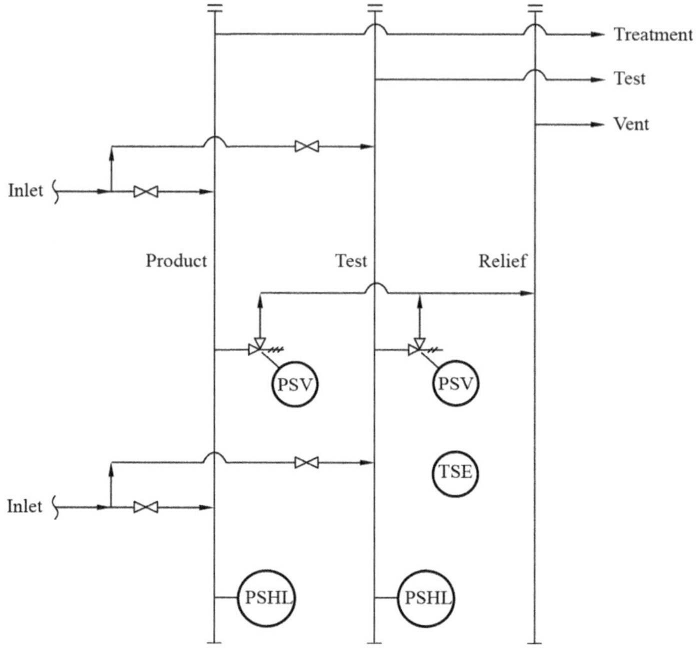

Fig. 3.9 Safety devices: manifold

Table 3.11 Safety Analysis Table (SAT): manifold

Accidental event	Causes	Detectable abnormal conditions of equipment
Overpressure	Outlet pipeline blockage or throttling Hydrate blockage Upstream control device failure Inlet flow too large	High pressure
Leakage	Pipe material deterioration Erosion Corrosion Impact damage Vibration	Low pressure

vessel is equipped with a submersible heating element that can withstand overheating, an LSLL sensor should be installed to cut off the heat source. The LSLL sensor should cut off the inlet fluid to prevent leakage, unless the liquid level in the vessel is not automatically maintained. When the normal input of the liquid may interfere with the sensor's leak detection, it is better to use a leak collection and discharge system to collect the leaked liquid than a low liquid level sensor.

Table 3.12 Safety Analysis Checklist (SAC): manifold

Serial number	Facility	Number	Optional item	Corresponding protection in PID	SAFE noted as
A3 Manifold					
a	PSH	1	Install PSH		
		2	Each input source is equipped with PSH, and the set value of PSH is less than the MAWP of the manifold		
		3	The manifold is protected by the downstream PSH and cannot be isolated from the manifold		
		4	The manifold is used for torches, pressure relief, venting, and other atmospheric purposes, and there is no valve on the outlet pipeline		
b	PSL	1	Install PSL		
		2	Each input source is protected by PSL, and there is no pressure control or throttling device between PSL and the manifold		
		3	The manifold is used for torches, pressure relief, venting, and other atmospheric purposes		
c	PSV	1	Install PSV		
		2	The MAWP of the manifold is greater than the shut-in pressure of any well		

(continued)

Table 3.12 (continued)

Serial number	Facility	Number	Optional item	Corresponding protection in PID	SAFE noted as
		3	Each input source provides pressure relief protection, and the maximum shut-off pressure of the input source is greater than the MAWP of the manifold MAWP		
		4	The manifold is protected by the downstream PSV, and the PSV cannot be separated from the manifold		
		5	The manifold is used for torches, pressure relief, venting, and other atmospheric purposes, and there is no valve on the outlet pipeline		
		6	The input source is the wellhead, the wellhead pressure is greater than the MAWP of the manifold, the wellhead is equipped with two independent PSH-controlled SDVs (one of which can be an SSV), and the PSH is connected to the isolation relay and test point. Other input sources with pressure greater than the MAWP of the manifold are protected by PSV		
		7	When the wellhead pressure is higher than the WAMP of the manifold, it is protected by HIPPS (High-Pressure Protection System)		

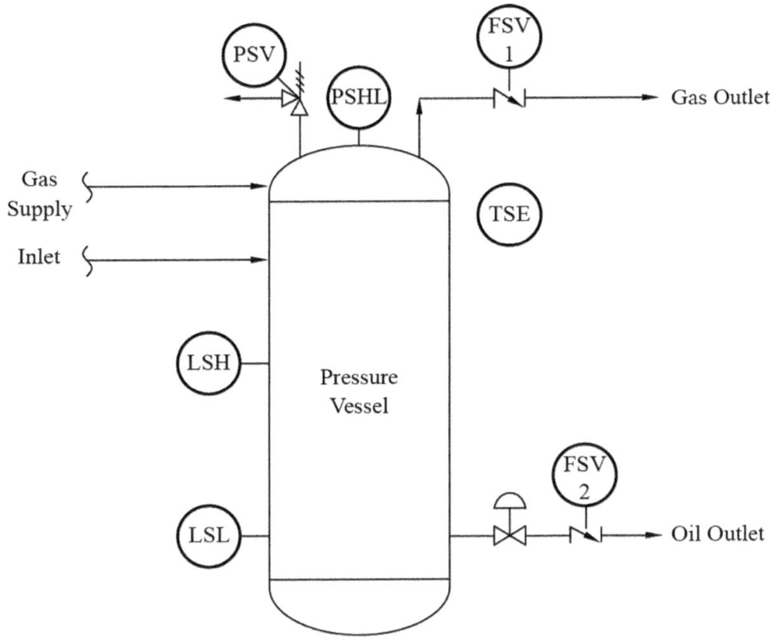

Fig. 3.10 Safety devices: pressure vessels. *Note 1* TSE is a symbol and does not reflect the actual position and quantity. *Note 2* If the pressure vessel is a high-temperature vessel, a TSH should be installed. *Note 3* If the pressure vessel is a low-temperature vessel, a TSL should be installed. *Note 4* The number of safety devices in the figure is for reference only and should be determined according to actual needs

3.4.5.3 Temperature Safety Devices (TSH)

If the atmospheric vessel needs to be heated, a TSH sensor should be installed, which can cut off the heat source when the process fluid overheats. The recommended safety devices for atmospheric vessels are shown in Fig. 3.11, and the safety analysis table and safety analysis checklist are shown in Tables 3.15 and 3.16.

3.4.6 Heating Equipment

Heating equipment is used to process and heat hydrocarbon substances. This section describes the protective measures required for the heating section of naturally ventilated heating equipment. The protective measures required for the process part of the heating equipment are described in the corresponding equipment.

The recommended safety devices for heating equipment are shown in Figs. 3.12, 3.13 and 3.14. The safety analysis table and safety analysis checklist are shown in Tables 3.17 3.18 3.19 and 3.20.

Table 3.13 Safety Analysis Table (SAT): pressure vessels

Accidental event	Causes	Detectable abnormal conditions of equipment
Overpressure	Outlet pipeline blockage or throttling; inlet flow exceeds outlet flow Gas breakthrough (upstream equipment); pressure control system failure Thermal expansion Excessive heat input Fire	High pressure
Negative pressure (vacuum)	Outlet flow exceeds inlet flow Cold shrinkage Outlet opening is too large Pressure control system failure	Low pressure
Overflow	Inflow of liquid exceeds outflow of liquid; Segment blockage Liquid outlet blockage or throttling; Liquid level control system failure	High liquid level
Gas breakthrough	Liquid outlet flow exceeds inlet flow; Liquid outlet opening is too large Liquid level control system failure	Low liquid level
Leakage	Pipe material deterioration Erosion Corrosion Impact damage Vibration	Low pressure Low liquid level
Overtemperature (high temperature)	Temperature control system failure Inlet temperature too high	High temperature
Overtemperature (low temperature)	Temperature control system failure Inlet temperature too low Ambient temperature too low Release or rapid depressurization	Low temperature

3.4.7 Pumps

Pumps are used to transport fluids within the production process system and feed them into the export pipeline. Export pumps are used to transport hydrocarbons produced from the process system into the pipeline. Occasionally, pumps that transport small amounts of hydrocarbons from auxiliary equipment (such as suction tanks, waste oil tanks, etc.) into the pipeline cannot be considered as export pumps. The pipeline mainly transports large quantities of hydrocarbons from other sources.

Table 3.14 Safety Analysis Checklist (SAC): pressure vessels

Serial number	Facility	Number	Optional item	Corresponding protection in PID	SAFE noted as
A4 Pressure vessel					
a	PSH	1	Install PSH		
		2	The input source comes from a pump or compressor, and the pressure they generate cannot exceed the MAWP of the container		
		3	The input source is not a wellhead oil pipeline, production manifold or pipeline, and each input source has a PSH protection, which also protects the container		
		4	The gas outlet is connected to downstream equipment with a suitable pipeline without a cutoff valve or regulating valve, which is protected by a PSH that also protects the upstream equipment		
		5	The container is the last stage gas scrubber of the torch, venting or venting system, and its design pressure can withstand the maximum allowable gathering pressure		

(continued)

Table 3.14 (continued)

Serial number	Facility	Number	Optional item	Corresponding protection in PID	SAFE noted as
		6	The container operates at atmospheric pressure and has an appropriate venting system		
b	PSL	1	Install PSL		
		2	The minimum pressure during operation is atmospheric pressure		
		3	Each input source is protected by PSL, and there is no control device or throttling device between PSL and the container		
		4	The container is a gas scrubber or small separator rather than process equipment and is provided with sufficient protection by the downstream PSL or improved design function (such as the container is a gas scrubber of the pneumatic safety system or the last stage gas scrubber of the torch, venting or venting)		

(continued)

Table 3.14 (continued)

Serial number	Facility	Number	Optional item	Corresponding protection in PID	SAFE noted as
		5	The gas outlet is connected to downstream equipment with a suitable pipeline without a cutoff valve or regulating valve, which is protected by a PSL that also protects the upstream equipment		
c	PSV	1	Install PSV		
		2	Each input source is protected by a PSV, whose set pressure is not higher than the MAWP of the container, and a PSV is installed on the container to protect against fire heat and thermal expansion		
		3	Each input source is protected by a PSV, whose set pressure is not higher than the MAWP of the container, and at least one of them will not be isolated from the container		
		4	The PSV on the downstream equipment can meet the venting requirements of the container and will not be cut off from the container		

(continued)

Table 3.14 (continued)

Serial number	Facility	Number	Optional item	Corresponding protection in PID	SAFE noted as
		5	The container is the last stage gas scrubber of the torch, venting or venting system, and its design pressure can withstand the maximum allowable gathering pressure, and there are no internal or external flow obstructions, such as fogging gas, back pressure valve or fire arrester		
		6	The container is the last stage gas scrubber of the torch, venting or venting system, and its design pressure can withstand the maximum allowable gathering pressure, and a rupture disc or safety cap (PSE) is installed to bypass internal or external flow obstructions, such as fogging gas, back pressure valve or fire arrester		

(continued)

Table 3.14 (continued)

Serial number	Facility	Number	Optional item	Corresponding protection in PID	SAFE noted as
		7	When there is a flow upstream with a pressure higher than the WAMP of the container, it is protected by HIPPS (High Integrity Pressure Protection System); Overpressure caused by other reasons is protected by PSV		
d	LSH (oil)	1	Install LSH		
		2	The downstream equipment of the gas outlet is not a torch or venting system, and can safely handle the maximum liquid carryover		
		3	The characteristics of the container do not require the handling of separated liquid phase		
		4	The container is a small catcher, and its internal liquid can be manually discharged		
e	LSL	1	Install LSL		

(continued)

Table 3.14 (continued)

Serial number	Facility	Number	Optional item	Corresponding protection in PID	SAFE noted as
		2	The liquid level in the container is not automatically maintained, and the container does not have any submersible heating elements and gas phase heating elements that can cause overheating		
		3	For containers that need to control the gas/liquid interface, the downstream equipment of the liquid outlet can safely handle the maximum gas flow that can be discharged through the liquid outlet, and there are no submersible heating elements in the container that can cause overheating; throttling facilities on the discharge pipeline can be used to limit gas flow		

(continued)

Table 3.14 (continued)

Serial number	Facility	Number	Optional item	Corresponding protection in PID	SAFE noted as
		4	For containers that need to control the oil/water interface, the downstream equipment of the liquid outlet can safely handle the maximum gas/liquid flow that can be discharged through the liquid outlet, and there are no submersible heating elements in the container that can cause overheating		
f	FSV	1	Install FSV on each outlet		
		2	The maximum amount of hydrocarbon material flowing back from downstream equipment can be ignored		
		3	The control device on the pipeline can effectively reduce backflow		
g	TSH	Note	TSH is only used when the container has a heat source		
		1	Install TSH		
		2	Heat sources can cause overheating		
h	TSL	Note	Use TSH only when the container may be cooled		

(continued)

Table 3.14 (continued)

Serial number	Facility	Number	Optional item	Corresponding protection in PID	SAFE noted as
		1	Install TSL		
		2	Materials need to adapt to low temperature conditions under normal and abnormal conditions		

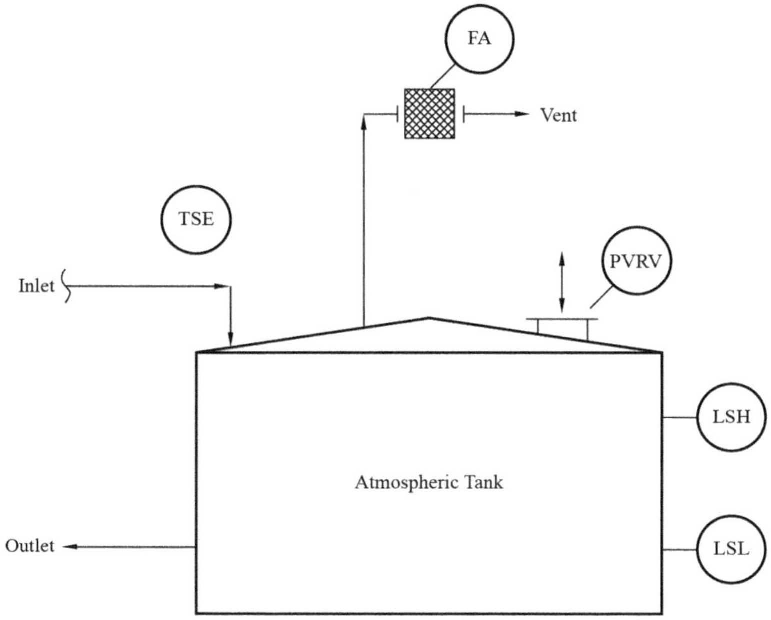

Fig. 3.11 Recommended safety devices for atmospheric vessels. *Note 1* TSE is a symbol, not reflecting the actual position and quantity. *Note 2* If the pressure vessel needs to be heated, TSE must be installed. *Note 3* Pressure and/or vacuum relief devices can be set up on the venting pipeline. *Note 4* A second venting pipeline can be installed to replace the pressure/vacuum relief device

3.4.7.1 Pressure Safety Devices (PSHH, PSLL, and PSV)

All short export pump discharge pipelines should be equipped with PSHH and PSLL sensors to cut off the inflow and shut down the pump. Other types of pump discharge pipelines should also be equipped with PSHH sensors for shutting down the pump, unless the pump's maximum discharge pressure does not exceed 70% of the maximum allowable working pressure of the discharge pipeline, or the pump is

Table 3.15 Safety Analysis Table (SAT): atmospheric vessels

Accident event	Cause	Equipment detectable abnormal state
Overpressure	Outlet pipeline blockage or throttling; Inlet flow exceeds outlet flow Gas breakthrough (upstream equipment); Pressure control system failure Thermal expansion; excessive heat input; fire	High pressure
Underpressure (vacuum)	Outlet flow exceeds inlet flow Cold shrinkage Pressure control system failure	Low pressure
Overflow	Inflow of liquid exceeds outflow of liquid Liquid outlet blockage or throttling level control system failure	High liquid level
Leak	Pipe material deterioration Erosion Corrosion Impact damage Vibration Vacuum rupture	Low liquid level
Overheating	Temperature control system failure Inlet temperature too high	High temperature

manually operated and under continuous surveillance. All export pump discharge pipelines should be equipped with PSV, unless it is a kinetic pump like a centrifugal pump, and it is impossible to generate a head pressure exceeding the maximum allowable working pressure of the discharge pipeline. Other types of pump discharge pipelines should also be equipped with PSV, unless the pump's maximum discharge pressure is lower than the pipeline's maximum allowable working pressure, or the pump has the ability to release internal pressure.

3.4.7.2 Check Valve (FSV)

A check valve (FSV) should be installed on the pump discharge pipeline to minimize backflow.

The recommended pump safety devices are shown in Figs. 3.15, 3.16, 3.17, 3.18 and 3.19. The safety analysis table and safety analysis checklist are shown in Tables 3.21 and 3.22.

Table 3.16 Safety Analysis Checklist (SAC): atmospheric vessels

Serial number	Facility	Number	Optional item	Corresponding protection in PID	SAFE noted as
A.5 Atmospheric vessels					
a	Venting system	1	Install venting system		
		2	The vessel should be protected by HIPPS (High Integrity Pressure Protection System)		
b	Pressure, vacuum relief device	1	Install PSV		
		2	The vessel has a second-stage venting system capable of handling the maximum gas volume		
		3	The equipment is a pressure vessel that cannot be crushed, operates at atmospheric pressure, and is equipped with a properly sized venting system		
		4	The container has no pressure source (filling gas and manual discharge trigger) and is equipped with a venting system suitable for its size		

(continued)

Table 3.16 (continued)

Serial number	Facility	Number	Optional item	Corresponding protection in PID	SAFE noted as
		5	The container should be protected by HIPPS (High Integrity Pressure Protection System)		
c	LSH	1	Install LSH		
		2	Filling operations are continuously monitored by operators		
		3	Overflow can be input or stored in other process equipment		
d	LSL	1	Install LSL		
		2	Install an appropriate leak collection and discharge system		
		3	The liquid level in the container is not automatically maintained, and there are no immersion heating elements in the container that can cause overheating		

(continued)

Table 3.16 (continued)

Serial number	Facility	Number	Optional item	Corresponding protection in PID	SAFE noted as
		4	The equipment is the final container of the leak collection and discharge system, and the design of the leak collection and discharge system is used to collect and discharge hydrocarbon liquids to a safe place		
e	TSH	Note	Use TSH only when the container has a heat source		
		1	Install TSH		
		2	The heat source can cause overheating		

3.4.8 Compressor

The compressor unit is used to transport gaseous hydrocarbons within the production process system and into the export pipeline from the platform.

SAT analyzed the cylinders and casings of the compressor unit, as well as the suction pipelines, discharge pipelines, and fuel gas pipelines of the compressor unit. Hydrocarbon handling equipment related to the compressor, other than the compressor cylinders and casings, should be protected according to the relevant sections of this book. Compressors and prime movers are usually equipped with devices to prevent mechanical damage. Overpressure, leakage, and overheating are accidental events that affect the normal operation of the compressor.

3.4.8.1 Pressure Safety Devices (PSH, PSL, and PSV)

Each suction pipeline of the compressor unit should be equipped with PSH and PSL sensors, unless each input source has PSH, PSL sensors protect the compressor, and each compressor discharge line should also be equipped with PSH and PSL sensors.

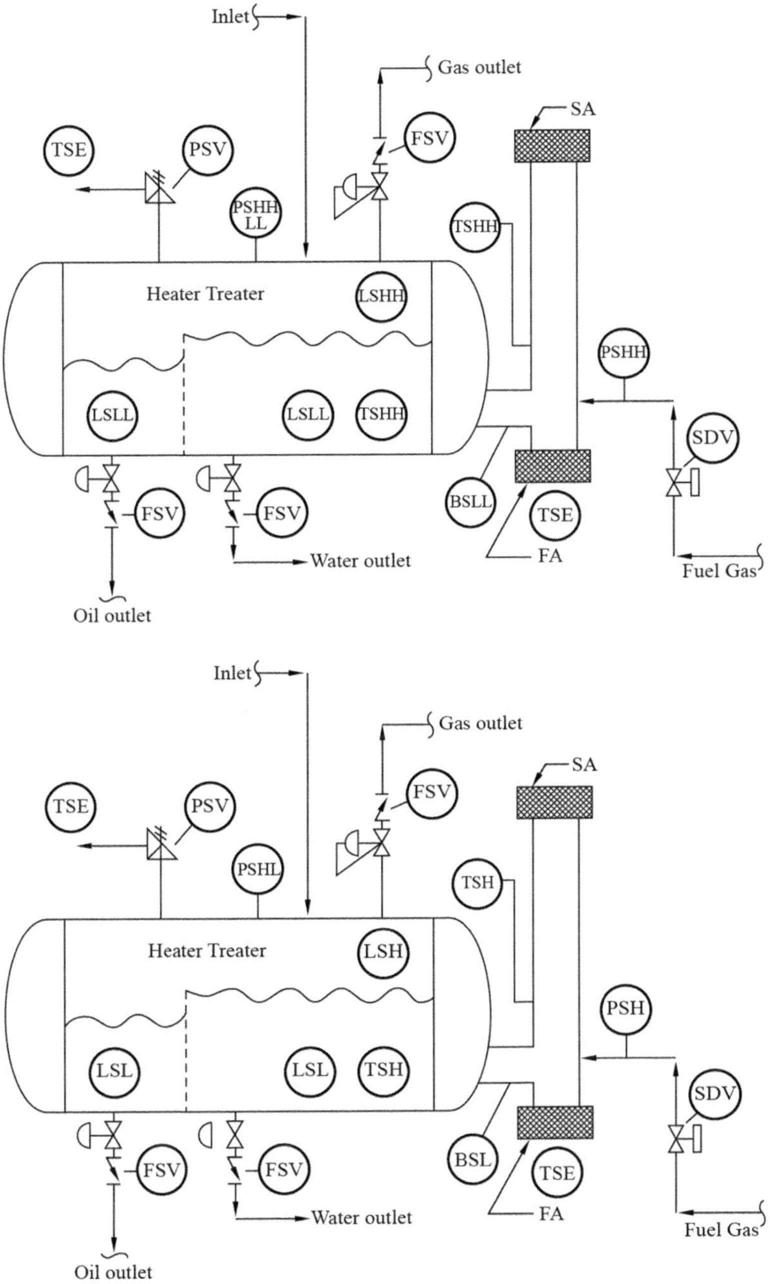

Fig. 3.12 Safety device: combustion container (natural ventilation). *Note 1* TSE is a symbol and does not reflect the actual position and quantity. *Note 2* The container part should be determined based on the previous corresponding analysis

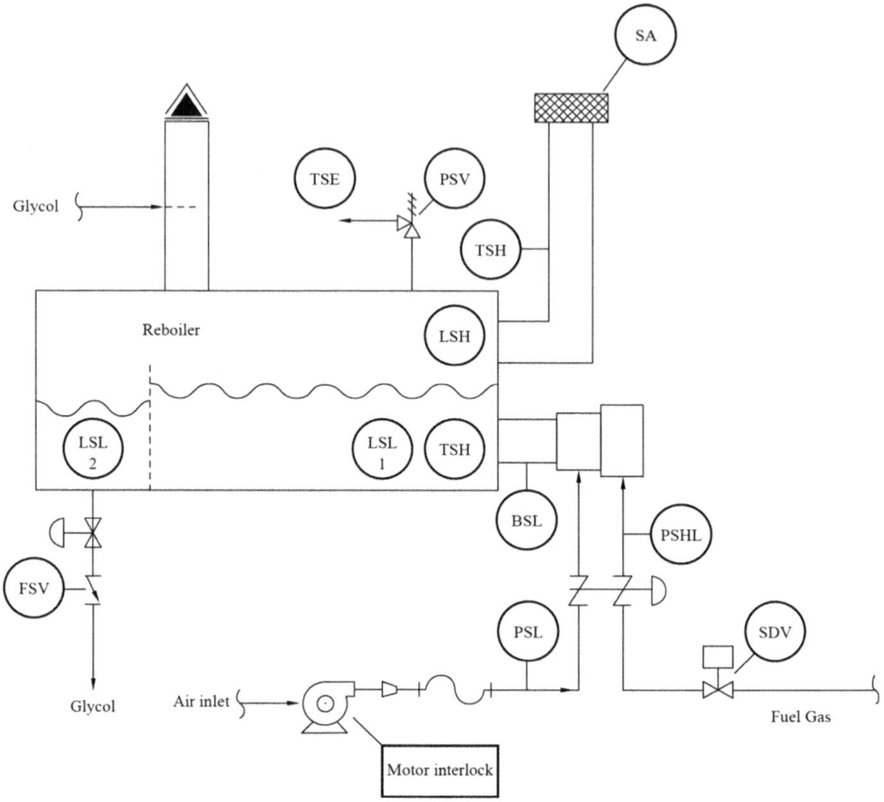

Fig. 3.13 Safety device: combustion container (forced ventilation)

PSH and PSL sensors should cut off all process input lines and fuel gas lines of the compressor. Each compressor suction line should be equipped with a PSV, unless each input source has PSV protection and can also protect the compressor. Each compressor discharge line should also be equipped with a PSV. If the compressor is kinetic and cannot generate pressure exceeding the maximum allowable working pressure of the compressor or discharge line, there is no need to install a PSV at the compressor outlet.

3.4.8.2 Flow Safety Device (SV)

Each final discharge line should be equipped with a check valve (FSV) to minimize backflow.

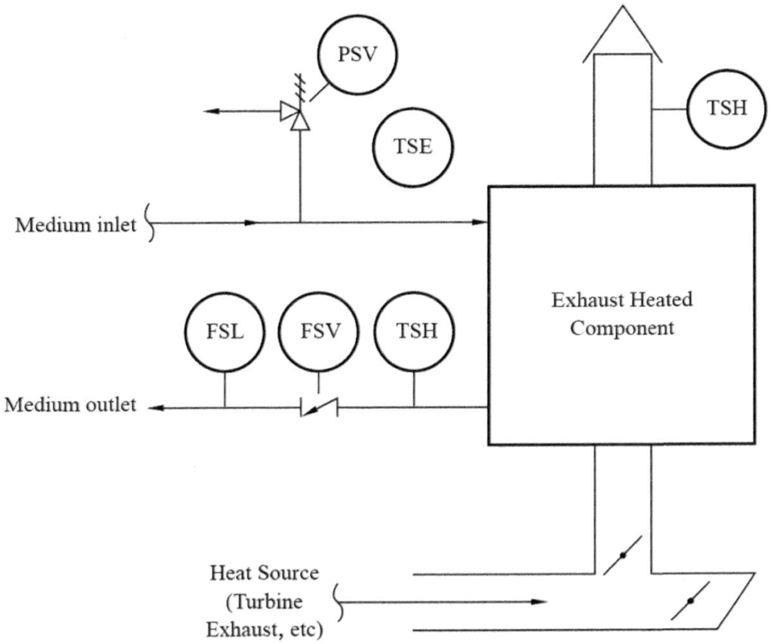

Fig. 3.14 Safety device: exhaust—heating element

Table 3.17 Safety Analysis Table (SAT): heating equipment (natural ventilation)

Accidental event	Cause	Equipment detectable abnormal state
Overheating	Failure of temperature control system Low medium flow rate Heat transfer is restricted Medium leaking into the combustion chamber is ignited Surface of heat transfer system exposed	High temperature (process) High temperature (flue) Low flow rate Low liquid level
Direct ignition source	Air inlet flame ejection Flue gas sparks Flue temperature too high Heat source surface exposed	Fire High temperature (flue)
Excessive combustible vapor in the combustion chamber	Failure of fuel control system	Flame extinguished High pressure (fuel) Low pressure (fuel)

Table 3.18 Safety Analysis Table (SAT): heating equipment (forced ventilation)

Accidental event	Cause	Equipment detectable abnormal state
Overheating	Failure of temperature control system Low medium flow rate; heat transfer is restricted Medium leaking into the combustion chamber is ignited Surface of heat transfer system exposed	High temperature (process) High temperature (flue) Low flow rate Low liquid level
Direct ignition source	Air inlet flame ejection Flue gas sparks Flue temperature too high Heat source surface exposed	Fire High temperature (flue)
Excessive combustible vapor in the combustion chamber	Failure of fuel control system; failure of air supply control system Air inlet blocked Fan failure	Low pressure (air) Flame extinguished High pressure (fuel) Low pressure (fuel) Low flow rate (air)

Table 3.19 Safety Analysis Table (SAT): exhaust—heating element

Accidental event	Cause	Equipment detectable abnormal state
Overheating	Failure of temperature control system Low medium flow rate Heat transfer is restricted Medium leaking into the combustion chamber is ignited Surface of heat transfer system exposed	High temperature (process medium) High temperature (flue) Low flow rate Low liquid level
Direct ignition source	Exhaust pipe sparks Flue temperature too high Heat source surface exposed	Fire High temperature (flue)

3.4.8.3 Gas Detection Device (ASH)

If the compressor unit is installed in a poorly ventilated building or within a wall, a gas detection device (ASH) should be installed to cut off all process input lines and fuel gas lines and vent the compressor.

Table 3.20 Safety Analysis Checklist (SAC): fire and flue gas heated equipment

Serial number	Facility	Number	Optional item	Corresponding protection in PID	SAFE noted as
A.6 Fire and flue gas heated equipment					
a	TSH (medium or process fluid)	1	Install TSH		
		2	The equipment is a steam boiler protected by PSH. If on fire, the equipment needs to be protected by LSL		
		3	The equipment is an indirect water bath heater working under normal pressure protected by LSL		
b	TSH (flue)	1	Install TSH		
		2	The equipment is isolated and does not process any combustible medium or process fluid other than fuel		
		3	The equipment has no fire for auxiliary heating, only uses flue gas for heating, and the medium is non-combustible		
c	PSL (air supply)	1	Install PSL		
		2	The equipment is a natural ventilation burner		
		3	Forced ventilation burner equipped with other types of low air supply sensors		
		4	The equipment has no fire for auxiliary heating, only uses flue gas for heating		
d	PSH (fuel supply)	1	Install PSH		

(continued)

Table 3.20 (continued)

Serial number	Facility	Number	Optional item	Corresponding protection in PID	SAFE noted as
		2	The equipment has no fire for auxiliary heating, only uses flue gas for heating		
e	PSL (fuel supply)	1	Install PSL		
		2	The equipment is a natural ventilation burner		
		3	The equipment has no fire for auxiliary heating, only uses flue gas for heating		
f	BSL	1	Install BSL (flame failure sensor)		
		2	The equipment has no fire for auxiliary heating, only uses flue gas for heating		
g	FSL (heat medium)	1	Install FSL (low flow sensor)		
		2	The equipment is not a closed heat transfer type; in closed heat transfer type equipment, combustible medium circulates in the coil in the combustion chamber or flue gas heating chamber		
h	Motor interlock device (forced ventilation)	1	Install Motor Interlock Device		
		2	The equipment is a natural ventilation burner		

(continued)

Table 3.20 (continued)

Serial number	Facility	Number	Optional item	Corresponding protection in PID	SAFE noted as
		3	The equipment has no fire for auxiliary heating, only uses flue gas for heating		
i	Fire damper (air intake)	1	Install Fire Damper		
		2	The equipment is a forced ventilation burner		
		3	The equipment is isolated and does not process any combustible medium or process fluid other than fuel		
		4	The equipment has no fire for auxiliary heating, only uses flue gas for heating		
j	Flue fire damper	1	Install flue fire damper		
		2	The equipment is a forced ventilation burner and meets: the heated fluid is non-combustible or at the outlet of the heat exchange part, the burner's discharge pressure is higher than the fluid's pressure (pressure head)		
		3	The equipment is isolated, so the process fluid will not contact the flue discharge		
		4	The equipment has no fire for auxiliary heating, only uses flue gas for heating		

(continued)

Table 3.20 (continued)

Serial number	Facility	Number	Optional item	Corresponding protection in PID	SAFE noted as
k	PSV heat medium circulation coil	1	Install PSV		
		2	The equipment is not a coil heater		
		3	PSV installed on another equipment will provide effective protection for this equipment, And the PSV is always connected to the coil		
l	FSV heat medium circulation coil	1	Each outlet is equipped with FSV		
		2	The maximum volume of combustible medium that may flow back from downstream equipment is very small, or the medium is non-combustible		
		3	The equipment is not a coil heater		

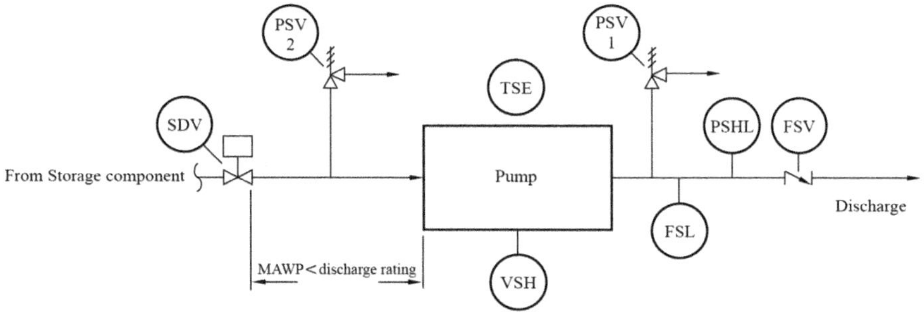

Fig. 3.15 Safety device: pipeline booster pump. *Note* TSE is a symbol and does not reflect the actual position and quantity

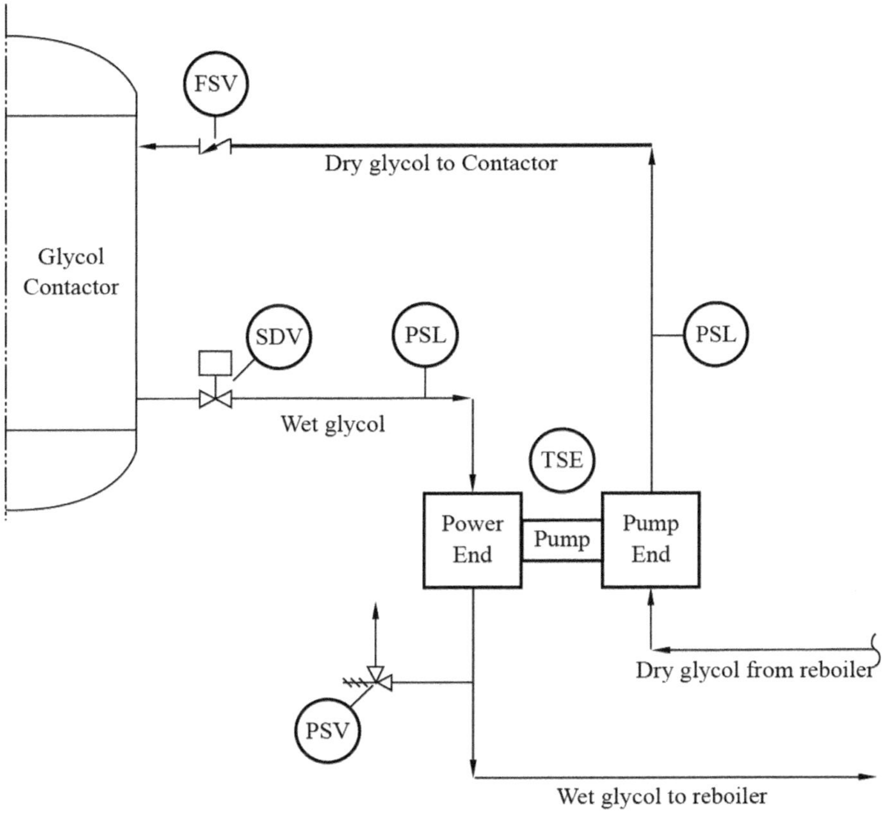

Fig. 3.16 Safety device: ethylene glycol booster pump

3.4.8.4 Temperature Safety Device (TSH)

TSH sensors should be installed to protect all compressor cylinders and casings. TSH sensors should cut off all process input lines and fuel gas lines.

The recommended compressor safety devices are shown in Fig. 3.20, and the safety analysis table and safety analysis checklist are shown in Tables 3.23 and 3.24.

3.4.9 Pipeline

Offshore pipelines are used to transport liquids and gases between platforms or between platforms and land. Depending on the flow direction, pipelines are divided into inlet and outlet pipelines and bidirectional pipelines. The inflow pipeline delivers fluid to the platform, the outflow pipeline outputs fluid from the platform. Bidirectional pipelines can transport fluid in any direction. The recommended safety devices

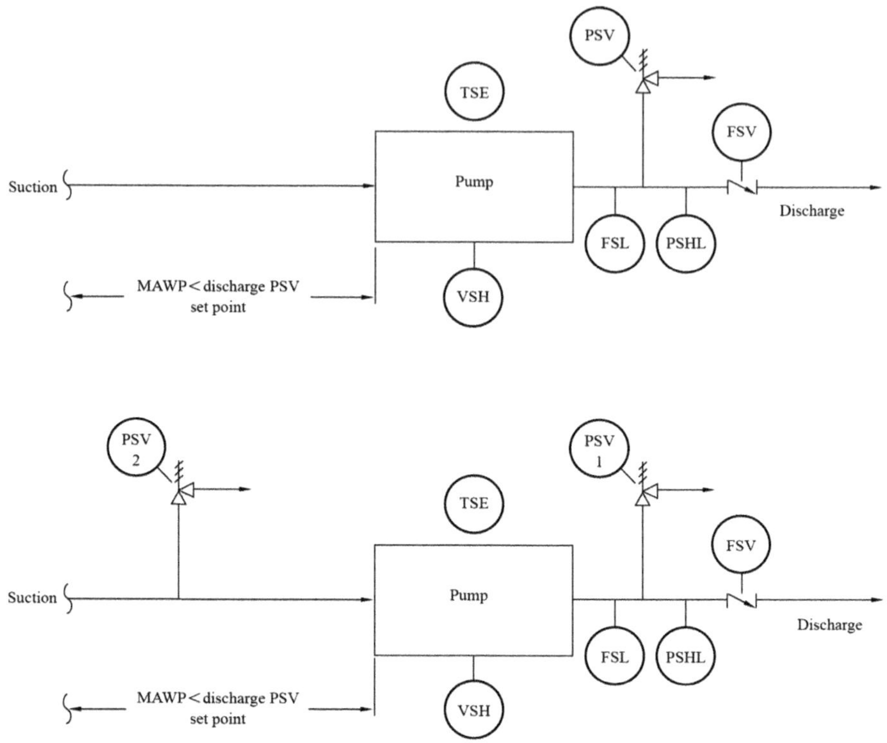

Fig. 3.17 Safety device: other pumps

for offshore pipelines are shown in Fig. 3.21, and the safety analysis table and safety analysis checklist are shown in Tables 3.25 and 3.26.

3.4.10 Heat Exchanger

Heat exchangers are used to transfer heat between two isolated fluids. When analyzing the pressure safety devices of the heat exchanger, the two parts of the heat exchanger (the heated part and the heating part) should be analyzed separately, because the designs of these two parts may be different, and the requirements for operating pressure may also be different. The recommended safety devices for heat exchangers are shown in Fig. 3.22, and the safety analysis table and safety analysis checklist are shown in Tables 3.27 and 3.28.

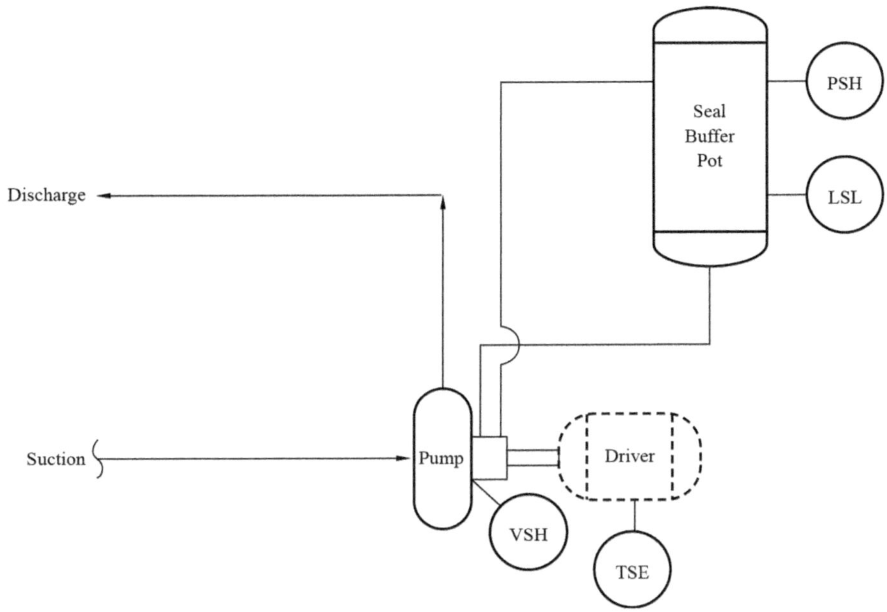

Fig. 3.18 Safety device: cantilever centrifugal pump sealing system

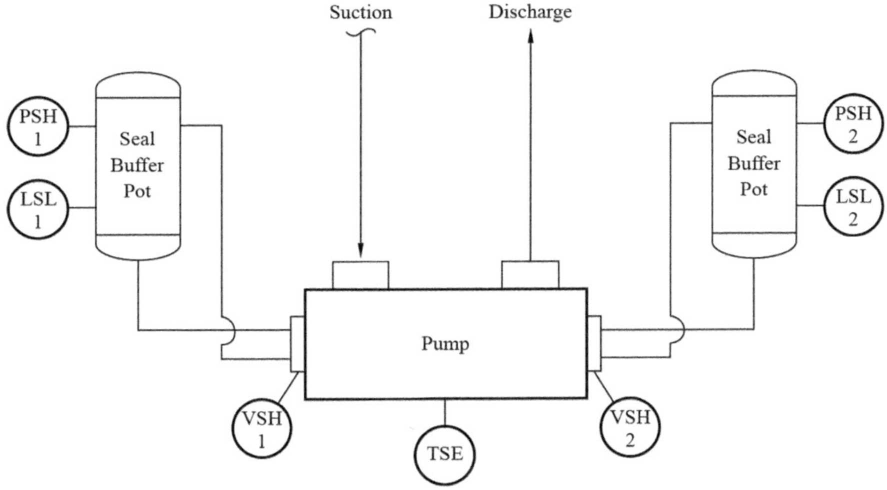

Fig. 3.19 Safety device: bearing type centrifugal pump sealing system

Table 3.21 Safety Analysis Table (SAT): pump

Accident event	Cause	Equipment detectable abnormal state
Overpressure	Discharge pipeline blockage or throttling High back pressure High inlet pressure (centrifugal) Over-speed operation Fluid density increase Backflow	High pressure Low flow rate
Leakage	Pipe material deterioration Erosion Corrosion Impact damage	Low pressure; vibration Low flow rate

3.5 HSE Analysis

3.5.1 Hazard Identification (HAZID)

HAZID (Hazard Identification) is a team-based brainstorming technique, usually guided by a hazard checklist or guidelines, benefiting from the collective knowledge and experience formed by the workshop team members, to identify potential health, safety, and environmental hazards [4].

Hazard identification is a purposeful technique used to identify relevant potential hazards and threats early on, so as to eliminate or reduce hazards at the earliest feasible stage of development or investment projects. Early detection of hazards helps to implement risk reduction measures sooner, reducing the impact on cost and schedule compared to later implementation of risk reduction measures.

3.5.2 Environmental Impact Identification (ENVID)

Environmental Impact Identification (ENVID) is a team-based brainstorming technique, usually conducted under the guidance of checklists and prompts, and uses the collective knowledge and experience of the workshop team to identify potential health, safety, and environmental hazards. The purpose of environmental impact identification is to systematically identify potential environmental factors and impacts related to the project. Once aspects and impacts related to environmental hazards are identified, they can be assessed and, if necessary, avoided, prevented, or controlled.

Table 3.22 Safety Analysis Checklist (SAC): pump

Serial number	Facility	Number	Optional item	Corresponding protection in PID	SAFE noted as
a	PSH export pump	1	Install PSH		
b	PSH other pumps	1	Install PSH		
		2	The pump's maximum discharge pressure does not exceed 70% of the MAWP of the discharge pipeline		
		3	The pump is manually operated and under continuous surveillance		
		4	The pump is a small low-displacement pump, such as a chemical injection pump		
		5	The pump discharges into an atmospheric pressure container		
		6	Ethylene glycol pump powered by ethylene glycol		
c	PSL export pump	1	Install PSL		
		2	The pump does not transport hydrocarbons		
d	PSL other pumps	1	Install PSL		
		2	The pump is manually operated and under continuous surveillance		
		3	Has a good leak collection and emission system		

(continued)

Table 3.22 (continued)

Serial number	Facility	Number	Optional item	Corresponding protection in PID	SAFE noted as
		4	The pump is a small, low-displacement pump, such as a reagent injection pump		
		5	The pump discharges into a normal pressure container		
e	PSV external pump	1	Install PSV		
		2	The pump is kinetic and cannot generate a head pressure exceeding the pipeline's MAWP		
f	PSV other pumps	1	Install PSV		
		2	The pump's maximum discharge pressure is lower than the pipeline's MAWP		
		3	The pump has the ability to internally relieve pressure		
		4	The pump is an ethylene glycol pump powered by ethylene glycol, and the rated pressure of the rich ethylene glycol low-pressure discharge pipeline is higher than the maximum discharge pressure		

(continued)

Table 3.22 (continued)

Serial number	Facility	Number	Optional item	Corresponding protection in PID	SAFE noted as
		5	The pump is an ethylene glycol pump powered by ethylene glycol, and the rich ethylene glycol low-pressure discharge pipeline is protected by a PSV on the downstream equipment that is always connected to the pump		
g	FSV	1	Install FSV check valve		
h	PSV (pump outlet)	1	Install PSV		
		2	The MAWP of the pump outlet pipeline is higher than the set point of the outlet PSV		
		3	The pressure rating of the pump outlet pipeline is not higher than the inlet pipeline, with no inlet flow pressure exceeding the MAWP of the inlet pipeline		
		4	The pump inlet pipeline is protected by the PSV of the upstream component, which cannot be isolated from the pump		
		5	The pump is an ethanol booster pump		
i	FSL	1	Install FSL		
		2	The pump is a volumetric pump		

(continued)

Table 3.22 (continued)

Serial number	Facility	Number	Optional item	Corresponding protection in PID	SAFE noted as
		3	The pump is manually operated and continuously supervised		
		4	Small capacity pump		
		5	There is no continuous low flow condition (blockage or throttling)		
		6	A reasonable reflux system is installed		
		7	PSH and/or PSL have a shutdown design point to detect flow loss		
j	VSH	1	Install VSH		
		2	The pump motor is 745.7 kW (1000hp) or less		
		3	The pump is manually operated and continuously supervised		
k	LSL centrifugal pump seal buffer tank	1	Install LSL		
		2	The pump motor is 745.7 kW (1000hp) or less		
		3	The pump is manually operated and continuously supervised		
		4	The pump has a secondary seal that can detect faults and stop the pump		
		5	No seal buffer tank installed		
i	PSH	1	Install PSH		
		2	The pump motor is 745.7 kW (1000hp) or less		

(continued)

Table 3.22 (continued)

Serial number	Facility	Number	Optional item	Corresponding protection in PID	SAFE noted as
		3	The pump is manually operated and continuously supervised		
		4	The pump has a secondary seal that can detect faults and stop the pump		
		5	No seal buffer tank installed		

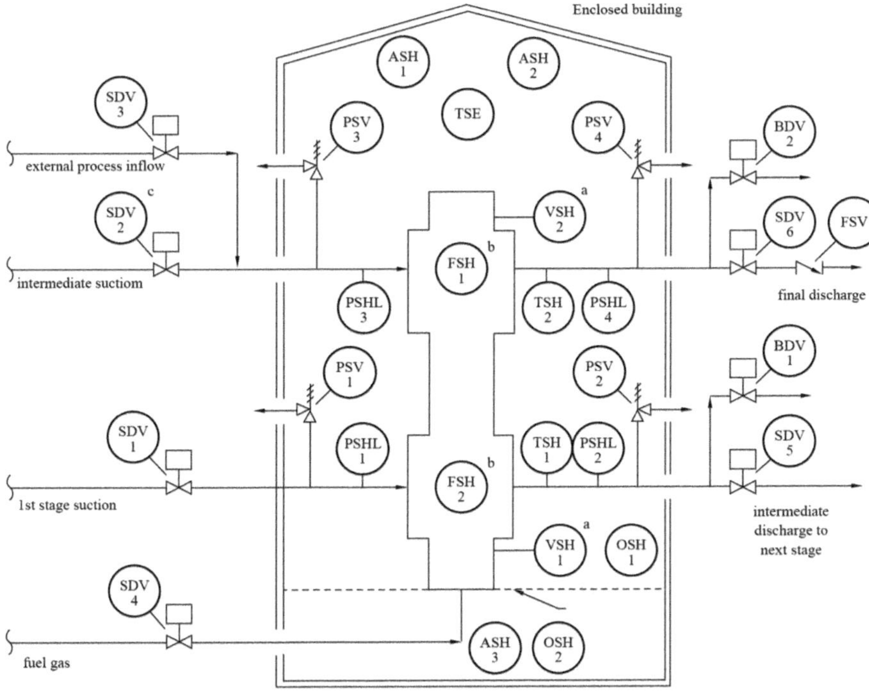

Fig. 3.20 Safety device: compressor. *Note 1* TSE is a symbol and does not reflect the actual position and quantity. *Note 2* If the compressor is not installed in a closed building, there is no need to install ASH 1, ASH 2, ASH 3, and OSH 1, OSH 2. *Note 3* If the compressor has no underground pipelines and potential gas leak sources, there is no need to install ASH 3 and OSH 2; pressure vessels are low-temperature vessels and should be equipped with TSL. *Note 4* Some equipment in the figure is not drawn (such as washers, air coolers, etc.); the number of safety devices in the figure is for reference only and should be determined according to actual needs. a—Placement of vibration sensors depends on compressor installation. b—For centrifugal or screw compressors, FSH is used to detect seal failures. c—Non-essential device

Table 3.23 Safety Analysis Table (SAT): compressor

Accidental event	Cause	Equipment detectable abnormal state
Overpressure (inlet)	Excessive inlet flow Suction pressure control system failure; compressor or driver failure Backflow Equal pressure	High pressure
Overpressure (outlet)	Discharge line blocked or throttled Back pressure too high Inlet pressure too high Overspeed operation	High pressure
High pressure	High pressure pipe material deterioration Erosion Corrosion Impact damage Vibration Suction of liquid phase causing damage Filling/sealing failure	Low pressure High gas concentration (inside building) Vibration Excessive flow (seal gas)
Overtemperature	Compressor valve failure; cooler failure; overcompression ratio Low flow	High temperature

3.5.3 *Occupational Health Hazard Identification (OHID)*

Occupational Health Hazard Identification (OHID) is a team-based brainstorming technique, usually conducted under the guidance of checklists and prompts, and uses the collective knowledge and experience of the workshop team to identify and determine potential health hazards. The purpose of occupational health hazard identification is to systematically analyze potential occupational health hazards caused by project facilities.

The Occupational Health Hazard Identification qualitative workshop should discuss the daily causes or planned events that may lead to occupational exposure (short-term or long-term). Occupational Health Hazard Identification should cover all facilities, workplaces, and related activities performed by employees and contractors that have potential hazards, and should at least include the following:

(1) Daily operational activities and tasks, including activities and tasks performed in an office environment.
(2) Planned maintenance activities.
(3) Emergency response activities to control the accidental release of chemicals, hazardous substances, or energy.
(4) Cleaning activities.
(5) Non-routine activities, such as shutdowns or unplanned maintenance.

Table 3.24 Safety Analysis Checklist (SAC): compressor

Serial number	Facility	Number	Optional item	Corresponding protection in PID	SAFE noted as
A8 Compressor					
a	PSH suction	1	Install PSH		
		2	Each input source has PSH protection, and it can also protect the compressor		
b	PSH discharge	1	Install PSH		
		2	The compressor is protected by a PSH installed downstream of it and upstream of any cooler, and this PSH is always connected to the compressor		
c	PSL suction	1	Install PSL		
		2	Each input source has PSL protection, and it can also protect the compressor		
d	PSL discharge	1	Install PSL		
		2	The compressor is protected by a PSL installed downstream of it, and this PSL is always connected to the compressor		
e	PSV inhale	1	Install PSV		
		2	Each input source has PSV protection, which can also protect the compressor		
f	PSV discharge	1	Install PSV		

(continued)

Table 3.24 (continued)

Serial number	Facility	Number	Optional item	Corresponding protection in PID	SAFE noted as
		2	The compressor is protected by a PSV installed downstream, upstream of any cooler, and the PSV is always connected to the compressor		
		3	The compressor is kinetic and will not generate pressure exceeding the compressor's MAWP		
g	FSV	1	Install FSV		
		2	FSV is installed at the final exhaust port, the compressor is volumetric		
h	TSV	1	Install TSV		
i	VSH	1	Install VSH		
		2	The compressor is manually operated and under continuous monitoring		
j	Secondary seal (main seal vent with FSH—centrifugal, screw compressor)	1	Compressor is less than 745.7 kW (1000hp) and has no recirculation system		
		2	The compressor is manually operated and under continuous monitoring		
		3	The secondary seal has a fault detection and shutdown system		
		4	The compressor has no dry gas seal		

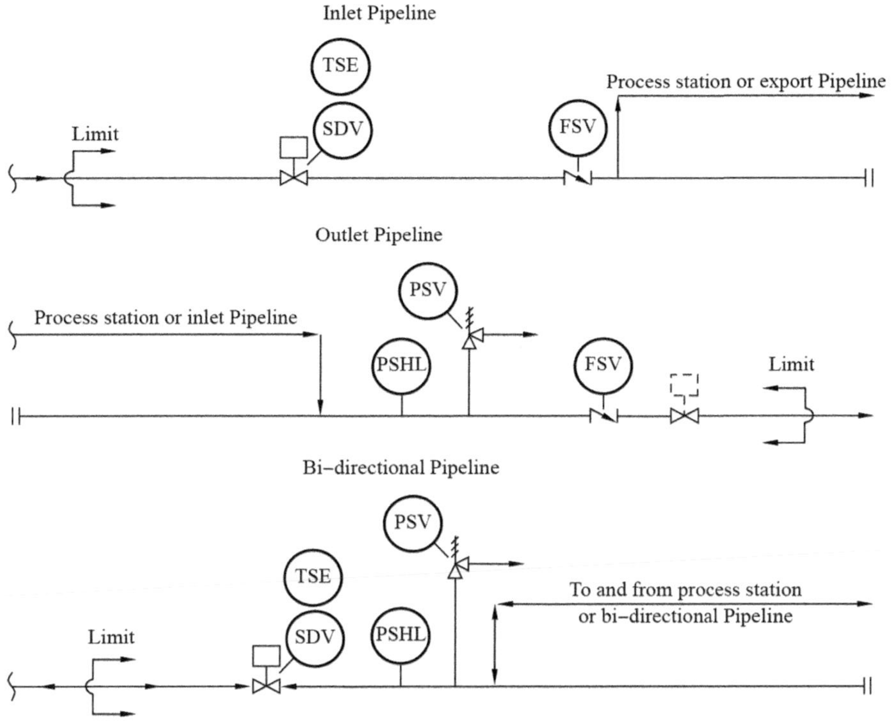

Fig. 3.21 Safety device: offshore pipeline

Table 3.25 Safety Analysis Table (SAT): pipeline

Accidental event	Cause	Equipment detectable abnormal state
Overpressure	Pipeline blockage or throttling Thermal expansion Inlet flow exceeds outlet flow	High pressure
Leakage	Pipe material deterioration; Erosion; corrosion Impact damage Vibration	Low pressure

(6) Waste handling and disposal tasks.
(7) Laboratories, as well as quality assurance and quality control activities.
(8) Other planned or unplanned activities.

Table 3.26 Safety Analysis Checklist (SAC): pipeline

Serial number	Facility	Number	Optional Item	Corresponding protection in PID	SAFE noted as
A9 Pipeline					
a	PSH	1	Install PSH		
		2	The output pipeline is protected by a PSH installed on its upstream equipment		
		3	Each input source has PSH protection, and this PSH can also protect the output pipeline or bidirectional pipeline		
		4	The pipeline is protected by a PSH installed on a parallel component		
b	PSL	1	Install PSL		
		2	The output pipeline is protected by a PSL installed on its upstream equipment		
		3	Each input source has PSL protection, and this PSL can also protect the output pipeline or bidirectional pipeline		
		4	The pipeline is protected by a PSL installed on a parallel component		
c	PSV	1	Install PSV		
		2	The maximum allowable operating pressure of the pipeline is higher than the maximum pressure of all input sources		

(continued)

Table 3.26 (continued)

Serial number	Facility	Number	Optional Item	Corresponding protection in PID	SAFE noted as
		3	Each input source with a pressure higher than the maximum allowable operating pressure of the pipeline is protected by a set of PSV devices, and the maximum allowable operating pressure of this PSV device is not higher than the maximum allowable operating pressure of the pipeline		
		4	The pipeline does not use the platform process system as its input source		
		5	The input source is a production well with a pressure higher than the maximum allowable operating pressure of the pipeline, equipped with two SDVs (one of which can be an SSV), controlled by independent PSHs connected to isolation relays and detection points. Other types of input sources with pressure higher than the maximum allowable operating pressure of the pipeline are protected by PSV		
d	FSV	1	Install FSV		
		2	SDV controlled by PSL installed on the output pipeline		

(continued)

Table 3.26 (continued)

Serial number	Facility	Number	Optional Item	Corresponding protection in PID	SAFE noted as
		3	All input sources on the pipeline are protected by FSV, and their location can prevent backflow in all effective sections of the pipeline		
		4	The pipeline is used for bidirectional flow		

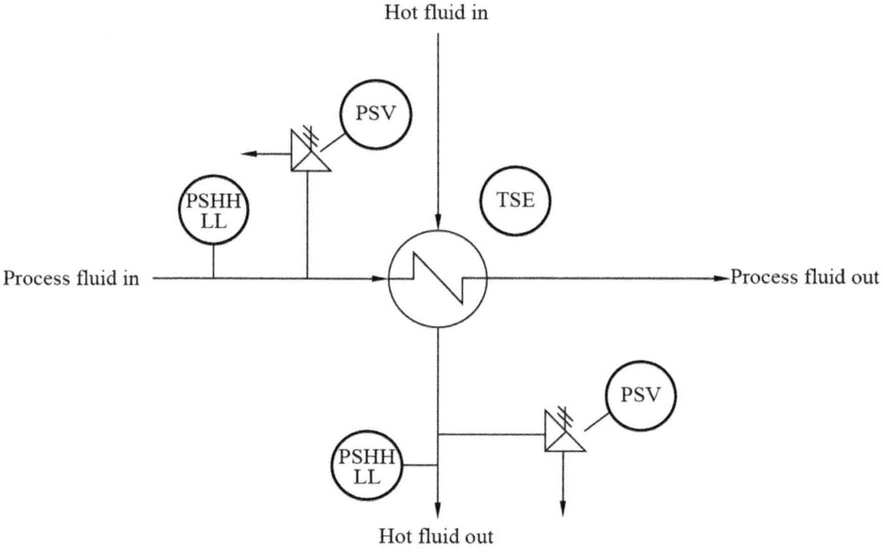

Fig. 3.22 Heat exchanger safety device. *Note* TSE is only a symbol and does not reflect the actual position and quantity

3.5.4 Safety Integrity Level (SIL) Analysis

There are various risks in oil and natural gas process stations, and no matter what technology and management are used, it is impossible to completely eliminate risks, but only to reduce risks to an acceptable range through certain measures. The Safety Integrity Level (SIL) can be simply understood as the quantitative level of risk reduction through system safety design. According to IEC standards, SIL levels are divided into four levels, namely SIL 1, SIL 2, SIL 3, and SIL 4, with SIL 4 being the highest

Table 3.27 Safety Analysis Table (SAT): heat exchanger

Accident event	Cause	Detectable abnormal state of equipment
Overpressure	Blockage or throttling; inflow exceeds outflow Thermal expansion Pipe rupture; vaporization	High pressure
Leakage	Pipe material deterioration Erosion Corrosion Impact damage; vibration	Low pressure
Overtemperature	Control failure Process pipeline outlet blockage; Inlet temperature rise	High temperature
Low temperature	Control failure JT effect or liquid phase vaporization Inlet temperature drop	Low temperature

level and SIL 1 being the lowest level. The higher the Safety Integrity Level (SIL), the higher the probability that the selected equipment will correctly execute the safety design and the higher the reliability. SIL analysis identifies all instrument safety functions (SIFs) that need to be considered in the system, determines all hazardous scenarios that need to be evaluated, and provides reliable design for all instrument safety functions (SIFs). The Safety Instrument System (SIS) is an instrument system used to implement one or several instrument safety functions. The purpose of the Safety Integrity Level (SIL) analysis is to determine the risk reduction level required by the Safety Instrument System (SIS), so that the identified hazards (including explosions, toxic gas diffusion, container rupture, etc.) are reduced to a reasonably acceptable risk level [5].

3.5.5 Quantitative Risk Analysis (QRA)

Quantitative Risk Analysis (QRA) is a method of quantitatively evaluating risk by considering the consequences and frequency of hazards. The main objectives of QRA for oilfield surface engineering are:

(1) Identify hazard sources, hazard escalation scenarios, and mitigation plans for each scenario.
(2) Quantify the overall risk of oilfield facilities or related activities.
(3) Identify the main risks of comprehensive risk.
(4) Identify and analyze effective evaluations, preventive measures, and plans to reduce comprehensive risk.

Table 3.28 Safety Analysis Checklist (SAC): shell and tube heat exchanger serial number

Serial number	Facility	Number	Optional item	Corresponding protection in PID	SAFE noted as
A.10 Shell and tube heat exchanger					
a	PSH	1	Install PSH		
		2	The input source of this part of the heat exchanger will not generate pressure higher than the MAWP of this part of the heat exchanger		
		3	Each input source is protected by PSH, and this PSH also protects this part of the heat exchanger		
		4	Downstream equipment of the heat exchanger is equipped with PSH, and this PSH is not isolated from this part of the heat exchanger due to cut-off valves or control valves		
b	PSL	1	Install PSL		
		2	The minimum operating pressure during operation is atmospheric pressure		

(continued)

Table 3.28 (continued)

Serial number	Facility	Number	Optional item	Corresponding protection in PID	SAFE noted as
		3	PSL installed on other equipment will provide protection for this part of the heat exchanger, and it is always connected with this part when the heat exchanger is working		
c	PRD (PSV or PSE)	1	Install PRD		
		2	Each input source is protected by PRD device, its pressure does not exceed the MAWP of this part of the heat exchanger, and it releases the pressure generated by thermal expansion and fire on this part of the heat exchanger		
		3	Each input source is protected by PRD device, its pressure does not exceed the MAWP of this part of the heat exchanger, and each input source is always connected with the heat exchanger		

(continued)

Table 3.28 (continued)

Serial number	Facility	Number	Optional item	Corresponding protection in PID	SAFE noted as
		4	The PRD on downstream equipment can meet the release requirements of the heat exchanger, and it is always connected with the heat exchanger		
		5	The input source of this part of the heat exchanger will not generate pressure higher than the MAWP of this part of the heat exchanger, and this part of the heat exchanger will not be overpressured due to the temperature or pressure of other parts		
		6	Each input source is protected by PRD or HIPPS, the set value is not higher than the MAWP of the heat exchanger, and the heat exchanger cannot be overpressured due to changes in temperature or pressure of a certain part		
d	TSH	1	Install TSH		

(continued)

Table 3.28 (continued)

Serial number	Facility	Number	Optional item	Corresponding protection in PID	SAFE noted as
		2	The input source temperature of the heat exchanger cannot exceed the maximum allowable working temperature of the heat exchanger		
e	TSL	1	Install TSL		
		2	The input source temperature of the heat exchanger cannot be lower than the minimum allowable working temperature of the heat exchanger		

(5) Determine the safety nature and cost-effectiveness of the proposed preventive measures and plans.
(6) Demonstrate compliance with national risk standards, and demonstrate and implement the As Low As Reasonably Practicable (ALARP) principle.
(7) Facilitate employees and third parties to understand the risks and impacts they will face.
(8) Comply with legal/company requirements.

3.5.6 *Fire Safety Assessment (FSA)*

Fire Safety Assessment (FSA) is a systematic and structured safety analysis that identifies hazards and makes comprehensive and integrated analysis based on quantitative analysis results. Considering factors such as economy, environment, reliability, and safety of the process system, appropriate risk management procedures are formulated to help system operators and managers make safety decisions.

Fire safety assessment mainly calculates risk results based on hazards such as fire, explosion, and toxic gas diffusion, and measures whether the fire partition in the project design process is reasonable, whether the active fire protection design meets the requirements, whether the fire protection of equipment/structures meets the requirements, and whether the escape route and assembly point are reasonable. For design considerations that are not comprehensive or unreasonable, reasonable optimization suggestions and recommended practices are given based on the results of quantitative analysis, and provide effective input data/information for future active fire protection (AFP), passive fire protection (PFP), escape, evacuation and rescue analysis (EERA), emergency system survivability analysis.

(ESSA) and other safety analyses.

3.5.7 Engineering Facility Layout Analysis

Engineering facility layout analysis is a method of analyzing whether the site layout is reasonable and whether the distances between areas meet the requirements based on the results of fire and explosion quantitative simulation calculations for the project site, and finally giving conclusions and suggestions. The scope of this analysis generally includes all process equipment and units that process, transport, and store flammable media within the plant area. The engineering facility layout analysis reviews whether the process unit, storage unit, public equipment unit, manned and unmanned buildings, and fire protection facilities will spread to neighboring units in the event of a fire, explosion, and other accidents, causing the situation to expand.

3.5.8 Bow-Tie Analysis

Bow-tie analysis is a method for identifying and managing major hazards during the project lifecycle, with the aim of identifying the causes and consequences of major hazards losing control, proposing corresponding preventive protection and mitigation barriers, and providing systematic risk management methods. This method dynamically identifies, records, and manages hazards during the project lifecycle, ultimately forming part of the health, safety, and environmental system case.

Bow-tie analysis uses diagrams to analyze identified hazards. It can be understood as a combination of accident tree analysis and event tree analysis. Bow-tie analysis can clearly display the relationship between the causes and possible consequences of hazards, as well as preventive measures to prevent hazards and mitigating measures to limit consequences. It also lists the escalation factors that may prevent these preventive and mitigating measures, showing the effectiveness of various barriers and mitigation measures, making people more clearly aware of deficiencies and areas for improvement, facilitating risk control and tracking, and ensuring their long-term effectiveness.

3.5.9 Emergency Systems Survivability Analysis

Emergency Systems Survivability Analysis (ESSA) is used to assess whether all reasonably practicable measures have been taken to ensure the existence of emergency systems and to analyze the existence of emergency systems.

3.5.10 Escape, Evacuation, and Rescue Analysis

Escape, Evacuation, and Rescue Analysis (EERA) is an HSE risk assessment and management technique used to assess the implementation of facility emergency measures and arrangements. It includes a review of the implementation of escape, evacuation, and rescue (EER) measures and arrangements for dealing with the most incredible hazardous scenarios. The main purposes of conducting escape, evacuation, and rescue analysis are:

(1) Determine the escape route for personnel from the workplace to the assembly point in an emergency.
(2) Considering the nature of the dangers that personnel may face, determine one or more assembly points for evacuation in an emergency.
(3) Ensure that personnel can safely and orderly evacuate the site in an emergency.
(4) Ensure that personnel evacuating the site in an emergency are reasonably arranged and transported to a safe place.
(5) The choice of emergency measures is determined by assessing the hazards of the facility. Based on the complexity of the production process and the facility, the type of equipment, the level of personnel staffing, and the environmental conditions of the operation site, the methods used are evaluated and determined and suggestions are made.

3.5.11 Environmental Impact Assessment (EIA) Analysis

The Environmental Impact Assessment (EIA) is a systematic procedure used to identify potential environmental impacts of any project. It is a preventive method that can enhance environmental management, predict potential adverse impacts of the project, and make appropriate design modifications. It also identifies the potential environmental benefits of the project. Its main value lies in its ability to predict adverse consequences and beneficial impacts early in the project, introducing improvement measures (mitigation measures) into the design process to eliminate or minimize identified adverse impacts, while taking appropriate action to enhance the project's positive impact on the environment. Introducing mitigation measures at the design stage is always more cost-effective than introducing control devices after commissioning, so for design engineers and managers, ensuring appropriate environmental impact assessment at an early stage of project design can bring significant economic benefits. The output of the environmental impact assessment process is provided in the form of an environmental impact assessment report, which provides all the analyses and assessments carried out during the environmental impact assessment, the methodologies and assumptions used, and relevant information on predicted significant impacts and their mitigation measures, and provides recommendations for subsequent environmental management and monitoring.

3.5.12 Occupational Health Risk Analysis

Occupational Health Risk Assessment (OHRA) is a process of demonstrating and assessing all identified high-risk occupational health hazards, and proposing or formulating appropriate controls, mitigation, and recovery measures (for existing facilities and operations) to reduce any remaining HSE risks as much as possible.

The purpose of occupational health management is to:

(1) Protect employees, contractors, and other personnel from potential health hazards related to the work environment.
(2) Promote employee health. The basis for the first purpose is to conduct an Occupational Health Risk Assessment (OHRA), which reviews a particular activity through an occupational health risk assessment to identify potential health hazards and create a hazard list; assesses health risks associated with exposure to these hazards based on specific screening criteria; determines the required control measures (if any) to eliminate or reduce health risks to the lowest reasonable and feasible level. At the same time, emergency (recovery) measures should be considered in case control measures fail, to mitigate acute and/or chronic health impacts.

3.5.13 *Fire and Gas Detector Layout Analysis*

Fire and Gas Detector Mapping Study is a safety analysis that scientifically determines the reasonableness of the design scheme or existing fire and gas detector coverage based on the diffusion analysis model of combustible and toxic gases, using a reasonable methodology and software as a tool. It can also be used as a means to verify whether the layout of fire and gas detectors has achieved the design purpose.

3.5.14 *Project Safety, Environmental Protection, and Occupational Health Review (PHSER)*

The Project HSE Review (PHSER) is generally chaired by a third party independent of the project team. It is a qualitative analysis hearing organized for the project HSE execution plan and its ongoing series of activities, to systematically identify HSE high-risk areas and supervise and control identified risks in the design and operation process of key projects. The PHSER review is a tool to ensure that the project execution and commissioning operations comply with the owner company's HSE regulations, local laws, and regulations.

The main purposes of the PHSER review include:

(1) To provide assurance to senior management, the owner's operations department, and the design management department that all identified HSE risks have been appropriately addressed at each review stage.
(2) To ensure that all company regulations, rules, etc., regarding risk have been reflected in the project design documents.
(3) To ensure that all regulations regarding HSEIA in the project have been completed.
(4) To increase the value from the conceptual design stage (PHSER -I) to the project retirement and demolition stage (PHSER-VII).

References

1. API Standard 521 Pressure-relieving and depressuring systems, Seventh Edition, June 2020.
2. API RP 14E Recommended practice for design and installation of offshore production platform piping systems, Fifth Edition, March 2007.
3. API RP 14C Recommended practice for analysis, design, installation, and testing of basic surface safety systems for onshore production platforms.
4. SY/T Analysis, Design, Installation, and Testing of Safety Systems for Basic Upper Facilities of the 10033 Offshore Production Platform: Recommended Practices.
5. Compiled by Sinopec Shanghai Engineering Co., Ltd. Chemical Process Design Manual. Beijing: Chemical Industry Press, 2018.

Chapter 4
Safety Protection Scheme for Oilfield Surface Engineering Gathering System

Well fluids flow from the oil tree into surface facilities and need to be transported to the central processing station for deep processing through the gathering system. The gathering system mainly includes well site facilities, single well pipelines, metering stations, gathering trunk lines, oil transfer stations, and transfer pipelines, etc. The main facilities vary depending on the station layout. According to the number of station levels from the oil wellhead to the central processing station, the station layout can be divided into first-level, second-level, and third-level layouts.

(1) First-level layout: The process only includes a central processing station.
(2) Second-level layout: The process includes a metering station and a central processing station.
(3) Third-level layout: The process includes a metering station, an oil transfer station, and a central processing station. In recent years, with the continuous improvement of oilfield automation, a "one-and-a-half-level layout process", also known as the well selection point process, has emerged. This process arranges several well selection points outside the central processing station. The well selection points only set up well selection valve groups for well metering, and do not design separators and metering instruments. The well selection point has two pipelines leading to the central processing station, one is a pipeline for oil well metering, connected to the metering separator set up in the central processing station; the other is a gathering pipeline for non-metered oil well flow. China's Ninghai oilfield and Sudan's Mugrande oilfield have adopted the one-and-a-half-level layout process.

K. Fang, *Onshore Oilfield Surface Facilities: Process Safety Engineering Guide*,
https://doi.org/10.1007/978-981-96-7830-3_4

4.1 Safety Protection Scheme for Wellhead and Single Well Pipeline

The typical gathering system process is shown in Fig. 4.1. The wellhead is equipped with wellhead control devices (manual and automatic) and a collection and discharge system for well fluid leaks, and provides a downhole entrance for well repair. The accidental events affecting the single well pipeline are overpressure and leakage. The safety analysis table and safety analysis checklist applicable to the single well pipeline are seen in Chap. 3, Sect. 3.4.

Since the wellhead is the main pressure source, a PSH and PSL sensor should be provided on each oil outlet pipe. These sensors detect abnormal high-pressure conditions. At the same time, the check valve (FSV) is only needed at the end of the single well pipeline to reduce fluid backflow. The following introduces common single well pipeline safety protection device setting schemes.

4.1.1 Full Pressure Design Scheme

The full pressure design scheme is a recommended scheme for single well pipeline safety protection. For overpressure accidents, the design pressure of the single well pipeline can be considered to be increased so that it exceeds the wellbore shut-in pressure. In this way, the pipeline will not burst even if it is overpressured, because the maximum operating pressure will not exceed the wellbore shut-in pressure. The design, installation, operation, and maintenance of the full pressure design scheme are relatively convenient, and the specific configuration is shown in Fig. 4.2.

In the full pressure design scheme, the design pressure of the entire section of the single well pipeline is greater than the wellbore shut-in pressure, and the pressure boundary is downstream of the manual ball valve in the metering station. When the pipeline is overpressured, there is no shut-off and venting action, and the design pressure of the pipeline itself can ensure the safe operation of the single well pipeline under overpressure conditions.

For the accidental event of leakage, there are generally two solutions: one is to realize it through the low-pressure protection system on the oil tree, that is, a low-pressure detection device (PSL) is set up on the oil tree, and once it is lower than the low-pressure set value, a certain valve on the oil tree will act to cut off the oil tree; the other is to set up a PSL alarm device near the wellhead and upload the alarm signal to the control room, and take corresponding measures after manual on-site confirmation; if the wellhead signal cannot be uploaded to the control room, it can rely on manual well patrol to discover the leakage and take measures.

The full pressure design scheme is simple and reliable, but the investment is higher. At the same time, in some situations such as project renovation, the full pressure design scheme lacks flexibility. In this case, alternative scheme one or alternative scheme two can be considered.

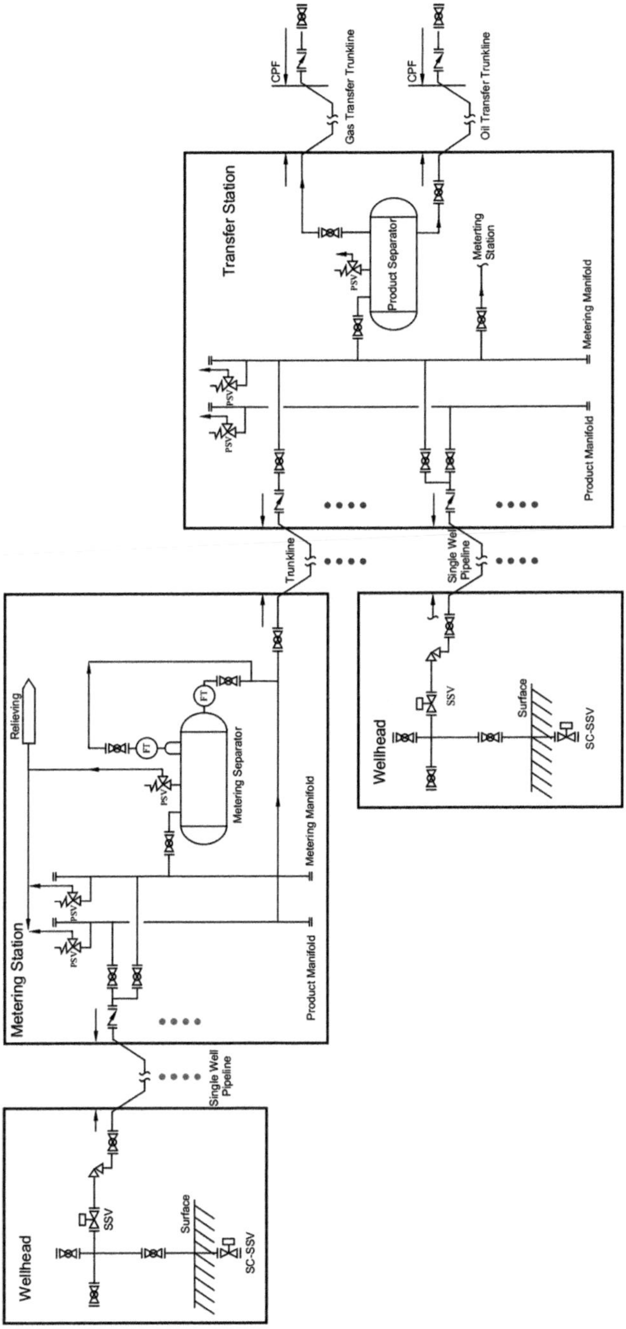

Fig. 4.1 Schematic diagram of the typical gathering system process. *Note* The oil tree in the figure is a typical configuration and does not represent the oil tree configuration of every project

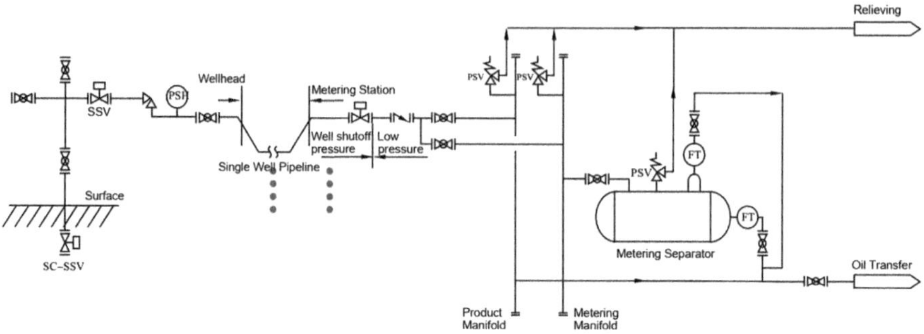

Fig. 4.2 Configuration diagram of the recommended scheme for single well pipeline overpressure protection

4.1.1.1 Alternative Scheme One

For overpressure accidents, alternative scheme one has three protections. The first and second protections are both realized by two SDVs set after the oil nozzle. The third protection is the PSV set at the end of the pipeline. Among them, the first and second SDVs should each have a separate PSH, and the PSH set value of the first SDV should be lower than that of the second SDV, so that step-by-step cut-off can be carried out, and the use of the second protection can be avoided as much as possible. The PSV is used for emergency venting when the SDV fails (such as leakage), so the venting volume of the PSV only needs to consider the leakage volume of the SDV or SSV, and a very small PSV can meet the requirements [1]. The specific configuration is shown in Fig. 4.3.

The prerequisite conditions for using this scheme are as follows:

(1) The volume of the upstream pipeline section of the cut-off valve at the end of the single well pipeline is sufficient to allow the SDV to have enough time to cut off the wellhead before the system pressure exceeds the maximum allowable working pressure.

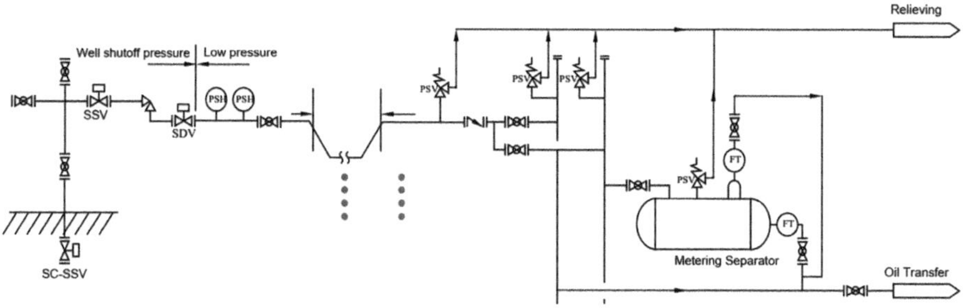

Fig. 4.3 Configuration diagram of alternative overpressure protection scheme for single well pipeline

(2) Both SDVs should be switch-type valves and comply with API 14D standard requirements.
(3) Both SDVs should be regularly inspected on-site to ensure that no leaks occur. In addition, if the single well pipeline can be protected by the PSV of the downstream equipment, and the downstream equipment is not isolated from the single well pipeline, and there is no throttling device and restrictions between the single well pipeline and the PSV, the PSV at the end of the pipeline can be cancelled. For leak accidents, a PSL can be set up at the same time as the PSH, and the SDV can be closed when the set value of the PSL is reached. One of the SDVs in this scheme can be replaced by an SSV on the oil tree, so that one less SDV can be installed on the ground pipeline, but this needs to be negotiated with the reservoir department. In addition, when determining the pressure set values of SDV, SSV and PSV, the pressure drop from the wellhead to the metering station of the single well pipeline should be considered.

In summary, if alternative scheme one is chosen, close cooperation with the reservoir department is required, as signals such as PSH and PSL on the ground pipeline need to be connected to the control panel of the oil tree.

4.1.1.2 Alternative Scheme Two

Alternative Scheme Two is developed based on Alternative Scheme One. The specific configuration is shown in Fig. 4.4.

Compared with Alternative Scheme One, Alternative Scheme Two has the following differences:

(1) One SDV is cancelled, leaving only a single SDV and PSV for protection.
(2) When the SDV fails, it relies entirely on the PSV's venting for overpressure protection. The venting volume should be considered based on the total well flow volume, and the venting time depends on the SDV's repair time.
(3) The PSV's action frequency is higher than that of Alternative Scheme One, and it also poses a greater potential hazard to the environment.

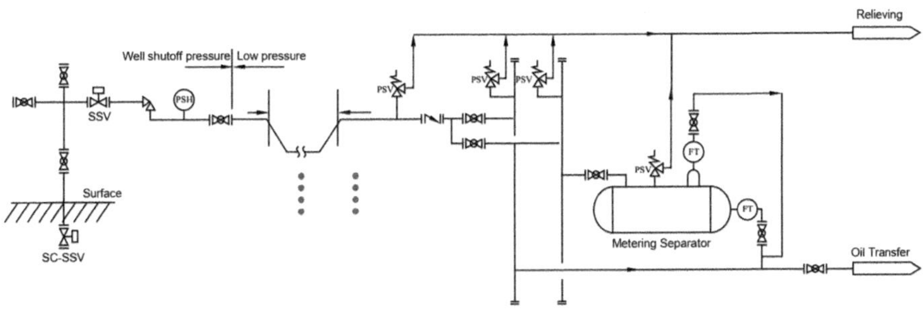

Fig. 4.4 Configuration diagram of alternative overpressure protection scheme two for single well pipeline

(4) The PSV at the end of the pipeline in this scheme cannot be cancelled. When determining the pressure set values of the SSV and PSV, the pressure drop from the wellhead to the metering station of the single well pipeline should also be considered.
(5) For leak accidents, a PSL can be set up at the same time as the PSH, and the SDV can be closed when the set value of the PSL is reached.

4.1.2 High-Pressure Integrity Protection Scheme (HIPPS)

4.1.2.1 Introduction to HIPPS

The High-Integrity Pressure Protection System (HIPPS) is a safety instrument system that complies with SIL3 certification. HIPPS generally consists of pressure detection components, controllers, and execution components. The pressure detection components are two out of three voting, and the execution components consist of two series of shut-off valves. The schematic diagram is shown in Fig. 4.5.

It has the following main features:

(1) To meet the requirement of cut-off time ($\leq$2s), the cut-off valve is generally a SIL3 certified straight-stroke hydraulic fail-safe cut-off valve.
(2) The controller is a SIL4 certified solid-state logic controller, which can be installed outdoors and powered by solar energy.
(3) After using HIPPS, the subsequent overpressure venting system can be cancelled to avoid adverse effects on the environment and personnel caused by venting.

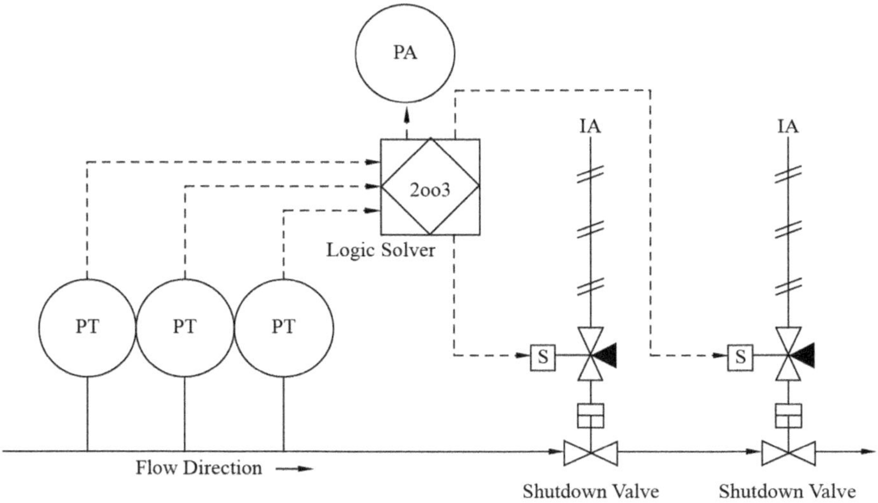

Fig. 4.5 HIPPS system schematic

The high integrity and reliability of HIPPS itself enable it to achieve 100% cut-off in the event of an accident, thus protecting the safety of pipelines and downstream equipment. It is based on this that the price of HIPPS is relatively high, and there are few suppliers, the famous ones being Germany's HIMA and Japan's Yokogawa.

The design, commissioning, operation, and maintenance of the HIPPS system should strictly follow the following standards:

(1) IEC 61,508 "Functional Safety of Electrical/Electronic/Programmable Electronic Safety-related Systems".
(2) IEC 61,511 "Functional Safety—Safety Instrumented Systems for the Process Industry Sector".
(3) GB/T 20,438 "Functional Safety of Electrical/Electronic/Programmable Electronic Safety-related Systems".

4.1.2.2 Application of HIPPS

Traditional overpressure protection is achieved through a venting system. A venting system is installed on the pipeline in advance, and a certain venting pressure is set. When the system pressure rises to the set value, the venting system automatically starts, venting the excess pressure in the system (usually to a flare or vent pipe), thus achieving the purpose of protecting the system. The traditional venting method discharges a large amount of combustible, toxic fluids or their combustion products into the atmosphere, causing environmental damage.

The HIPPS system can timely cut off the upstream fluid to avoid the discharge of overpressure fluid, which is more environmentally friendly. However, HIPPS is mainly used on offshore platforms at present. The process design standard of the Norwegian Petroleum Standardization Organization (NORSOK) mentions: "HIPPS system should only be used when the discharge device is impractical."

In addition, due to the high safety of the HIPPS system, minor faults in the field may trigger the shutdown of the HIPPS system, leading to the shutdown of the entire oil field. This indicates that the HIPPS system may cause a lot of trouble for on-site production management, especially during the initial production of the oil field or during the unstable production period. Frequent shutdowns will not only affect the output of the oil field, but also bring a large workload to the operators, increasing the requirements for oil field operation management personnel.

4.2 Safety Protection Plan for Metering Stations

A metering station is a place where the oil and gas water production of each oil well in the oil area is measured. The metering station does not require high measurement accuracy, and the measurement error of oil, gas, and water is allowed to be within $\pm$ 10%. The metering station measures each oil well in turn, and the continuous measurement time for each well is 4–8 h. The number of oil wells under the jurisdiction of the metering station is generally less than 20. When there are more than 20, a second set of metering equipment should be added to the station.

The main facilities in a typical metering station include a ball receiver and sender, an inlet manifold, a metering separator, and a closed drain tank. The valve group in the metering station directs the well flow of the test oil well into the metering separator. The metering separator separates the well flow into gas and liquid phases, measures the flow rate of the gas and liquid phases separately, and determines the water content in the liquid phase of the crude oil, thus obtaining the oil, gas, and water production of the oil well. The gas after measurement either enters the gas gathering network of the oil field, or is recombined with the liquid phase after measurement, and is sent to the central processing station after being combined with the well flow of other oil wells. Some metering stations mix the well flow of oil wells that are not measured in the valve group, introduce it into the production separator to measure the gas and liquid flow, and the sum of the gas and liquid flow of the test oil well can obtain the oil, gas, and water production of the oil area under the jurisdiction of the metering station.

Since the late 1990s, after many years of research, several types of multiphase flow meters have been developed at home and abroad that can simultaneously measure the oil, gas, and water flow in the pipeline, without the need to separate the well flow into gas and liquid phases for measurement. As a new type of instrument, the adaptability and measurement accuracy of multiphase flow meters to complex well flow still need to be improved, and the price also needs to be reduced. Multiphase flow meters are currently only used in extremely harsh natural environments such as marine and desert oil fields. The following is a description of its safety protection plan.

4.2.1 Manifold

The manifold in the metering station receives the fluid produced from two or more wellheads and distributes these produced fluids to the required process systems, i.e., test separation facilities and production pipelines. The accidental events affecting the manifold are overpressure and leakage. The safety analysis table and safety analysis checklist of the manifold refer to Chap. 3, Sect. 3.4. Normally, a pressure safety device PSH and PSL sensor or PSV should be installed where the entire manifold pressure can be detected. If the pressure of each section of the manifold is different, each section should have a protective device. If the manifold is equipped with PSH and PSL sensors, the sensor signals should be able to shut off all input sources entering the manifold.

(1) If one of the following conditions is met, PSH is not required to be installed:

   ① Each input source has installed PSH, and the set pressure of PSH is less than the maximum allowable working pressure of the manifold.

② The manifold is protected by downstream PSH and cannot be isolated from the manifold.

(2) If the following condition is met, PSL is not required to be installed:

Each input source is protected by PSL, and there is no pressure control device or throttle device between PSL and the manifold.

(3) If one of the following conditions is met, PSV is not required to be installed:

① The maximum allowable working pressure of the manifold is greater than the maximum shut-in pressure of any connected well.
② Each input source has pressure relief protection, and the maximum shut-off pressure of the input source is less than the maximum allowable working pressure of the manifold.
③ The manifold is protected by downstream PSV, and PSV cannot be isolated from the manifold.
④ The input source is the wellhead, the wellhead pressure is greater than the maximum allowable working pressure of the manifold, and the wellhead is equipped with two SDVs controlled by independent PSHs (one of which can be SSV), and PSH is connected to the isolation relay and detection point. Other input sources with pressure greater than the maximum allowable working pressure of the manifold are protected by PSV (Fig. 4.6).

4.2.2 Metering Separator

The metering separator is the main equipment within the metering station. The unexpected events that affect the metering separator are overpressure, negative pressure, overflow, gas breakthrough, leakage, and overtemperature (if the container is heated). The safety analysis table and safety analysis checklist of the metering separator can refer to the pressure vessel or atmospheric vessel in Sect. 3.4 of Chap. 3. Figures 4.7 and 4.8 are the recommended safety devices for typical metering separators.

4.2.2.1 Pressure Protection

The metering separator receives pressure fluid from the wellhead, and the PSH sensor provides overpressure protection, timely cutting off the separator feed. When the leakage of the metering separator is large enough to depressurize the metering separator, the PSL sensor detects the set value and cuts off the separator feed. If the PSH is under negative pressure, the metering separator should install a gas supplement system to maintain the appropriate pressure state of the metering separator. The installation positions of PSH and PSL sensors and PSV should be able to detect or release the pressure of the gas in the metering separator, which is usually at or near the top of the metering separator. If the pressure drop from the metering separator to

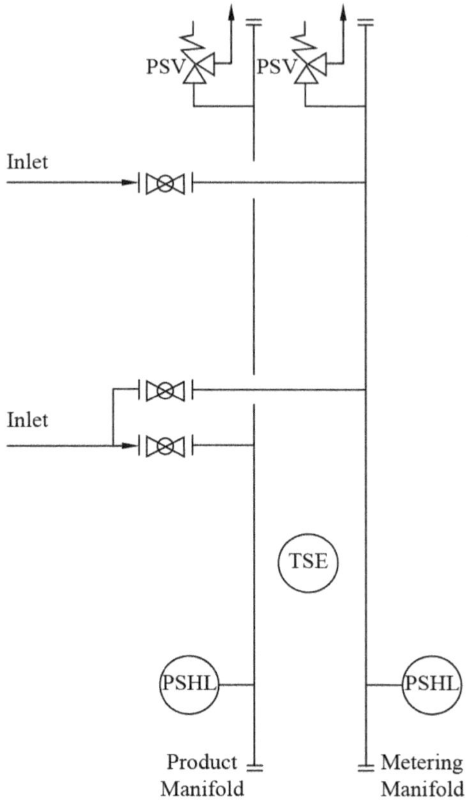

Fig. 4.6 Recommended safety devices for a typical manifold. *Note* TSE is a symbol and does not reflect the actual position and quantity

the monitoring point can be ignored, and these safety devices cannot be isolated from the metering separator, these devices can be installed at the outlet of the gas pipeline. Some external (such as the closure of the gas outlet valve) or internal (such as the blockage of the demister) reasons may isolate the safety device and the metering separator [2].

(1) If one of the following conditions is met, PSH does not need to be installed:

① The maximum pressure generated by the upstream fluid of the metering separator cannot be greater than the maximum allowable working pressure of the metering separator.
② Each input source is equipped with a PSH, and the set pressure of the PSH is less than the maximum allowable working pressure of the metering separator.
③ The gas outlet is connected to the downstream equipment with a suitable pipeline without a cut-off valve or regulating valve, and this downstream equipment is protected by a PSH that also protects the metering separator.
④ The metering separator operates at atmospheric pressure and has an appropriate venting system.

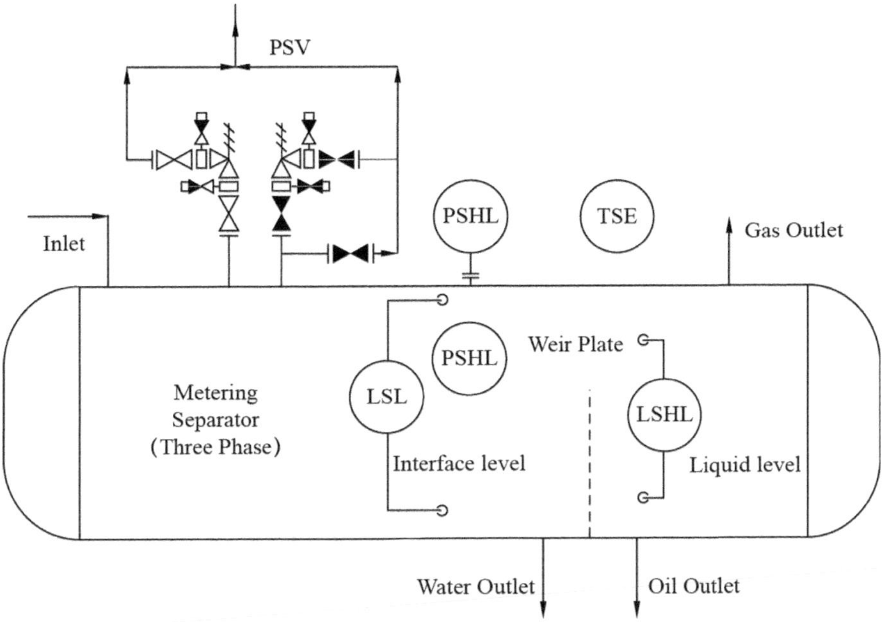

Fig. 4.7 Recommended safety devices for typical metering separators (three-phase). *Note 1* TSE is a symbol, not reflecting the actual position and quantity. *Note 2* If the metering separator needs to be heated, TSH must be installed

(2) If the following conditions are met, PSL does not need to be installed:

① Each input source is protected by PSL, and there is no pressure control device or throttling device between PSL and the manifold.
② The minimum operating pressure is atmospheric pressure during operation.
③ The gas outlet is connected to the downstream equipment with a suitable pipeline without a cut-off valve or regulating valve, and this equipment is protected by a PSL that also protects the metering separator.

(3) If one of the following conditions is met, PSV does not need to be installed:

① Each input source is protected by a PSV, whose set pressure is not higher than the maximum allowable working pressure of the metering separator, and a PSV is installed on the metering separator to protect against heat from fire and thermal expansion.
② Each input source is protected by a PSV, whose set pressure is not higher than the maximum allowable working pressure of the metering separator, and at least one of them will not be isolated from the metering separator.
③ The PSV on the downstream equipment can meet the pressure relief requirements of the metering separator and will not be isolated from the metering separator.

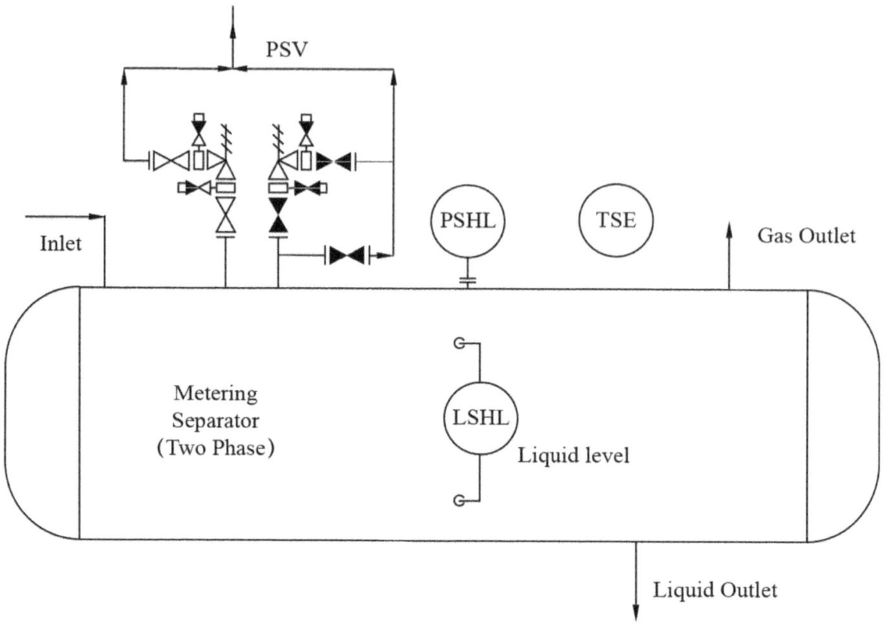

Fig. 4.8 Recommended safety devices for typical metering separators (two-phase). *Note 1* TS is a symbol, not reflecting the actual position and quantity. *Note 2* If the metering separator needs to be heated, TSH must be installed

4.2.2.2 Overflow and Leakage Protection

LSH can cut off the separator feed to prevent the metering separator from overflowing. The LSL sensor can cut off the metering separator feed or close the liquid outlet to protect it from the impact of gas breakthrough. If the metering separator has an immersed heating element, the LSL sensor should also be able to cut off the fuel supply.

The position of the LSH sensor should be high enough above the highest operating liquid level to prevent mis-cutoff, but there should be appropriate container space above the LSH sensor to prevent overflow before the cutoff takes effect. The position of the LSL sensor should be low enough below the lowest operating liquid level to prevent mis-cutoff, but there should be appropriate liquid space between the LSL sensor and the liquid outlet to prevent gas breakthrough before the cutoff takes effect. In fire-tube heating equipment, LSH should be above the fire tube. It is best to install LSH and LSL sensors outside the container, which can be isolated from the container. In this way, these devices can be tested without interrupting the process system.

(1) If one of the following conditions is met, LSH does not need to be installed:

① The downstream equipment of the gas outlet is not a torch or venting system, and can safely handle the maximum liquid carryover.

② The container characteristics do not require the handling of the separated liquid phase.

(2) The LSL does not need to be installed if one of the following conditions is met:

① The liquid level in the container is not automatically maintained, and the container does not have an immersion heating element that can withstand overheating.
② The downstream equipment of the liquid outlet can safely handle the maximum gas flow that can be discharged through the liquid outlet, and there is no immersion heating element in the meter separator that can withstand overheating. The throttling device on the discharge pipeline can be used to limit the gas flow.

(3) Over-temperature protection

If the meter separator needs to be heated, a TSH sensor should be installed to cut off the heat source when the process fluid is overheated. If the heat source cannot cause overheating, the TSH does not need to be installed.

(4) Backflow protection

If a large amount of fluid may flow back from the downstream equipment in the event of a leak, a check valve (FSV) should be installed on each gas and liquid outlet pipeline. If one of the following conditions is met, the FSV does not need to be installed:

(1) The maximum amount of hydrocarbon substances that may flow back from the downstream equipment can be ignored.
(2) The control device on the pipeline can effectively reduce backflow.

4.2.3 Closed Drain Tank

The closed drain tank is a device used in the metering station to collect sewage liquid, which can be understood as the last stage of gas scrubbing for discharge. In general, the closed drain tank is buried underground and collects liquid by gravity.

The accidental events affecting the closed drain tank are overpressure, negative pressure, overflow, gas breakthrough, leakage, and overheating (if the container is heated). The safety analysis table and safety analysis checklist of the closed drain tank can refer to the pressure vessel or atmospheric vessel in Sect. 3.4 of Chap. 3. Figure 4.9 shows the recommended safety devices for a typical closed drain tank.

4.2.3.1 Pressure Protection

(1) The closed drain tank does not need to install a PSH under any of the following conditions:

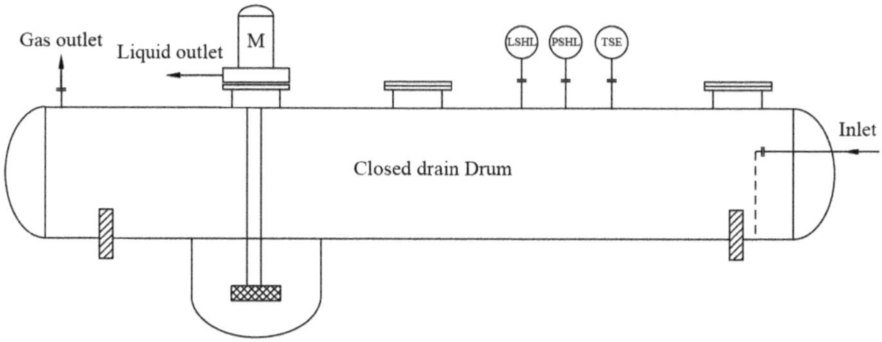

Fig. 4.9 Recommended safety devices for a typical closed drain tank

① Its design pressure can withstand the maximum accumulated pressure.
② The closed drain tank is equipped with an appropriate venting system, which is guaranteed not to be cut off.

(2) The closed drain tank usually operates under normal pressure, slight positive pressure, or changes from slight positive pressure to normal pressure. A locking device is usually provided at the entrance of the closed drain tank to ensure that the entrance is not cut off, so the PSL does not need to be installed.
(3) If the design pressure can withstand the maximum accumulated pressure, and there are no obstructions inside the tank or at the gas outlet, such as mist catchers, back pressure valves, and fire extinguishers, the closed drain tank does not need to install a PSV.

The positions where the PSH sensor and PSV are installed should be able to detect or release the pressure of the gas in the closed drain tank, which is usually at or near the top of the meter separator. However, if the pressure drop from the closed drain tank to the monitoring point can be ignored, and these safety devices cannot be isolated from the closed drain tank, these devices can be installed at the outlet of the gas pipeline. The isolation of these safety devices from the closed drain tank may be caused by external reasons (such as the closure of the gas outlet valve) or internal reasons (such as blockage of the mist catcher).

4.2.3.2 Overflow and Leakage Protection

The LSH can start the closed drain pump in time to avoid overflow of the closed drain tank, which would prevent the upstream equipment from properly discharging sewage, affecting the normal operation of the upstream equipment, or causing liquid to overflow to the downstream equipment. The LSL sensor can cut off the closed drain pump to prevent the pump from cavitation. If there is an immersion heating element in the closed drain tank, the LSL sensor should also be able to cut off the energy source to prevent the heating element from dry burning.

The position of the LSH sensor should be a sufficient distance below the top of the container to prevent the tank from being full and overflowing. The position of the LSL sensor should be a sufficient distance below the lowest operating liquid level to prevent too much liquid from being discharged, thereby preventing the closed drain pump from cavitation. If there is an immersion heating element in the closed drain tank, the position of the LSL sensor should also be higher than the heating element to avoid dry burning. Generally, both the LSH and LSL sensors are installed outside the closed drain tank and can be isolated from the tank. This allows these devices to be tested without interrupting the process system.

(1) The LSH does not need to be installed if one of the following conditions is met:

   ① The downstream equipment of the gas outlet is not a torch or venting system, and it can safely handle the maximum liquid carryover without affecting the normal discharge of the upstream equipment.
   ② The liquid in the tank can be manually discharged without affecting the normal discharge of the upstream equipment.

(2) The LSL does not need to be installed if one of the following conditions is met:

   ① The liquid level in the container is not automatically maintained, and the container does not have an immersion heating element that can withstand overheating.
   ② The downstream equipment of the liquid outlet can safely handle the maximum gas flow that can be discharged through the liquid outlet, and there are no immersion heating elements in the tank that can withstand overheating. The throttling device on the discharge pipeline can be used to limit the gas flow.

4.2.3.3 Over-Temperature Protection

If the closed tank needs to be heated, a TSH sensor should be installed so that it can cut off the heat source when the process fluid is overheated. If the heat source cannot cause overheating, there is no need to install a TSH.

4.2.3.4 Backflow Protection

If a large amount of fluid may flow back from the downstream equipment in the event of a leak, a check valve (FSV) should be installed on each gas and liquid outlet pipeline. If one of the following conditions is met, there is no need to install an FSV:

(1) The maximum amount of hydrocarbon substances that may flow back from the downstream equipment can be ignored.
(2) The control device on the pipeline can effectively reduce backflow.

4.3 Safety Protection Scheme for Oil Transfer Stations

The main facilities in a typical oil transfer station include receiving and sending ball cylinders, station inlet manifolds, metering separators, production separators, external transfer pumps, open and closed tanks, and torches or vent pipes. The following is a description of its safety protection scheme.

4.3.1 Station Inlet Manifold

The manifold in the oil transfer station mainly receives fluids from the metering station, and often also receives fluids from nearby oil wells, and distributes these produced fluids to the required process systems, i.e., metering facilities and production facilities. The accidental events affecting the station inlet manifold are overpressure and leakage. The safety analysis table and safety analysis checklist for the manifold can be found in Sect. 3.4 of Chap. 3.

Normally, pressure safety devices PSH and PSL sensors or PSV should be installed where the entire manifold pressure can be detected. If the pressure of each section of the manifold is different, each section should have the required protection device. If the manifold needs PSH and PSL sensors, the signals from the sensors should be able to shut off all input sources entering the manifold.

(1) If one of the following conditions is met, there is no need to install a PSH:

  ① Each input source is equipped with a PSH, and the set pressure of the PSH is less than the maximum allowable working pressure of the manifold.
  ② The manifold is protected by a downstream PSH and cannot be isolated from the manifold.

(2) If each input source is protected by a PSL, and there is no pressure control device or throttling device between the PSL and the manifold, there is no need to install a PSL.
(3) If one of the following conditions is met, there is no need to install a PSV:

  ① The maximum allowable working pressure of the manifold is greater than the maximum shut-in pressure of any connected well.
  ② Each input source provides pressure relief protection, and the maximum inlet shut-off pressure of the input source is less than the maximum allowable working pressure of the manifold.
  ③ The manifold is protected by a downstream PSV, and the PSV cannot be isolated from the manifold.
  ④ The input source is a wellhead, the wellhead pressure is greater than the maximum allowable working pressure of the manifold, and the wellhead is equipped with two SDVs controlled by independent PSHs (one of its inlets can be an SSV), and the PSH is connected to the isolation relay and detection

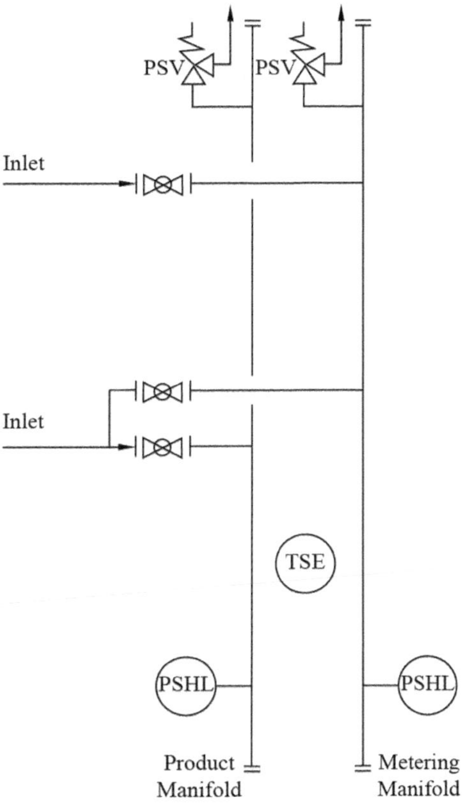

Fig. 4.10 Recommended safety devices for typical manifolds. *Note* TSE is a symbol and does not reflect the actual position and quantity

point. Other input sources with pressure greater than the maximum allowable working pressure of the manifold are protected by a PSV (Fig. 4.10).

If the oil transfer station also receives liquid from nearby oil wells, a metering separator needs to be set up. The accidental events affecting the metering separator are overpressure, negative pressure, overflow, gas breakthrough, leakage, and over-temperature (if the vessel is heated). The safety analysis table and safety analysis checklist for the metering separator can refer to the pressure vessel or atmospheric vessel in Sect. 3.4 of Chap. 3.

4.3.1.1 Pressure Protection

The metering separator receives pressure fluid from the wellhead. If it causes overpressure, it is protected by a PSH sensor, which timely cuts off the inlet flow of the vessel. When the leak of the metering separator is large enough to depressurize the metering separator, the PSL sensor detects the set value and cuts off the inlet fluid of the vessel. The PSV depressurizes the metering separator when the PSH fails.

If the metering separator may be subjected to negative pressure, the metering separator should be equipped with a gas replenishment system to maintain the metering separator in an appropriate pressure state.

The locations where PSH and PSL sensors and PSV are installed should be able to detect or release the pressure of the gas in the metering separator, which is usually at or near the top of the metering separator. However, if the pressure drop from the metering separator to the monitoring point is negligible, and these safety devices cannot be isolated from the metering separator, these devices can be installed on the outlet of the gas pipeline. The isolation of these safety devices from the metering separator may be caused by external factors (such as the closure of the gas outlet valve) or internal factors (such as blockage of the mist catcher).

(1) PSH does not need to be installed if one of the following conditions is met:

① The maximum pressure generated by the upstream fluid of the metering separator cannot exceed the maximum allowable working pressure of the metering separator.
② Each input source has a PSH installed, and the set pressure of the PSH is less than the maximum allowable working pressure of the metering separator.
③ The gas outlet is connected to the downstream equipment by a suitable pipeline without a cut-off valve or regulating valve, and this equipment is protected by a PSH that also protects the metering separator.
④ The metering separator operates at atmospheric pressure and has an appropriate venting system.

(2) PSL does not need to be installed if the following conditions are met:

① Each input source is protected by a PSL, and there is no pressure control device or throttling device between the PSL and the manifold.
② The minimum operating pressure is atmospheric pressure during operation.
③ The gas outlet is connected to the downstream equipment by a suitable pipeline without a cut-off valve or regulating valve, and this equipment is protected by a PSL that also protects the metering separator.

(3) PSV does not need to be installed if one of the following conditions is met:

① Each input source is protected by a PSV with a set pressure not higher than the maximum allowable working pressure of the metering separator, and a PSV is installed on the metering separator to protect against heat from fire and thermal expansion.
② Each input source is protected by a PSV with a set pressure not higher than the maximum allowable working pressure of the metering separator, at least one of which will not be cut off from the metering separator.
③ The PSV on the downstream equipment can meet the pressure relief requirements of the metering separator and will not be cut off from the metering separator.

4.3.1.2 Overflow and Leakage Protection

LSH can cut off the inlet fluid of the container to protect the metering separator and prevent overflow. The LSL sensor can cut off the inlet fluid of the metering separator or close the liquid outlet to protect it from gas breakthrough. If the metering separator has an immersed heating element, the LSL sensor should also be able to cut off the fuel supply.

The position of the LSH sensor should be high enough above the highest operating liquid level to prevent accidental shutdown, but there should be adequate container space above the LSH sensor to prevent overflow before the shutdown function takes effect. The position of the LSL sensor should be low enough below the lowest operating liquid level to prevent accidental shutdown, but there should be adequate liquid space between the LSL sensor and the liquid outlet to prevent gas breakthrough before the shutdown function takes effect. In fire tube heating equipment, the LSH should be above the fire tube. It is best to install LSH and LSL sensors outside the container, where they can be isolated from the container. This allows these devices to be tested without interrupting the process system.

(1) LSH does not need to be installed if one of the following conditions is met:

    ① The downstream equipment of the gas outlet is not a torch or venting system, and it can safely handle the maximum liquid carryover.
    ② The characteristics of the container do not require the handling of the separated liquid phase.

(2) LSL does not need to be installed if one of the following conditions is met:

    ① The liquid level in the container is not automatically maintained, and the container does not have an immersed heating element that can withstand overheating.
    ② The downstream equipment of the liquid outlet can safely handle the maximum gas flow that can be discharged through the liquid outlet, and there is no immersed heating element in the metering separator that can withstand overheating. The throttling device on the discharge pipeline can be used to limit the gas flow.

4.3.1.3 Overtemperature Protection

If the metering separator needs to be heated, a TSH sensor should be installed so that it can cut off the heat source when the process fluid is overheated. If the heat source cannot cause overheating, TSH does not need to be installed.

4.3.1.4 Backflow Protection

(1) If a large amount of fluid may flow back from the downstream equipment in the event of a leakage accident, a check valve (FSV) should be installed on each gas and liquid outlet pipeline.
(2) FSV does not need to be installed if one of the following conditions is met:

    ① The maximum amount of hydrocarbon substances that may flow back from the downstream equipment can be ignored.
    ② The control device on the pipeline can effectively reduce backflow.

Figures 4.11 and 4.12 show the recommended safety devices for typical metering separators.

4.3.2 Production Separators

Production separators are important equipment within the oil transfer station, typically used for the separation of gas–liquid two-phase flow. The accidental events affecting the production separator are overpressure, negative pressure, overflow, gas

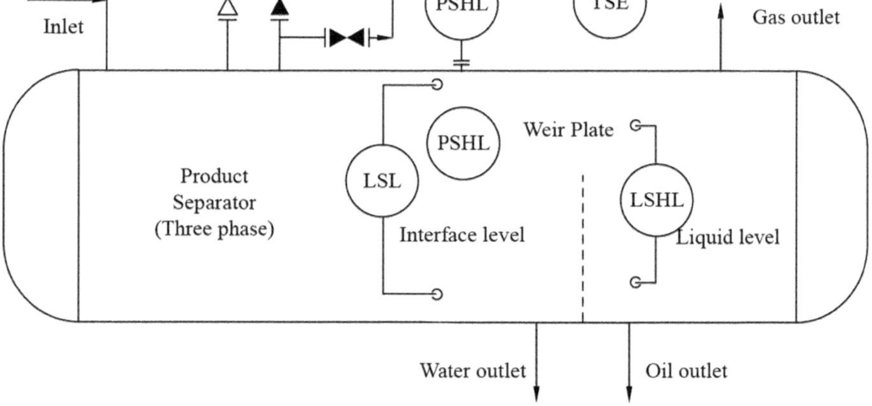

Fig. 4.11 Recommended safety devices for typical metering separators (three-phase). *Note 1* TSE is a symbol, not reflecting the actual position and quantity. *Note 2* If the metering separator needs to be heated, TSH must be installed

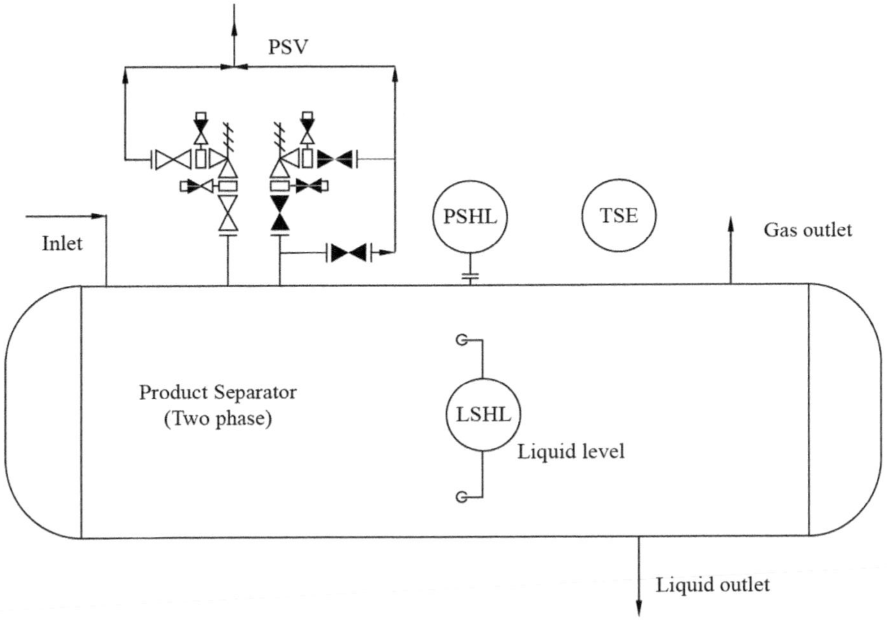

Fig. 4.12 Recommended safety devices for typical metering separators (two-phase). *Note 1* TSE is a symbol, not reflecting the actual position and quantity. *Note 2* If the metering separator needs to be heated, TSH must be installed

breakthrough, leakage, and overtemperature (if the vessel is heated). The safety analysis table and safety analysis checklist for the production separator can refer to the pressure vessel or atmospheric vessel in Sect. 3.4 of Chap. 3.

The production separator receives pressure fluids from the metering station and/or wellhead. If overpressure is caused, the PS sensor provides protection, timely cutting off the inlet flow of the vessel. When the leakage of the production separator is large enough to depressurize the production separator, the PSL sensor detects the set value and cuts off the inlet fluid of the vessel. The PSV depressurizes the production separator when the PSH fails. If the production separator may withstand a negative pressure that could crush it, the production separator should install a gas replenishment system, which can maintain the appropriate pressure state of the production separator.

4.3.2.1 Pressure Protection

The PSH and PSL sensors and PSV should be installed in a position that can detect or release the pressure of the production separator gas, which is usually at or near the top of the production separator. However, if the pressure drop from the production separator to the monitoring point can be ignored, and these safety devices cannot be isolated from the production separator, these devices can be installed at the outlet of the gas pipeline. The isolation of these safety devices from the production separator may be caused by external (such as the closure of the gas outlet valve) or internal (such as the blockage of the mist catcher) reasons.

(1) If one of the following conditions is met, PSH does not need to be installed:

① The maximum pressure generated by the upstream fluid of the production separator cannot be greater than the maximum allowable working pressure of the production separator.
② Each input source is equipped with a PSH, and the set pressure of the PSH is less than the maximum allowable working pressure of the production separator.
③ The gas outlet is connected to the downstream equipment with a suitable pipeline without a cut-off valve or regulating valve, which is protected by a PSH that also protects the production separator.
④ The production separator operates at atmospheric pressure and has an appropriate venting system.

(2) If the following conditions are met, PSL does not need to be installed:

① Each input source is protected by PSL, and there is no pressure control device or throttling device between the PSL and the manifold.
② The minimum operating pressure is atmospheric pressure during operation.
③ The gas outlet is connected to the downstream equipment with a suitable pipeline without a cut-off valve or regulating valve, which is protected by a PSL that also protects the metering separator.

(3) If one of the following conditions is met, PSV does not need to be installed:

① Each input source is protected by a PSV, whose set pressure does not exceed the maximum allowable working pressure of the production separator, and a PSV is installed on the production separator to protect against fire heating and thermal expansion.
② Each input source is protected by a PSV, whose set pressure does not exceed the maximum allowable working pressure of the production separator, at least one of which will not be isolated from the production separator.
③ The PSV on the downstream equipment can meet the depressurization requirements of the production separator and will not be isolated from the production separator.

4.3.2.2 Overflow and Leakage Protection

The LSH can cut off the inlet fluid of the vessel to protect the production separator and prevent overflow. The LSL sensor can cut off the inlet fluid of the production separator or close the liquid outlet to protect it from the impact of gas breakthrough. If the production separator has an immersed heating element, the LSL sensor should also be able to cut off the fuel supply.

The position of the LSH sensor should be high enough above the highest operating liquid level to prevent mis-cutoff, but there should be appropriate vessel space above the LSH sensor to prevent overflow before the cutoff action takes effect. The position of the LSL sensor should be low enough below the lowest operating liquid level to prevent mis-cutoff, but there should be appropriate liquid space between the LSL sensor and the liquid outlet to prevent gas breakthrough before the cutoff action takes effect. In fire-tube heating equipment, the LSH should be above the fire tube. It is best to install the LSH and LSL sensors outside the vessel, where they can be isolated from the vessel. This allows these devices to be tested without interrupting the process system.

(1) If one of the following conditions is met, LSH does not need to be installed:

① The downstream equipment of the gas outlet is not a torch or venting system, and it can safely handle the maximum liquid carryover.

② The characteristics of the container do not require the handling of the separated liquid phase.

(2) If one of the following conditions is met, an LSL is not required to be installed:

① The liquid level in the container is not automatically maintained, and the container does not have an immersion heating element that can withstand over-temperature.

② The downstream equipment of the liquid outlet can safely handle the maximum gas flow that can be discharged through the liquid outlet, and there is no immersion heating element in the production separator that can withstand over-temperature. The throttling device on the discharge pipeline can be used to limit the gas flow.

4.3.2.3 Over-Temperature Protection

If the production separator needs to be heated, a TSH sensor should be installed so that the heat source can be cut off when the process fluid is over-temperature. If the heat source cannot cause over-temperature, a TSH is not required to be installed.

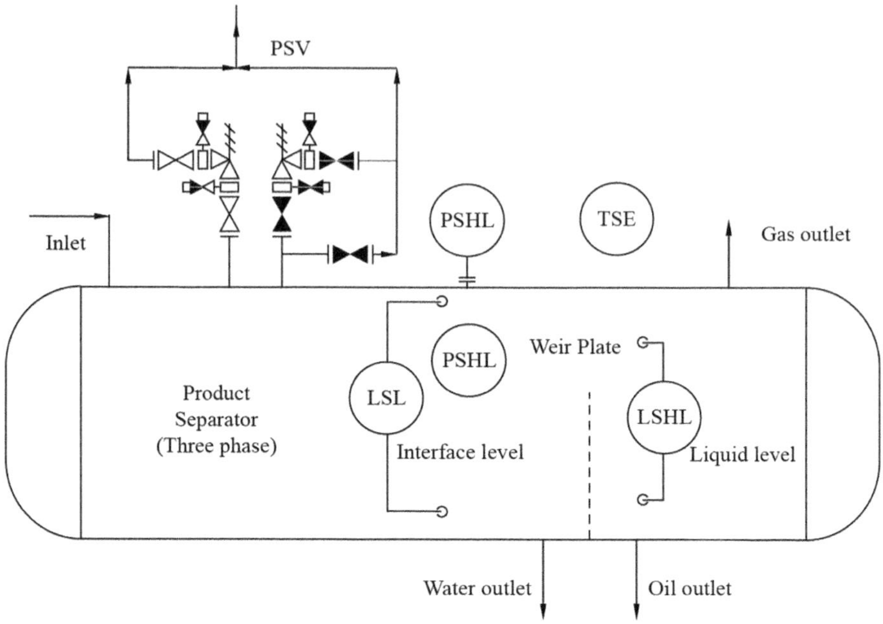

Fig. 4.13 Safety devices recommended for typical production separators (three-phase). *Note 1* TSE is a symbol and does not reflect the actual position and quantity. *Note 2* If the production separator needs to be heated, a TSH must be installed

4.3.2.4 Backflow Protection

(1) If a large amount of fluid may flow back from the downstream equipment in the event of a leak, a check valve (FSV) should be installed on each gas and liquid outlet pipeline.
(2) If one of the following conditions is met, an FSV is not required to be installed:

① The maximum amount of hydrocarbon substances that may flow back from the downstream equipment can be ignored.
② The control device on the pipeline can effectively reduce backflow.

Figures 4.13 and 4.14 are safety devices recommended for typical production separators.

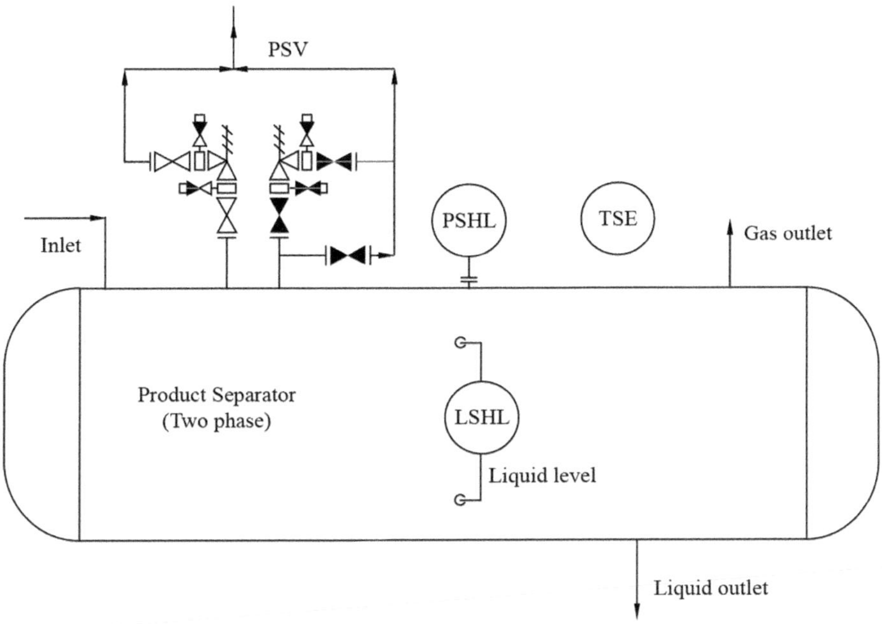

Fig. 4.14 Safety devices recommended for typical production separators (two-phase). *Note 1* TSE is a symbol and does not reflect the actual position and quantity. *Note 2* If the metering separator needs to be heated, a TSH must be installed

References

1. API RP 14E Recommended practice for design and installation of offshore production platform piping systems, Fifth Edition, March 2007.
2. API RP 14C Recommended practice for analysis, design, installation, and testing of basic surface safety systems for onshore production platforms.

Chapter 5
Central Processing Station Safety Protection System Scheme

The crude oil transported by the gathering and transportation system is deeply processed in the central processing station, mainly including crude oil separation, crude oil heating, crude oil electro-dehydration, crude oil electro-desalting, crude oil storage and qualified oil export systems. With the continuous improvement of environmental protection standards, the associated gas produced is generally pressurized and exported after treatment.

5.1 Inlet Manifold

The manifold in the metering station receives fluids produced from two or more wellheads and distributes these produced fluids to the required process systems, i.e., test separation facilities and production pipelines.

The accidental events affecting the manifold are overpressure and leakage. The safety analysis table of the manifold is seen in Sect. 3.4 of Chap. 3. Normally, pressure safety devices PSH and PSL sensors or PSV should be installed where the entire manifold pressure can be detected. If the pressures of the various sections of the manifold are different, each section should have the required protection devices. If the manifold needs PSH and PSL sensors, the signals from the sensors should be able to shut off all input sources entering the manifold.

(1) If one of the following conditions is met, a PSH is not required to be installed [1]:

① Each input source is equipped with a PSH, and the set pressure of the PSH is less than the maximum allowable working pressure of the manifold.

② The manifold is protected by a downstream PSH and cannot be isolated from the manifold.

K. Fang, *Onshore Oilfield Surface Facilities: Process Safety Engineering Guide*,
https://doi.org/10.1007/978-981-96-7830-3_5

(2) If each input source is protected by a PSL and there is no pressure control device or throttling device between the PSL and the manifold, a PSL is not required to be installed.
(3) If one of the following conditions is met, a PSV is not required to be installed:

① The maximum allowable working pressure of the manifold is greater than the maximum shut-in pressure of any connected well.
② Pressure relief protection is provided at each input source, and the maximum shut-off pressure of the input source is greater than the maximum allowable working pressure of the manifold.
③ The manifold is protected by a downstream PSV, and the PSV cannot be isolated from the manifold.
④ The input source is a wellhead, the wellhead pressure is greater than the maximum allowable working pressure of the manifold, and two SDVs controlled by independent PSHs are installed at the wellhead (one of which can be an SSV). The PSH is connected to the isolation relay and detection point. Other input sources with pressure greater than the maximum allowable working pressure of the manifold are protected by a PSV. Figure 5.1 shows the safety devices recommended for a typical manifold.

5.2 Production Separator

The production separator is an important piece of equipment within the oil transfer station, typically used for the separation of gas–liquid two-phase flow. The unexpected events that can affect the production separator include overpressure, negative pressure, overflow, gas breakthrough, leakage, and overheating (if the container is heated). The safety analysis table for the production separator can be found in Sect. 3. 4 of Chap. 3, under the pressure vessel section.

The production separator receives pressure fluids from the metering station and/ or wellhead. If overpressure occurs, the PSH sensor provides protection by promptly cutting off the container's inlet flow. When the leakage from the production separator is large enough to depressurize it, the PSL sensor detects the set value and cuts off the container's inlet fluid. The PSV depressurizes the production separator in case of PSH failure. If the production separator might be subjected to negative pressure that could crush it, a gas replenishment system should be installed, which can maintain the production separator at an appropriate pressure state [2].

5.2.1 *Pressure Protection*

The locations where PSH and PSL sensors and PSV are installed should be able to detect or release the pressure of the gas in the production separator, usually at or near the top of the production separator. However, if the pressure drop from the

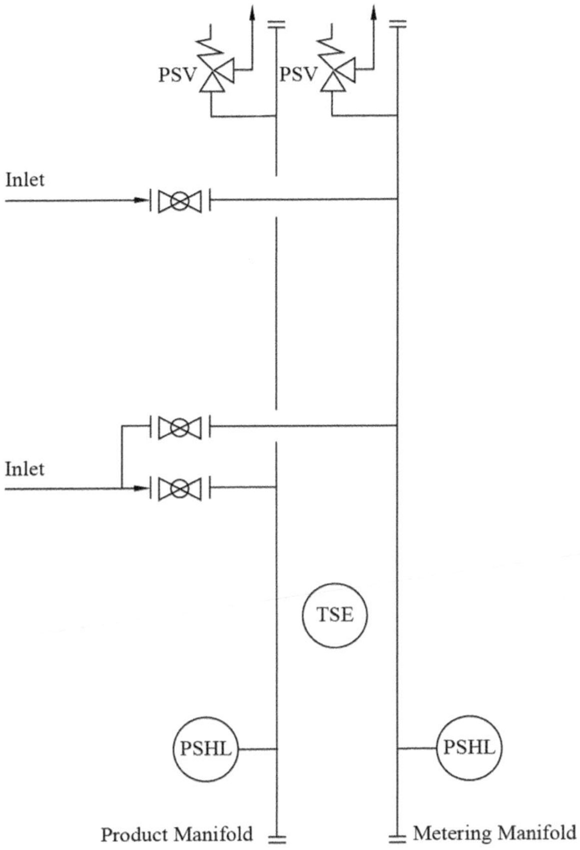

Fig. 5.1 Safety devices recommended for a typical manifold. *Note* TSE is a symbol and does not reflect the actual position and quantity

production separator to the monitoring point is negligible, and these safety devices cannot be isolated from the production separator, they can be installed at the outlet of the gas pipeline. The isolation of these safety devices from the production separator may be caused by external (for example, the gas outlet valve is closed) or internal (for example, the mist catcher is blocked) reasons.

(1) PSH does not need to be installed if one of the following conditions is met:

① The maximum pressure generated by the upstream fluid of the production separator cannot exceed the maximum allowable working pressure of the production separator.

② Each input source has a PSH installed, and the set pressure of the PSH is less than the maximum allowable working pressure of the production separator.

③ The gas outlet is connected to downstream equipment with a suitable pipeline without a cutoff valve or regulating valve, and this equipment is protected by a PSH that also protects the production separator.

④ The production separator operates at atmospheric pressure and has an appropriate venting system.

(2) PSL does not need to be installed if the following conditions are met:

① Each input source is protected by a PSL, and there is no pressure control device or throttling device between the PSL and the manifold.
② The minimum operating pressure during operation is atmospheric pressure.
③ The gas outlet is connected to downstream equipment with a suitable pipeline without a cutoff valve or regulating valve, and this equipment is protected by a PSL that also protects the metering separator.

(3) PSV does not need to be installed if one of the following conditions is met:

① Each input source is protected by a PSV, whose set pressure does not exceed the maximum allowable working pressure of the production separator, and a PSV is installed on the production separator to protect against heat from fire and thermal expansion.
② Each input source is protected by a PSV, whose set pressure does not exceed the maximum allowable working pressure of the production separator, and at least one of them will not be isolated from the production separator.
③ The PSV on the downstream equipment can meet the depressurization requirements of the production separator and will not be isolated from the production separator.

5.2.2 Overflow and Leakage Protection

The LSH can cut off the container's inlet fluid to protect the production separator from overflow. The LSL sensor can cut off the inlet fluid of the production separator or close the liquid outlet to protect it from gas breakthrough. If the production separator has an immersed heating element, the LSL sensor should also be able to cut off the fuel supply [3].

The position of the LSH sensor should be a sufficient distance above the highest operating liquid level to prevent accidental shutdown, but there should be adequate container space above the LSH sensor to prevent overflow before the shutdown function takes effect. The position of the LSL sensor should be a sufficient distance below the lowest operating liquid level to prevent accidental shutdown, but there should be adequate liquid space between the LSL sensor and the liquid outlet to prevent gas breakthrough before the shutdown function takes effect. In fire-tube heating equipment, the LSH should be above the fire tube. It is best to install the LSH and LSL sensors outside the container, where they can be isolated from the container. This allows these devices to be tested without interrupting the process system.

(1) LSH does not need to be installed if one of the following conditions is met:

① The downstream equipment of the gas outlet is not a torch or venting system, and it can safely handle the maximum liquid carryover.
② The characteristics of the container do not require the handling of the separated liquid phase.

(2) LSL does not need to be installed if one of the following conditions is met:

① The liquid level in the container is not automatically maintained, and the container does not have an immersed heating element that can withstand overheating.
② The downstream equipment of the liquid outlet can safely handle the maximum gas flow that can be discharged through the liquid outlet, and there are no over-temperature immersion heating elements in the production separator. The throttling device on the discharge pipeline can be used to limit the gas flow.

5.2.3 Over-Temperature Protection

If the production separator needs to be heated, a TSH sensor should be installed so that it can cut off the heat source when the process fluid is overheated. If the heat source cannot cause overheating, there is no need to install a TSH.

5.2.4 Backflow Protection

If a large amount of fluid may flow back from the downstream equipment in the event of a leak, a check valve (FSV) should be installed on each gas and liquid outlet pipeline. If one of the following conditions is met, there is no need to install an FSV:

(1) The maximum amount of hydrocarbon substances that may flow back from the downstream equipment can be ignored.
(2) The control device on the pipeline can effectively reduce backflow.

Figures 5.2 and 5.3 show the safety devices recommended for typical production separators.

5.3 Heat Exchanger (Tube-Shell Type)

The heat exchanger is used to transfer heat between two isolated fluids and is an important equipment in oilfield production, usually used for heat exchange between hot oil after treatment and cold oil entering the station. The accidental events affecting

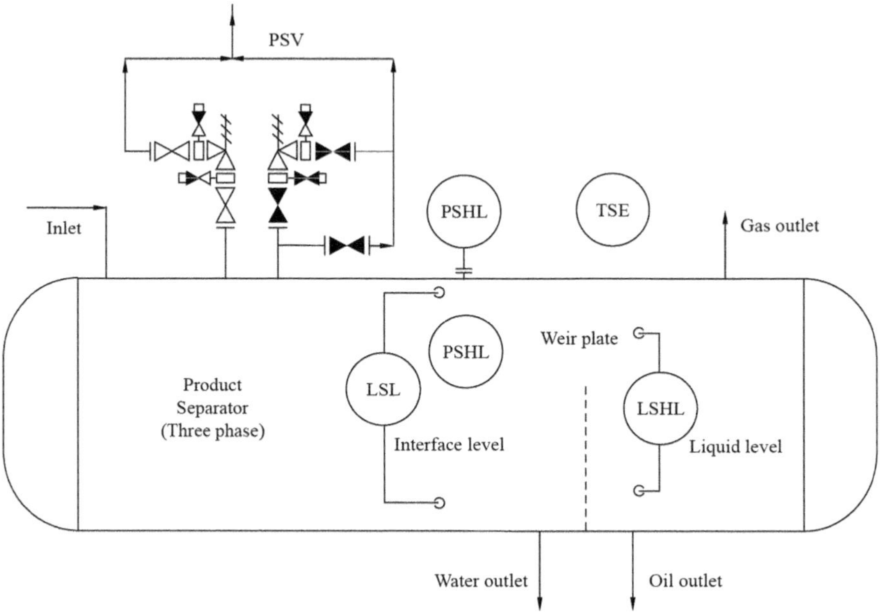

Fig. 5.2 Recommended safety devices for typical production separators (three-phase). *Note 1* TSE is a symbol and does not reflect the actual position and quantity. *Note 2* If the production separator needs to be heated, a TSH must be installed

the heat exchanger are overpressure and leakage. The safety analysis of the heat exchanger is seen in Sect. 3.4 of Chap. 3.

5.3.1 Pressure Protection

The heat exchanger has two pressurized parts, namely the heated part and the heating part. When analyzing the pressure safety device, they should be analyzed separately, because the design of these two parts may be different, and the requirements for operating pressure may also be different. If a certain input source may cause overpressure, the part of the heat exchanger that receives this input source should be protected by a PSH sensor to cut off the inflow of overpressure fluid. In addition, if a part of the heat exchanger will be overpressured due to the rupture or leakage of another part, the part that may cause overpressure should be protected by a PSV to cut off the inflow of overpressure fluid. The part of the heat exchanger that contains hydrocarbons should be equipped with a PSL sensor to cut off the inflow when the leakage is large enough to reduce the pressure.

PSH, PSL sensors and PSV should be located in positions where they can detect and relieve pressure for each part of the heat exchanger. If the pressure drop from the heat exchanger part to the detection point is small, and these safety devices are

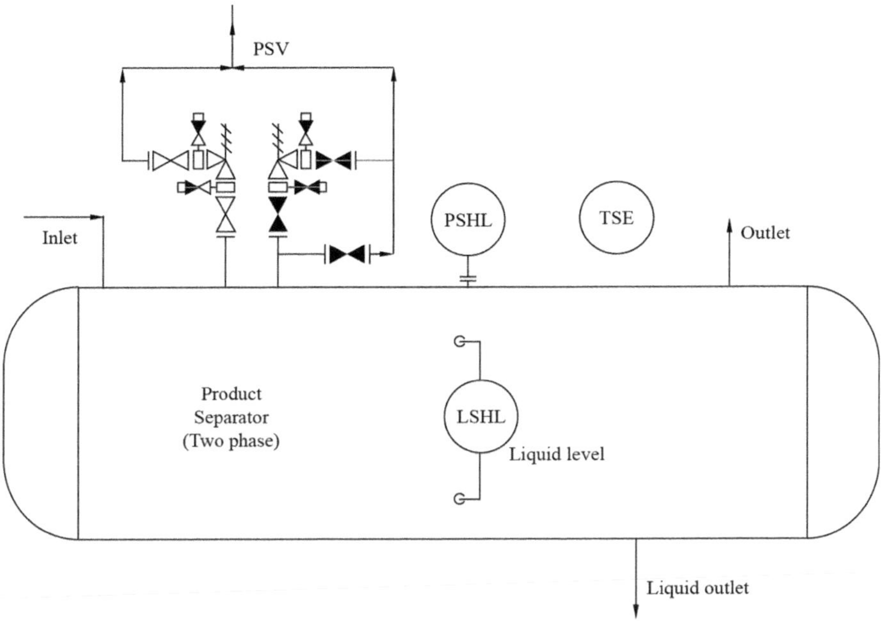

Fig. 5.3 Recommended safety devices for typical production separators (two-phase). *Note 1* TSE is a symbol and does not reflect the actual position and quantity. *Note 2* If the metering separator needs to be heated, a TSH must be installed

always connected with the heat exchanger part, these safety devices can be installed on the inlet or outlet pipeline.

(1) If one of the following conditions is met, there is no need to install a PSH:

① The input source of this part of the heat exchanger will not generate pressure higher than the maximum allowable working pressure of this part of the heat exchanger.
② Each input source has PSH protection, and this PSH also protects this part of the heat exchanger.
③ A PSH is installed on the downstream equipment of the heat exchanger, and this PSH is not cut off from this part of the heat exchanger by a cut-off valve or a regulating valve.

(2) If the following conditions are met, there is no need to install a PSL:

① The minimum operating pressure during operation is atmospheric pressure.
② A PSL installed on other equipment will provide the necessary protection for this part of the heat exchanger, and it is always connected with this part when the heat exchanger is working.

(3) If one of the following conditions is met, there is no need to install a PSV:

① Each input source is protected by a PSV device, and its pressure does not exceed the maximum allowable operating pressure of this part of the heat exchanger; and a PSV is installed on this part of the heat exchanger to release the pressure generated by thermal expansion and fire.
② Each input source is protected by a PSV device, and its pressure does not exceed the maximum allowable operating pressure of this part of the heat exchanger, and each input source should always be connected with this part of the heat exchanger.
③ The PSV on the downstream equipment can meet the release requirements of this part of the heat exchanger, and it is always connected with this part of the heat exchanger.
④ The input source of this part of the heat exchanger will not generate pressure higher than the maximum allowable working pressure of this part of the heat exchanger, and this part of the heat exchanger will not be overpressured due to the temperature or pressure of other parts.
⑤ Each input source is protected by a PSV device, which ensures that the pressure does not exceed the maximum allowable operating pressure of the heat exchanger part, and this part of the heat exchanger will not be over-pressurized due to the temperature or pressure of other parts.

5.3.2 Over-temperature Protection

Since the design of a shell-and-tube heat exchanger usually makes both parts adapt to the highest temperature of the thermal medium, a TSH is generally not required on a shell-and-tube heat exchanger.

Figure 5.4 shows the safety devices recommended for typical heat exchangers.

5.4 Heating Equipment

Heating equipment is used to process and heat hydrocarbon substances. The safety analysis of heating equipment is covered in Sect. 3.4 of Chap. 3.

5.4.1 Temperature Safety Device (TSHH)

(1) The temperature of the medium or process fluid in the heating equipment should be monitored by a TSHH sensor to cut off the fuel supply and the flow of combustible fluid. Steam boilers protected by PSHH sensors and LSLL sensors do not need to install TSE sensors, because PSHH sensors can detect high pressure caused by high temperature, and LSLL sensors can detect low liquid

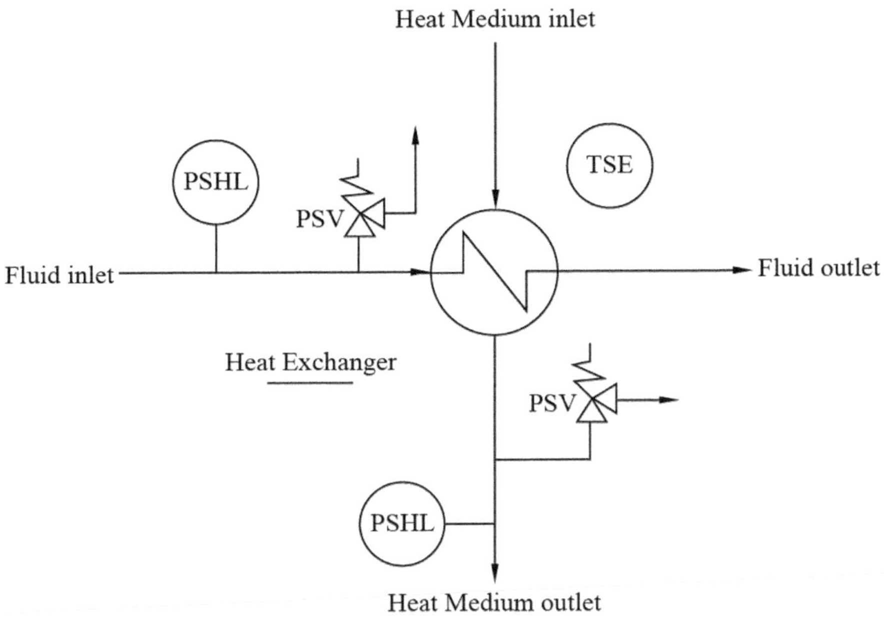

Fig. 5.4 Safety devices recommended for typical heat exchangers. *Note* TSE is a symbol and does not reflect the actual position and quantity

level conditions that can cause high temperature. For indirectly heated water bath heaters operating at atmospheric pressure, it is usually not necessary to set up a TSHH sensor to monitor the temperature of the medium or process fluid, because the boiling point of the bath water limits the maximum temperature of the medium.

(2) In a closed heat transfer system, the combustible medium circulates in the coil located in the combustion chamber, and the flow of the combustible medium cannot be cut off before the combustion chamber cools down. If an uncontrollable fire occurs in this area or the medium escapes from the closed system, the ESD system and the fusible plug circuit should immediately cut off the flow of the medium.

5.4.2 *Flow Safety Device (FSL and FSV)*

If the fuel medium circulates in the combustion chamber coil, the flow rate of the medium should be monitored by a sensor from FSL to cut off the fuel supply of the heating equipment. In this type of equipment, the medium may reach a high temperature state before being detected by the TSHH sensor (installed on the medium pipeline) located outside the heater. In other types of heating vessels, an FSL sensor is not required because the sensor (installed on the medium pipeline) is located in the medium section and can quickly check for high temperature conditions. A check valve (FSV) should be installed on the coil outlet pipeline to prevent the medium from flowing back into the combustion chamber or heating chamber when the coil ruptures.

5.4.3 *Pressure Safety Device (PSHH, PSLL and PSV)*

The pressure of the fuel supply pipeline should be monitored by a PSHH sensor to cut off the fuel supply of the burner. A PSLL sensor is not required in a naturally ventilated burner because the intake pressure is low. The fluid coil located in the tubular heater or flue gas heating chamber should be protected by a PSV to prevent overpressure caused by the expansion of the medium or process fluid.

5.4.4 *Ignition Safety Device*

(1) The air inlet of a natural ventilation burner should be equipped with a fire arrestor to prevent flames from entering the air inlet.
(2) A flue fire arrestor must be installed on the flue of a natural ventilation burner to prevent sparks from being ejected. If the heating equipment does not process combustible materials other than fuel and is located in an isolated area, a fire arrestor is not required. If the heated fluid is flammable or the discharge pressure of the burner is lower than the pressure of the fluid (pressure head) at the outlet of the heat exchange part, a flue fire arrestor is required.
(3) The flame in the combustion chamber should be monitored by a BSLL or TSL sensor, which ignites the combustible material entering the combustion chamber and cuts off the fuel supply immediately when it detects insufficient flame. The recommended safety devices for heating equipment are shown in Fig. 5.5.

5.5 Booster Pump

Due to constraints on formation pressure or considerations of crude oil production, often times, when well fluids enter the central processing station, the pressure is controlled at around 0.5 MPa (gauge pressure). Thus, after the crude oil has undergone two stages of separation, a booster pump needs to be set up to meet the downstream pressure requirements.

The unexpected events that affect the booster pump are overpressure and leakage. The safety analysis table for the booster pump can be found in Chap. 3, Sect. 3. 4. The booster pump is used for boosting between stages of crude oil treatment. If overpressure occurs, it is protected by a PSH sensor, which promptly cuts off

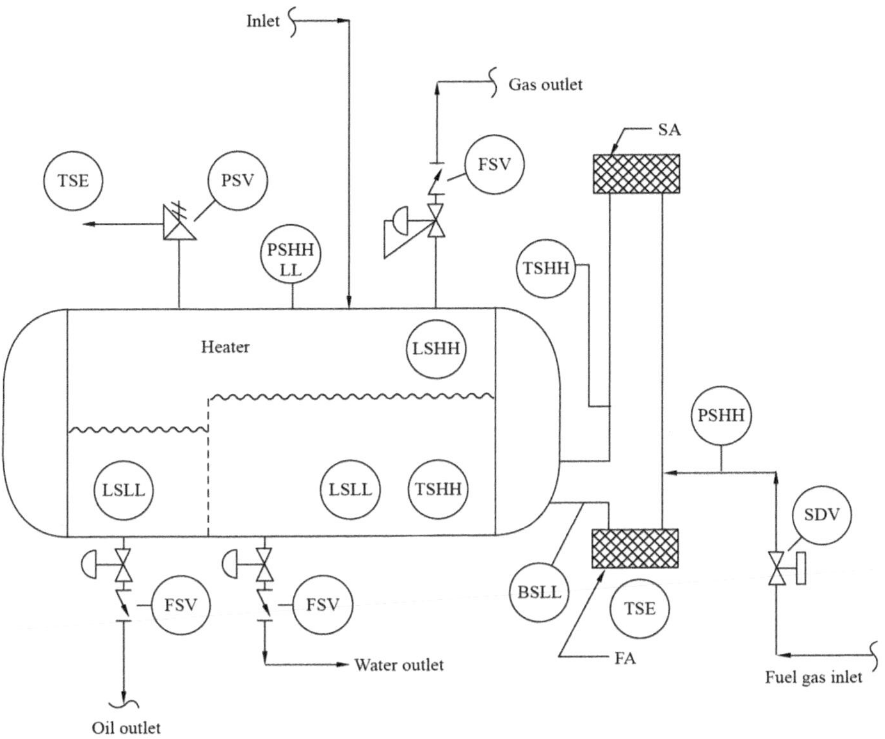

Fig. 5.5 Safety devices recommended for typical heating equipment. *Note 1* TSE is a symbol and does not reflect the actual position and quantity. *Note 2* The vessel part should be determined according to the previous corresponding analysis

the inflow and shuts down the pump. If a leak occurs in the downstream pipeline causing a drop in pressure, it is protected by a PSL sensor, which promptly shuts down the pump. The PSV is used to relieve pressure to protect the booster pump and downstream pipeline in case of PSH failure.

A check valve (FSV) should be installed on the pump discharge pipeline to minimize backflow.

The PSH and PSL sensors should be located upstream of the FSV or any shut-off valve on the pump discharge pipeline. The PSV should be located upstream of any shut-off valve on the pump discharge pipeline. The SDV should be located near the outlet of the separator to prevent crude oil from flowing into the downstream pipeline through the booster pump in case of a leak on the downstream pipeline.

(1) A PSH does not need to be installed if one of the following conditions is met:

① The maximum discharge pressure of the booster pump does not exceed 70% of the maximum allowable working pressure of the discharge pipeline.

② The booster pump is manually operated and under continuous surveillance.

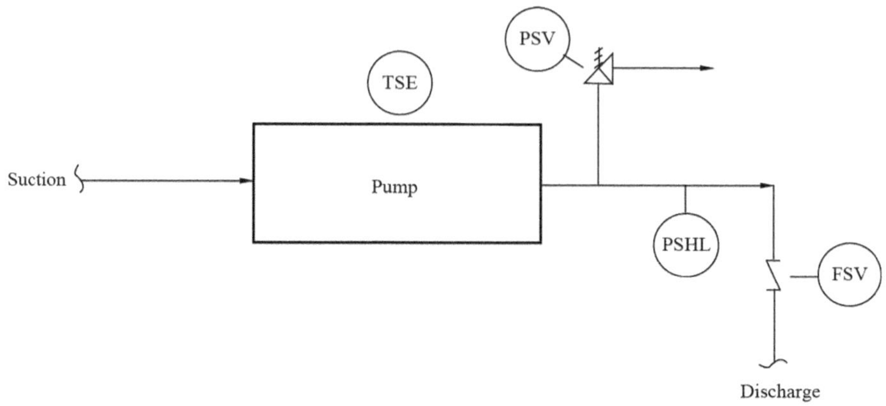

Fig. 5.6 Recommended safety devices for a typical booster pump. *Note* TSE is a symbol and does not reflect the actual position and quantity

(2) A PSL does not need to be installed if the following conditions are met:

① The booster pump is manually operated and under continuous surveillance.
② There is a good leak collection and discharge system.

(3) A PSV does not need to be installed if one of the following conditions is met:

① The maximum discharge pressure of the booster pump is lower than the maximum allowable working pressure of the discharge pipeline.
② The booster pump has the ability to relieve pressure internally. In any case, an FSV should be installed at the outlet of the booster pump. Figure 5.6 shows the recommended safety devices for a typical booster pump.

5.6 Electrolytic Dehydrator/Desalter

The electrolytic dehydrator (desalter) is a popular deep dehydration and desalting device internationally. After the crude oil has undergone primary dehydration through the first and second stage separators, it enters the two-stage electrolytic dehydrator (desalter) for deep desalting, so that the treated crude oil meets the qualified oil index. The electrolytic dehydrator (desalter) removes emulsified water from the crude oil under a certain temperature with the dual action of an externally applied high voltage electric field and a demulsifier. The electrolytic dehydrator (desalter) has electrode plates suspended inside, and a high and low voltage electric field is generated between and around the electrode plates by the externally applied current, causing the emulsified water to undergo electrophoresis and aggregation until separation. The electrolytic dehydrator (desalter) generally operates at full tank and is prohibited from separating gas inside. The unexpected events that affect the electrolytic dehydrator (desalter) are overpressure, underpressure, overflow, leakage, and overheating (if the vessel is heated). The safety analysis table for the electrolytic dehydrator (desalter) can be found in Chap. 3, Sect. 3.4 on pressure vessels.

5.6.1 Pressure Protection

In case of overpressure, the electrolytic dehydrator (desalter) is protected by a PSH sensor, which promptly cuts off the inflow of fluid to the vessel. When the leakage of the electrolytic dehydrator (desalter) is large enough to depressurize the production separator, the PSL sensor detects the set value and cuts off the inflow of fluid to the vessel. The PSV relieves pressure on the electrolytic dehydrator (desalter) in case of PSH failure.

The locations where the PSH and PSL sensors and PSV are installed should be able to detect or release the pressure of the liquid inside the electrolytic dehydrator (desalter), which is usually at or near the top of the electrolytic dehydrator (desalter). However, if the pressure drop from the electrolytic dehydrator (desalter) to the monitoring point is negligible, and these safety devices cannot be isolated from the electrolytic dehydrator (desalter), these devices can be installed on the crude oil outlet pipeline. The isolation of these safety devices from the electrolytic dehydrator (desalter) may be caused by external reasons (for example: the crude oil outlet valve is closed).

(1) A PSH does not need to be installed if one of the following conditions is met:

   ① The maximum pressure generated by the upstream fluid of the electrolytic dehydrator (desalter) cannot possibly be greater than the maximum allowable working pressure of the electrolytic dehydrator (desalter).
   ② Each input source has a PSH installed, and the set pressure of the PSH is less than the maximum allowable working pressure of the electrolytic dehydrator (desalter).
   ③ The gas outlet is connected to downstream equipment with a suitable pipeline without a shut-off or regulating valve, and this equipment is protected by a PSH that also protects the electrolytic dehydrator (desalter).
   ④ The electrolytic dehydrator (desalter) operates at atmospheric pressure and has an appropriate venting system.

(2) PSL installation is not required if the following conditions are met:

   ① Each input source is protected by PSL, and there is no pressure control device or throttle device between PSL and the manifold.
   ② The minimum operating pressure during operation is atmospheric pressure.
   ③ The crude oil outlet is connected to the downstream equipment with an appropriate pipeline without a cut-off valve or regulating valve, which is protected by a PSL that also protects the electro-dehydrator (salt remover).

(3) PSV installation is not required if one of the following conditions is met:

① Each input source is protected by a PSV, whose set pressure is not higher than the maximum allowable working pressure of the electro-dehydrator (salt remover), and a PSV has been installed on the electro-dehydrator (salt remover) to protect against fire heating and thermal expansion.
② Each input source is protected by a PSV, whose set pressure is not higher than the maximum allowable working pressure of the electro-dehydrator (salt remover), at least one of which will not be cut off from the electro-dehydrator (salt remover).
③ The PSV on the downstream equipment can meet the pressure relief requirements of the electro-dehydrator (salt remover) and will not be cut off from the electro-dehydrator (salt remover).

5.6.2 Overflow and Leakage Protection

Unlike the production separator, the electro-dehydrator (salt remover) requires full tank operation, so there is no need to install an LSH. During operation, it requires the electrode to be completely submerged in the liquid above the oil–water interface, so an LSL needs to be installed to monitor the liquid level in real time. If the liquid level drops, the motor needs to be cut off to avoid puncturing the electrode.

The position of the LSL sensor should be below the top of the container by a certain distance, which needs to be determined in conjunction with the supplier. The LSL sensor is best installed outside the container and can be isolated from the container, so these devices can be tested without interrupting the process system.

5.6.3 Overtemperature Protection

If the electro-dehydrator (salt remover) needs to be heated, a TSH sensor should be installed to cut off the heat source when the process fluid is overheated. If the heat source cannot cause overheating, there is no need to install a TSH.

Figure 5.7 shows the safety devices recommended for a typical electro-dehydrator (salt remover).

5.7 Export Pumps

Export pumps are used to send treated qualified crude oil from the process system to the export pipeline. The accidental events affecting the export pump are overpressure and leakage. The safety analysis table for the export pump can be found in Sect. 3.4 of Chap. 3.

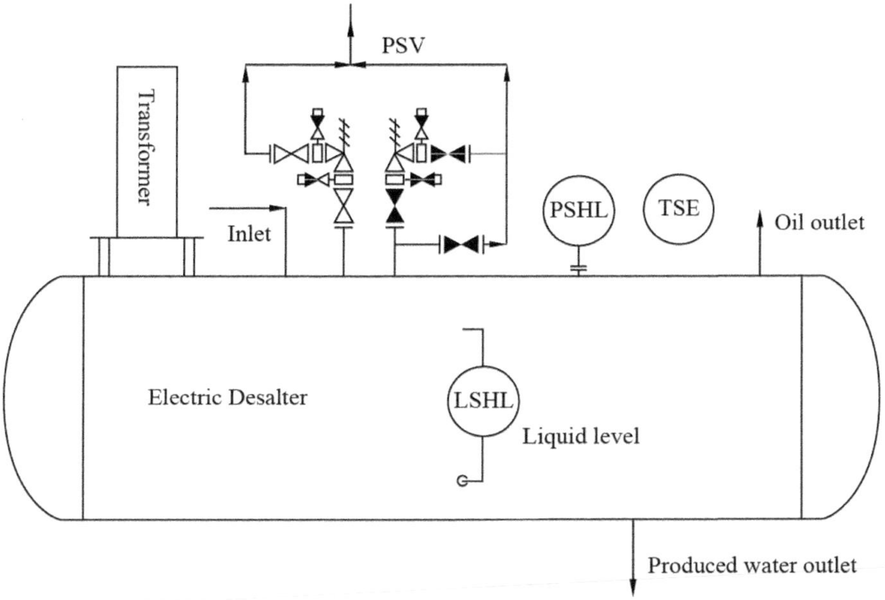

Fig. 5.7 Safety devices recommended for a typical electro-dehydrator (salt remover). *Note 1* TSE is a symbol and does not reflect the actual position and quantity. *Note 2* If the electro-dehydrator (salt remover) needs to be heated, a TSH must be installed

5.7.1 Pressure Protection

The export pump is used for crude oil export pressurization. If overpressure is caused, it is protected by a PSH sensor, which cuts off the inflow and shuts down the pump in time. If a leak in the pump's oil pipeline causes a pressure drop, it is protected by a PSL sensor, which shuts down the pump in time. The PSV relieves pressure to protect the export pump and downstream pipeline when the PSH fails.

A check valve (FSV) should be installed on the pump discharge pipeline to minimize backflow.

PSH and PSL sensors should be located upstream of the FSV or any cut-off valve on the pump discharge pipeline. The PSV should be located upstream of any cut-off valve on the pump discharge pipeline. On the storage equipment (such as tanks) supplying products to the export pump, the SDV should be set near its outlet to prevent crude oil from flowing into the export pipeline through the export pump in the event of a leak on the export pipeline.

(1) If one of the following conditions is met, there is no need to install a PSH:

① The maximum discharge pressure of the export pump does not exceed 70% of the maximum allowable working pressure of the discharge pipeline.

② The export pump is manually operated and under continuous surveillance.

(2) If the following conditions are met, there is no need to install a PSL:

① The export pump is manually operated and under continuous surveillance.
② There is a good leakage collection and discharge system.

(3) If one of the following conditions is met, there is no need to install a PSV:

① The maximum discharge pressure of the export pump is lower than the maximum allowable working pressure of the discharge pipeline.
② The export pump has the ability to relieve pressure internally.

5.7.2 *Backflow Protection*

In any case, an FSV should be installed at the outlet of the export pump. Figure 5.8 shows the safety devices recommended for a typical export pump.

5.8 Crude Oil Storage Tanks

Crude oil storage tanks are atmospheric vessels used to store qualified crude oil. Sometimes it is necessary to heat the crude oil storage tank to maintain a certain temperature. This discussion only covers the impact of heat input on the treatment section of the crude oil storage tank.

The accidental events that affect crude oil storage tanks are overpressure, negative pressure, overflow, leakage, and overheating (if the vessel is heated). The safety analysis table for crude oil storage tanks can be found in Sect. 3.4 of Chap. 3 on atmospheric vessels.

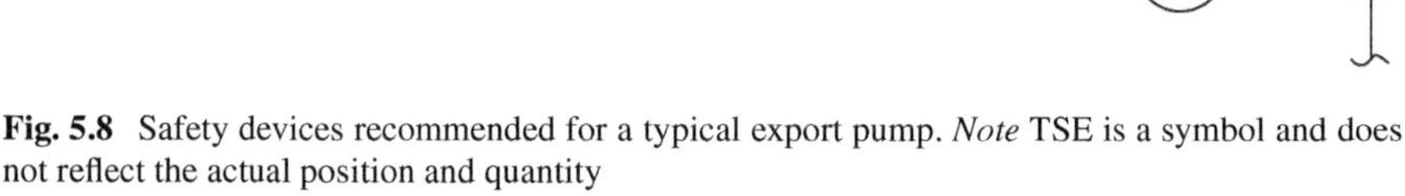

Fig. 5.8 Safety devices recommended for a typical export pump. *Note* TSE is a symbol and does not reflect the actual position and quantity

5.8.1 Pressure Protection

Crude oil storage tanks should be protected by a venting system of appropriate size to prevent overpressure and negative pressure. API Std 2000 Venting atmospheric and low-pressure storage tanks can be used as a guide for calculating the size of the venting system. A flame arrester should be installed in the venting system to prevent backfire. Once overpressure is formed, protection is provided by the PSH sensor, which promptly cuts off the inlet flow to the vessel. When the leakage from the crude oil storage tank is large enough to depressurize it, the PSL sensor detects the set value and cuts off the inlet fluid to the vessel. The PSV depressurizes the crude oil storage tank in the event of PSH failure.

The vent and PSV should be located at the top of the atmospheric vessel (at the highest possible position of the steam section). The PSV may not be installed if the following conditions are met:

(1) The vessel has a second-stage venting system capable of handling the maximum gas volume.
(2) The crude oil storage tank is designed as a pressure vessel that cannot be crushed, operates at atmospheric pressure, and has an appropriately sized venting system installed.
(3) The crude oil storage tank has no pressure source (except for gas replenishment and/or manual discharge) and has an appropriately sized venting system installed.

5.8.2 Overflow and Leakage Protection

Crude oil storage tanks should be protected by an LSH sensor to cut off the inlet fluid and prevent liquid overflow. If the crude oil storage tank is equipped with an immersion heating element that can withstand overheating, an LSL sensor should be installed to cut off the heat source. The inlet fluid should be cut off by the LSL sensor to prevent leakage, unless the liquid level in the crude oil storage tank is not automatically maintained. When the normal input of liquid may interfere with the sensor's leakage detection, it is better to use a leakage collection and discharge system to collect the leaked liquid than to use a low liquid level sensor.

The position of the LSH sensor should be high enough above the highest operating liquid level to prevent accidental shutdown, but there should be enough space in the vessel above the LSH sensor to accommodate the liquid flowing in during shutdown. The lowest position of the LSL sensor should be low enough below the lowest operating liquid level to avoid accidental shutdown. In equipment heated by fire tubes, the LSL should be above the fire tubes. LSH and LSL sensors are best installed on the outside of the vessel for easy testing without interrupting the process system operation. However, if solid deposition or foam causes external safety devices to become dirty or give false indications, the liquid level sensor can be directly installed inside the vessel.

(1) An LSH does not need to be installed if one of the following conditions is met:

① There is continuous monitoring by operators during liquid inflow.
② Overflow can be input or stored in other process equipment.

(2) An LSL does not need to be installed if one of the following conditions is met:

① A suitable leakage collection and discharge system has been installed.
② The liquid level in the crude oil storage tank is not automatically maintained, and there are no immersion heating elements in the tank that can withstand overheating.
③ The crude oil storage tank is the final container in the leakage collection and discharge system, and the design of the leakage collection and discharge system is used to collect crude oil and discharge it to a safe place.

5.8.3 Overheating Protection

If the atmospheric vessel needs to be heated, a TSH sensor should be installed to cut off the heat source when the process fluid overheats. Temperature sensors, except for fusible or surface contact types, should be installed in thermocouple sheaths for easy removal and testing. The thermocouple sheath should be installed in a place easily accessible to the operator and should be kept immersed in the hot fluid.

If the heat source cannot cause overheating, a TSH does not need to be installed. Figure 5.9 shows the recommended safety devices for a typical crude oil storage tank.

5.9 Compressors

Compressor units are used to transport gaseous hydrocarbons within the production process system. Compressors and prime movers are usually equipped with devices to prevent mechanical damage. Overpressure, leakage, and overheating are accidental

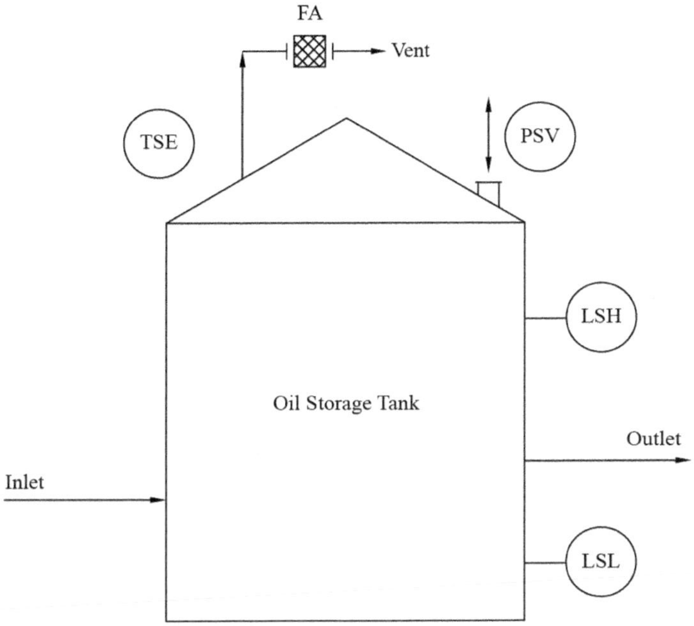

Fig. 5.9 Recommended safety devices for a typical crude oil storage tank. *Note 1* TSE is a symbol and does not reflect the actual position and quantity. *Note 2* Pressure and/or vacuum relief devices can be set up on the venting pipeline. *Note 3* A second venting pipeline can be installed to replace the pressure-vacuum relief device

events that affect the normal operation of the compressor. The safety analysis table for compressors can be found in Sect. 3.4 of Chap. 3.

(1) Pressure safety devices (PSH, PSL, and PSV). A PSH and PSL sensor should be installed on each suction pipeline of the compressor unit, unless each input source is protected by a PSH and PSL sensor and can also protect the compressor. A PSH and PSL sensor should also be installed on each discharge pipeline of the compressor. PSH and PSL sensors should cut off all process input pipelines and fuel gas pipelines of the compressor. A PSV should be installed on each suction pipeline of the compressor, unless each input source is protected by a PSV and can also protect the compressor. A PSV should also be installed on each outlet pipeline of the compressor. If the compressor is a kinetic type and cannot generate pressure exceeding the maximum allowable working pressure of the compressor or discharge pipeline, a PSV does not need to be installed at the compressor outlet.
(2) Flow safety devices (FSV). A check valve (FSV) should be installed on each final discharge pipeline to minimize backflow.

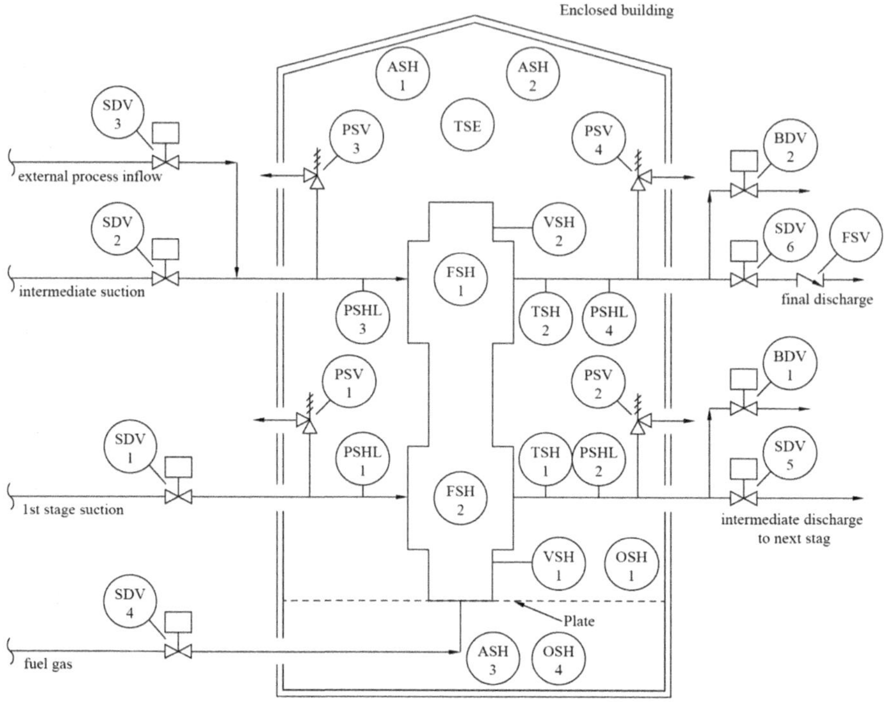

Fig. 5.10 Recommended safety devices for a typical associated gas compressor. *Note 1* TSE is only a symbol and does not reflect the actual position and quantity. *Note 2* If the compressor is not placed in a closed building, ASH1, ASH2, OSH1, and OSH2 are not required. *Note 3* If there are no pipelines or other potential leak sources under the solid base of the compressor, ASH3 is not required

(3) Gas detection devices (ASH). If the compressor unit is installed in a poorly ventilated building or within a wall, a gas detection device (ASH) should be installed to cut off all process input pipelines and fuel gas pipelines and vent the compressor.
(4) Temperature safety devices (TSH). A TSH sensor should be installed to protect all compressor cylinders and casings. TSH sensors should cut off all process input pipelines and fuel gas pipelines.

Figure 5.10 shows the recommended safety devices for a typical associated gas compressor.

References

1. API RP 14E Recommended practice for design and installation of offshore production platform piping systems, Fifth Edition, March 2007.
2. API RP 14C Recommended practice for analysis, design, installation, and testing of basic surface safety systems for onshore production platforms.
3. SY/T 10033 Recommended practice for analysis, design, installation, and testing of basic surface safety systems for offshore production platforms.

Chapter 6
Design and Calculation of Safety Valves

A safety valve is an automatic valve that discharges a specified amount of fluid using the force of the medium itself without any external force to prevent the pressure from exceeding the rated safety value. When the pressure returns to normal, the valve closes again to prevent the medium from continuing to flow out.

The description of safety valves internationally often follows the American ASME standard. In this standard, "safety valve" only refers to pressure relief facilities for steam or gas conditions, "relief valve" is mainly used for incompressible fluids, and "safety relief valve" refers to all pressure relief facilities including safety valves, relief valves, and safety relief valves. This section still follows the current national standard for naming, using safety valve to represent the entire meaning of ASME's safety relief valve.

The content of safety valve selection introduced in this section is only applicable to petrochemical enterprises, for the protection of pressure vessels and pipelines from overpressure accidents, and is not applicable to the protection of pressure vessels and pipelines in other industries.

The professional terms related to safety valves mainly include:

(1) Actual discharge area: The actual discharge area is the minimum net area that determines the valve flow rate measured in practice.
(2) Effective discharge area: The effective discharge area is different from the actual discharge area. It is the nominal or calculated area of the medium flowing through the safety valve, used in the flow calculation formula to determine the discharge capacity of the safety valve.
(3) Accumulation: During the discharge process of the safety valve, the value exceeding the maximum allowable working pressure of the container (when the design document of the pressure vessel does not give the maximum allowable working pressure, the design pressure of the container can be considered as the maximum allowable working pressure) is expressed in pressure units or percentages.

K. Fang, *Onshore Oilfield Surface Facilities: Process Safety Engineering Guide*,
https://doi.org/10.1007/978-981-96-7830-3_6

(4) Back pressure: It refers to the pressure at the outlet of the safety valve due to the pressure in the discharge system. Back pressure has two forms: fixed and variable. Back pressure is the sum of additional back pressure and accumulation back pressure.
(5) Additional back pressure: It refers to the static pressure at the outlet of the safety valve when it starts, which is caused by the pressure from the discharge of other valves.
(6) Accumulation back pressure: The pressure increase in the discharge main pipe due to flow after the relief valve opens.
(7) Set pressure: Also known as action pressure, it refers to the inlet pressure of the pressure relief facility when it opens under operating conditions.
(8) Maximum allowable working pressure: This refers to the maximum pressure that the top of the container is allowed to withstand at the design temperature. This pressure is based on the normal thickness, metal corrosion allowance, load, and pressure calculated in the equipment. The maximum allowable working pressure is the basis for setting the safety valve pressure to protect the equipment.
(9) Overpressure: The pressure that exceeds the set pressure of the safety valve, expressed in pressure units or percentages. It accumulates the same as the maximum allowable working pressure set for the container, assuming there is no pipeline loss at the safety valve inlet.
(10) Safety valve release pressure: The pressure at the valve inlet when the valve core of the safety valve rises to its maximum height. The release pressure equals the set pressure plus overpressure.
(11) Safety valve reseating pressure: After the safety valve pops, the pressure when the valve core returns to the valve seat as the pressure in the protected system decreases.

6.1 Classification of Safety Valves

The safety valves covered in this book are safety components in ASME BPVC Section VIII, mainly including conventional safety valves, bellows safety valves, and pilot-operated safety valves. Various safety valves are selected in oilfield surface engineering according to back pressure.

According to the recommendations of API Std 521 Pressure-relieving and depressuring systems, the maximum back pressure requirements for different types of safety valves are as follows:

(1) Conventional safety valve: The maximum back pressure is 10% of the set pressure.
(2) Balanced safety valve: The maximum back pressure is 30% of the set pressure.
(3) Pilot-operated safety valve: The maximum back pressure is 50% of the set pressure. The back pressure of the safety relief valve should be comprehensively calculated in the venting network according to the venting conditions [1].

6.1.1 Conventional Safety Valve

The conventional safety valve is a spring-actuated safety valve, and its set pressure is controlled by the spring. It consists of a sealing surface composed of a valve disc and a valve seat. The valve disc is connected to the valve stem, and the total displacement of the valve stem must meet the requirements of the valve from closed to fully open [2]. The set pressure of the safety valve is mainly adjusted by changing the spring pressure by adjusting the bolt. When the inlet pressure of the safety valve is less than the set pressure, the safety valve is closed; when the inlet pressure of the safety valve is greater than or equal to the set pressure, the system pressure can balance the force of the safety valve spring, and the valve disc begins to slowly open, and the medium begins to slowly release; as the valve disc opens, the pressure generated by the medium acts on a larger area of the valve disc, causing the valve disc to continue to rise until it is fully open; as the medium continues to be discharged, the pressure in the system gradually decreases, and the spring force will overcome the reaction force generated by the medium acting on the valve disc, thereby closing the safety valve (Fig. 6.1).

6.1.2 Balanced Safety Valve

The balanced safety valve is a spring-actuated safety valve, and its set pressure is controlled by the spring. A piston or bellows is used to reduce the impact of back pressure on the performance of the safety valve. The bellows safety valve refers to the addition of a bellows between the valve disc support and guide of the valve, mainly balancing the area of the back pressure acting on the valve disc, and the shape of the bellows is somewhat similar to the drain pipe of an old-fashioned washing machine. Of course, the function of the bellows is not only to balance the back pressure, but it can also prevent the medium from having a chance to enter the spring chamber, providing a better sealing effect between the valve body chamber and the spring chamber, and reducing the impact of back pressure (Fig. 6.2).

6.1.3 Pilot-Operated Safety Valve

The pilot-operated safety valve is a safety valve controlled by a pilot valve, and its set pressure is controlled by the pilot valve, and its operating characteristics are basically not affected by back pressure. Safety valves with pilot valves are divided into quick-opening (full-opening) and regulating (gradual-opening) types, and pilot valves are divided into flow and non-flow types. The pilot-operated safety valve is composed of a main valve, a pilot valve, a leak prevention valve, a joint box pipe, etc.

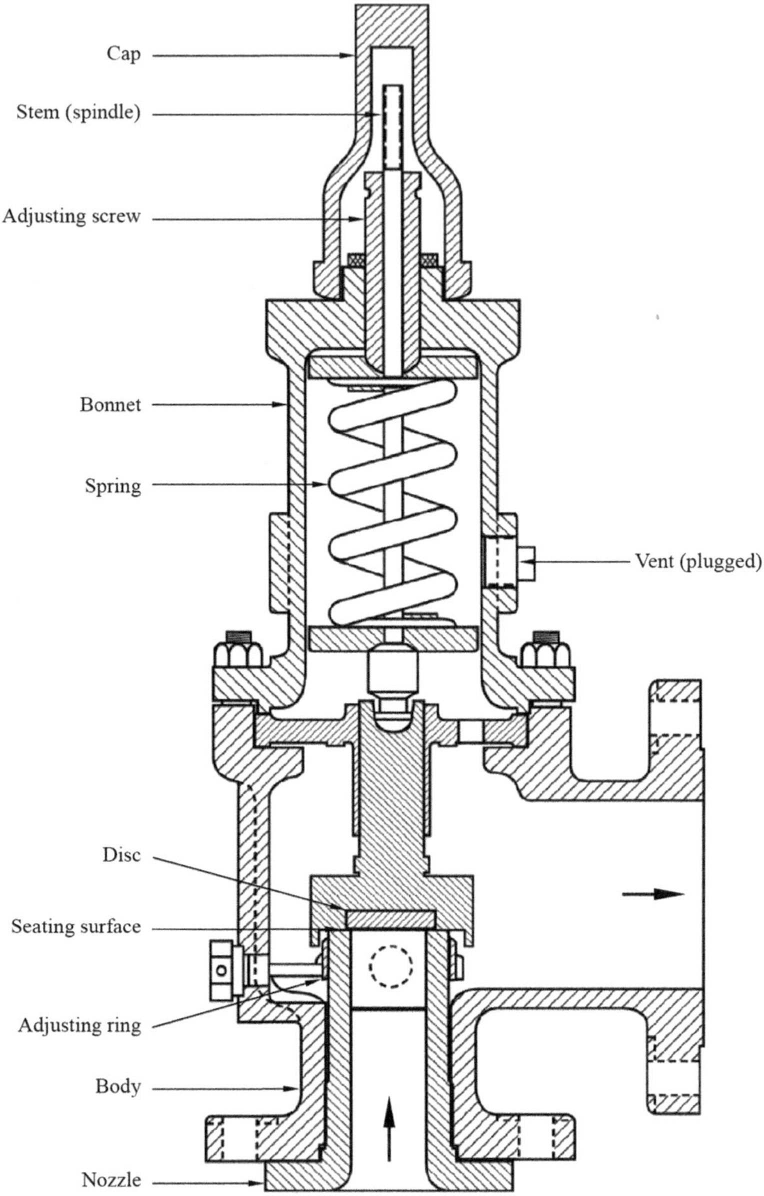

Fig. 6.1 Conventional safety valve

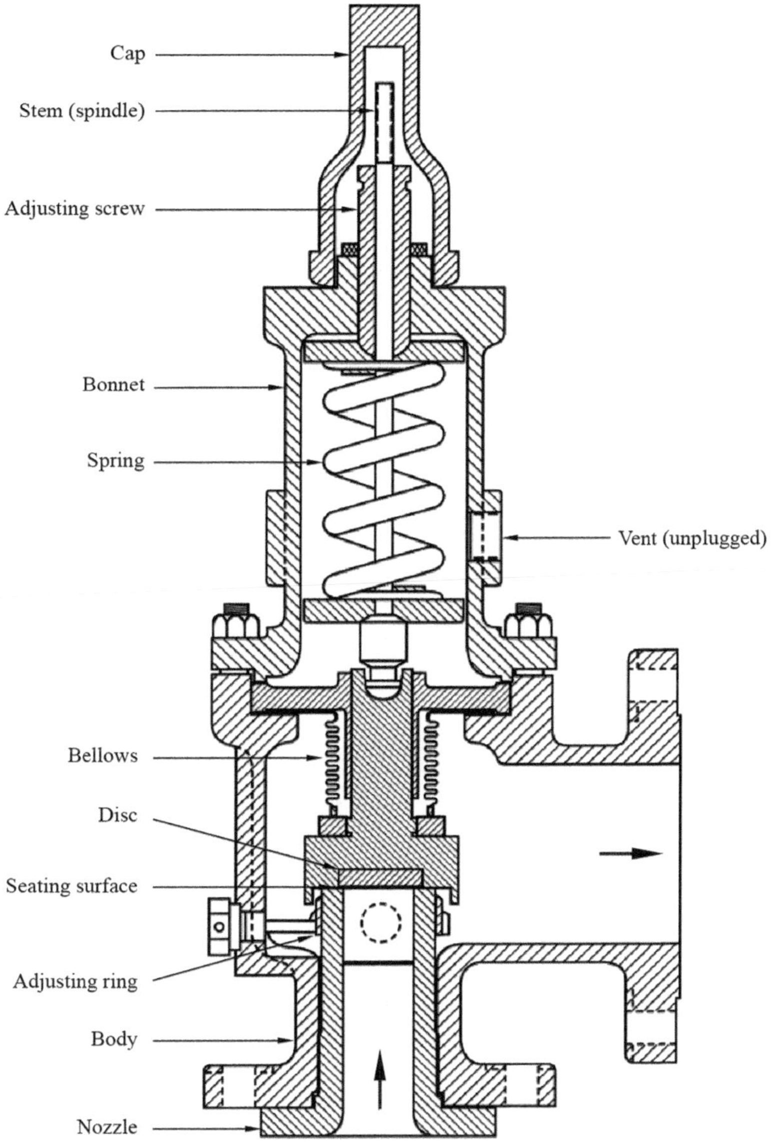

Fig. 6.2 Balanced safety valve

The pilot valve responds to the system medium pressure and controls the opening and closing action of the main valve by itself [3].

Under normal conditions, the system's medium pressure acts both below and above the main valve disc. Since the area of the medium acting on the upper part of the main valve disc is larger than the area of the medium acting on the lower part, the main valve disc is in a closed state under the action of the pressure difference. When the pipeline medium pressure is abnormal and reaches or exceeds the set pressure, the system pressure overcomes the spring force of the pilot valve, the pilot valve opens and the shuttle closes the air inlet of the pilot valve, and the pressure in the main valve chamber is released through the pilot valve. Because the system pressure is higher than the backseat pressure, the shuttle closes the air inlet, thereby blocking the flow of fluid in the pilot valve. At this time, the pressure in the main valve chamber drops sharply, the main valve disc is fully opened, and the fluid is released through the main valve to reduce the pressure. As the medium continues to be discharged, the pressure in the system gradually decreases. The spring force of the pilot valve will overcome the reaction force generated by the medium acting on the shuttle, the shuttle moves downward, and the medium can enter the main valve chamber through the pilot valve again. The main valve disc moves downward and the main valve closes (Fig. 6.3).

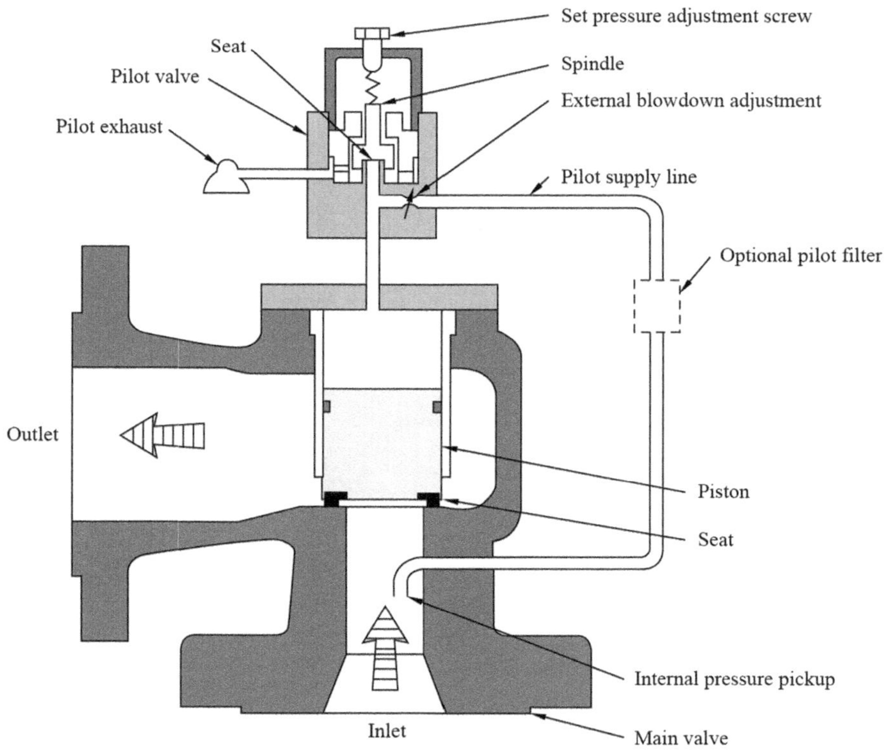

Fig. 6.3 Pilot-operated safety valve

6.2 Setting of Safety Valve

Safety valves are suitable for clean, non-particulate, low-viscosity fluids. In places where safety pressure relief devices must be installed and safety valves are not suitable, rupture discs or safety valves should be used in series with rupture discs. Any of the following situations must install a safety valve.

6.2.1 Pressure Vessel

(1) All independent pressure systems need to set up pressure relief facilities.
(2) When a safety valve is used to protect multiple pressure vessels, it must be ensured that there are no valves, regulating valves, etc. on the pipelines connecting the vessels, heat exchangers, and towers that can disconnect the equipment and the safety valve, and the size of the connected pipelines meets the pressure relief requirements. When the vessel is connected to the heat exchanger, the cut-off valve on the heat exchanger pipeline needs to be lead-sealed and kept open during normal operation.

6.2.2 Heat Exchanger

(1) If the tube side or shell side of the heat exchanger is designed to withstand the pressure when the pump outlet valve is closed, a safety valve may not be set on that side.
(2) If the low-temperature side of the heat exchanger has inlet and outlet valves, and the low-temperature side valves may be fully or partially closed during operation, a safety valve must be set on the low-temperature side for over-temperature protection.
(3) On the high-temperature side, if the vapor pressure of the condensed liquid at room temperature may exceed 110% of the equipment design pressure, a safety valve must be set.
(4) When the pressure difference between the two sides of the heat exchanger is large (exceeding 50%), a safety valve must be set on the low-pressure side to protect the heat exchange tube from rupture pressure.
(5) If the outlet valve of the cold side medium of the tube-shell heat exchanger is closed and the hot side medium is still in operation, the cold side may undergo vaporization causing overpressure. If the saturated vapor pressure of the cold side medium under the inlet temperature of the hot side medium does not exceed 1.3 times the design pressure of the cold side, this condition can be ignored. Otherwise, a safety valve needs to be set on the cold side.

6.2.3 Heating Furnace

If there is a cut-off valve or control valve on the outlet pipeline of the heating furnace, a safety valve should be set upstream of the valve, so that when the safety valve discharges, the medium must flow through the furnace tube, which can protect the furnace tube from overheating.

6.2.4 Pump

(1) If the pressure when the outlet valve of the reciprocating pump is closed may exceed the maximum pressure that the pump body can withstand, a safety valve must be set. Generally, the set pressure of the safety valve of the reciprocating pump is the maximum allowable working pressure of the pump body and does not exceed 121% of the maximum allowable working pressure of the downstream pipeline. The outlet of the pump safety valve is generally connected to the pump inlet.
(2) Under abnormal suction conditions, the pressure of the centrifugal pump may exceed the maximum pressure that the pump body can withstand, or it may be higher than 133% of the maximum allowable working pressure of the downstream pipeline, a safety valve must be set.
(3) In some cases, due to the leakage of the pump outlet check valve, a safety valve needs to be set on the inlet pipeline of the pump.

6.2.5 Compressor

Reciprocating compressors must have safety valves at each stage outlet, and they are discharged to the same stage inlet.

6.2.6 Pipeline System

(1) General pipelines within the device do not need to consider overpressure caused by liquid thermal expansion. For pipelines operating at temperatures below room temperature, when the valves at both ends may be cut off, and the vapor pressure of the medium at ambient temperature may exceed 133% of the maximum allowable working pressure of the pipeline, a safety valve needs to be set.
(2) For overhead liquid pipelines outside the device, when the diameter is greater than or equal to 200 mm, the length exceeds 30 m, and the cut-off valves at both ends may be cut off, a safety valve needs to be set to protect against liquid expansion pressure relief. If the cut-off valve has a bypass with a check valve,

when the pressure in the pipeline increases, the check valve can relieve pressure, and a safety valve does not need to be set.

6.2.7 Other Places Where Safety Valves Should be Set

When the container is exposed to a fire environment, due to radiation, convective heat transfer, and direct contact with flames, the substance stored in the container is heated, which can cause pressure to rise. According to the provisions of HG/T 20,570.2 "Setting and Selection of Safety Valves", containers within a height range of 7.5 m from the ground or above a large-area flame platform 7.5 m above the ground need to set safety valves [4].

6.2.8 Relief Pressure of Safety Valve

(1) For pressure vessel equipment, during non-fire condition relief, the maximum allowable relief pressure cannot exceed 10% of the maximum allowable working pressure (16% for multiple valves); during fire condition relief, the maximum allowable relief pressure cannot exceed 21% of the maximum allowable working pressure.
(2) For boiler equipment, the maximum allowable relief pressure cannot exceed 6% of the maximum allowable working pressure.
(3) The safety valve does not open only at the set pressure, it generally leaks at about 90% of the set pressure, so be careful when setting high pressure alarms. The pressure relationship related to the safety valve is shown in Fig. 6.4.

The set pressure of the safety valve should be the design pressure of the equipment and should be higher than the highest operating pressure of the equipment. The lower limit of the set pressure of the safety valve is determined according to the standard design principle (that is, the design with the minimum total cost of the equipment and pipeline system, safety relief device, and safety relief system). When the set pressure is less than this lower limit, the safety relief valve will have micro-leakage, and even cause deposits to accumulate on the valve seat and valve disc surfaces, which may cause the valve to stick and not open under the set pressure, causing an overpressure accident. ASME BPVC Section III proposes recommended values for the minimum operating pressure difference as shown in Table 6.1.

Each manufacturer has more detailed suggestions, and the minimum operating pressure difference of the safety valve is related to factors such as valve type, set pressure, and seal face structure.

The minimum operating pressure difference of the pilot-operated safety valve is relatively small. For the same type of safety relief valve, when the set pressure is higher, the required minimum operating pressure difference is smaller. Under the

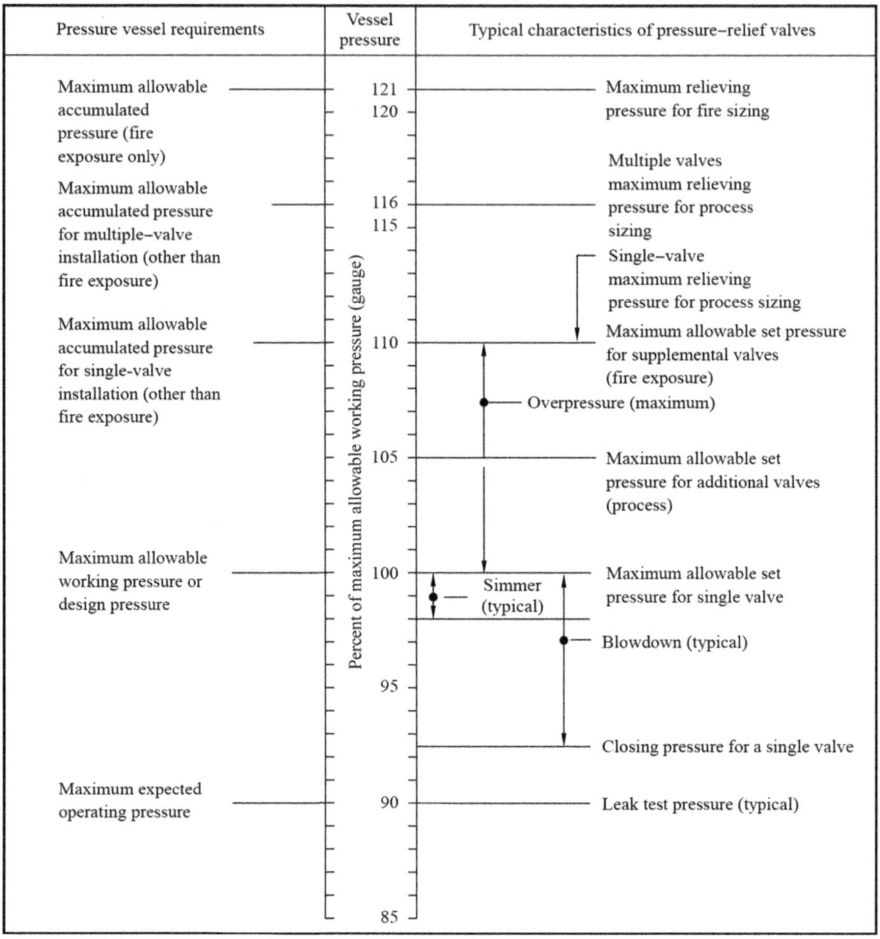

Fig. 6.4 Pressure relationship related to safety valve

Table 6.1 ASME BPVC Section III recommended minimum operating pressure difference for safety relief valve

Set pressure, MPa	Minimum operating pressure difference
≤0.485	35 kPa
0.490–6.90	10% set pressure
>6.90	7% set pressure

same set pressure, the minimum operating pressure difference required by the soft seal is smaller than that of the metal seal, and the smaller diameter valve seat requires a larger minimum operating pressure difference.

According to ASME BPVC Section VIII, when installing a safety valve, the upper limit of the set pressure is the maximum allowable working pressure; when installing

Table 6.2 Limitations of set pressure and accumulation pressure of pressure relief valves

Failure		Single valve device		Multi-valve device	
		Set pressure, %	Maximum accumulation pressure, %	Set pressure, %	Maximum accumulation pressure, %
Non-fire	First valve	100	110	100	116
	Additional valve	–	–	105	116
Fire	First valve	100	121	100	121
	Additional valve	–	–	105	121
	Auxiliary valve	–	–	110	121

Note The above data are percentages of the maximum allowable working pressure

multiple safety valves, at least one safety valve's set pressure must not exceed the maximum allowable working pressure, and the set pressure of the remaining additional safety valves must not exceed 5% of the maximum allowable working pressure; the set pressure of the auxiliary safety valve used for fire accidents must not exceed 10% of the maximum allowable working pressure (Table 6.2).

6.3 Analysis of Safety Valve Working Conditions

TSG 21 "Safety Technical Supervision Regulations for Fixed Pressure Vessels" has clear regulations on calculating the discharge capacity of safety valves under different working conditions. For content beyond the regulations, refer to the relevant parts of API Std 520 Sizing, selection, and installation of pressure-relieving devices and API Std 521 Pressure-relieving and depressuring systems. The methods introduced in this section consider the processing of the project and the recommended methods of relevant regulations in our country. Generally speaking, they are consistent with or safer than the methods recommended by API Std 520 Sizing, selection, and installation of pressure-relieving devices and API Std 521 Pressure-relieving and depressuring systems, and also meet the requirements of TSG 21 "Safety Technical Supervision Regulations for Fixed Pressure Vessels" in our country.

API Std 521 Pressure-relieving and depressuring systems standard explains in two parts how to determine the required processing capacity of a pressure relief system. First, it discusses the various causes that may lead to overpressure during design, and then provides guidance on how to determine the discharge volume under some common pressure situations. It also clearly states principles such as not considering the hazard of double accidents, needing to consider overpressure caused by operator errors, and generally not considering the protective effect of instruments on pressure relief. The working condition analysis method introduced in this section is based on API Std 521 Pressure-relieving and depressuring systems, and also refers to certain engineering experience.

6.3.1 Outlet Blockage

The outlet blockage condition may be caused by the following situations:

(1) Automatic control failure, closing the control valve.
(2) Accidental closure of the valve.
(3) Pump or compressor failure due to power loss, material flow loss, mechanical failure, etc.
(4) Loss of instrument air causes the valve to close.
(5) Incorrect installation, such as pipeline blind plate blinding, incorrect operation sequence, check valve installed in reverse, etc.
(6) Mechanical failure of the valve, such as the gate falling, the check valve baffle getting stuck (pay attention to the installation direction of the lead-sealed open valve to ensure that the failure will not cause the gate to fall to prevent flow).
(7) Blockage, freezing, formation of hydrates, polymerization, coking, salt deposition.
(8) Failure of internal components of the vessel [5].

6.3.1.1 Valve Blockage

To determine the pressure relief load, each valve blockage condition should be considered separately as completely closed, and the impact on downstream equipment should be analyzed. This may cause backflow and have adverse effects.

6.3.1.2 Liquid Phase Blockage

Liquid blockage will cause the liquid level in the upstream container to rise. If the design pressure of the container is higher than the maximum pressure of the upstream liquid, the container overflow may not cause a pressure relief situation. Liquid discharge should be avoided as much as possible to prevent liquid from being discharged into the gas phase or relief system. If liquid discharge cannot be avoided, sufficient volume of liquid treatment equipment must be provided in the relief system.

6.3.1.3 Heater Outlet Blockage

It is not recommended to install valves on the process pipes of heaters, unless they are equipped with lead-sealed valves or lock-open to ensure that they will not accidentally close. If necessary, pressure relief devices can be added.

6.3.1.4 Fog Catcher Blockage

The demister must ensure that the pressure relief path will not be blocked (such as blockage of the internal parts or blockage of the pressure relief valve caused by the falling off of the internal parts). If the demister is installed in a downstream vessel with a pressure reducing valve, the position of the pressure reducing valve should be based on the following criteria:

(1) If blockage is not allowed, the safety valve should be installed upstream of the demister.
(2) If there is a tendency for blockage or the tendency for blockage is unknown, a safety valve must be installed downstream of the demister element.
(3) In all cases, especially when the safety valve is located above the demister, the device must be securely installed to prevent displacement.

When there is only inlet flow, it should be ensured that the safety valve can be opened. The pressure relief flow is the flow of each inlet flow. For flows from centrifugal pumps, the pressure relief flow under higher discharge pressure should be determined according to the pump's characteristic curve. For volumetric pumps and compressors, the pressure relief flow should be determined according to their design capacity. For steam turbines, gas turbines, and hydraulic turbines, the pressure relief flow should be determined based on the maximum throughput.

The first consideration should be to increase the design pressure of the tank to the upstream pressure to cope with overflow conditions. If overflow will cause overpressure, a safety valve can be set to avoid certain conditions, such as the failure of the upstream high-pressure feed valve to open, which may cause liquid overflow and subsequently generate a large volume of vapor. A safety instrument system with a high enough SIL level can also be installed to reduce risk.

The weakest protection measure is operator intervention, which includes steps such as instrument detection of conditions, personnel recognition of conditions and determination of operations, deployment of personnel actions, etc., and will take a considerable amount of time.

6.3.2 *Failure or Backflow of Cooling Water*

The failure of cooling water is usually caused by other accident conditions. Cooling water failure should be considered as a local failure of a single heat exchanger or mechanical equipment.

6.3.2.1 Causes of Local Failure of Cooling Water

The local failure of cooling water may be caused by the following reasons:

(1) Accidental closure of the valve.
(2) Automatic control failure.
(3) Gas phase lock.

6.3.2.2 Causes of Cooling Water System Failure

The failure of the cooling water system may be due to the following fundamental reasons:

(1) Failure of the cooling medium pump due to power loss, steam loss (steam-driven pump), mechanical failure, etc.
(2) Power failure of the cooling tower fan.
(3) Accidental closure of the valve.
(4) Automatic control failure (loss of makeup water, high discharge flow rate).
(5) Insufficient cooling water supply leading to low liquid level in the cooling tower.

6.3.2.3 Consequences of Cooling Water System Failure

The failure of the cooling water system may lead to the following consequences:

(1) No cooling water supply for a certain process unit.
(2) No cooling water supply for a certain cooling water manifold.
(3) No cooling water supply for heat exchangers at higher altitudes.
(4) Complete loss of cooling water. In order to minimize the risk of cooling water system failure, various solutions should be considered in the design, such as setting up self-starting backup pumps; increasing the buffer capacity under the low liquid level alarm of the cooling tower to extend the response time of the operators; setting up low flow alarms for the makeup water supply pipeline and the off-site circulating cooling water manifold, etc.

6.3.2.4 Discharge Volume

The required discharge volume is determined by the heat balance and material balance at the discharge pressure of the system.

(1) Full condensation: The required discharge volume is the total steam inlet volume of the condenser, and the discharge volume should be recalculated according to the temperature under the new steam composition corresponding to the discharge condition and the heat inflow at the moment of discharge.
(2) Partial condensation: The required discharge volume is the difference in steam volume entering and leaving the condenser under the discharge condition.
(3) Air cooler fan failure: Due to the effect of natural convection, unless there is a significant change in the discharge condition, it is usually considered that 20–30% of the normal load of the air cooler is included in the partial condensation capacity, and the calculation of the required discharge volume is based on the remaining 70–80% of the normal load, depending on different situations, the actual effective load carried away by natural convection can also be calculated using HTRI or EDR, combined with the design parameters of the air cooler.
(4) Air cooler louvers closed: The closure of the air cooler louvers is considered to result in a complete loss of cooling effect. The reasons for the closure of the louvers may be a failure of the automatic control system, a failure of the mechanical linkage device, or a failure of the manually adjusted louvers due to destructive vibration.
(5) Top of the tower circulation reflux: In many cases, such as pump shutdown or valve closure causing top of the tower reflux failure, it will cause the top of the tower condenser to overflow, which is equivalent to the situation where the cooling function is completely lost. The component changes caused by reflux failure will produce different gas properties, which will affect the required discharge amount.
(6) Mid-section circulation reflux: The required discharge amount equals the vaporization amount produced by the heat generated by the heat removed by the mid-section circulation reflux. The latent heat of vaporization is equivalent to the latent heat of vaporization corresponding to the temperature and pressure at the time of discharge under discharge conditions.
(7) Top of the tower reflux failure. When a top of the tower reflux failure occurs, the initial discharge amount is the amount of vapor entering the top tray at normal operating temperature minus the condensation amount of the top condenser. Once the condenser is filled with liquid, the pressure relief discharge amount is the normal feed amount entering the top tray at normal temperature.
(8) Bottom of the tower discharge failure. When the bottom discharge pump stops running or the valve is closed, the amount of steam generated is equivalent to the steam generated by the heat taken away by the pump extraction flow.

6.3.3 Volatile Media Entering the System

Water or light hydrocarbons entering the hot oil system are potential sources of overpressure. Taking water as an example, if the amount of water that appears and the available heat in the process flow are known, then the size of the pressure relief facility can be determined by calculation like a steam valve. But the actual situation is that the amount of water is almost impossible to know, and even the size range is unknown; and when water changes from liquid to gas, the volume expands greatly (nearly 1:1400 under normal pressure), and the generation rate of steam is instantaneous, causing the pressure relief facility to be unable to open quickly enough to play a protective role. Therefore, pressure relief facilities are usually not provided for such accidents, and they should be avoided through reasonable process system design, such as the following designs:

(1) The operating pressure on the water side should be designed to be lower than that on the hot oil side.
(2) Avoid the occurrence of water (liquid) bags.
(3) Double cut-off valves and drain valves are set at the connection between the water pipeline and the thermal process pipeline.
(4) Set up interlocking devices to cut off the heat source when the raw material is contaminated with water.

6.3.4 Overflow

Many process vessels or buffer tanks have requirements for controlling the liquid level under normal operation, start-up or shutdown conditions. However, past experience has shown that under certain specific conditions, these devices may overflow. If the pressure at the source of the liquid feed and supply pipeline exceeds the set pressure of the relief facility and/or the design pressure of the equipment, then overflow should be specifically analyzed. System design options to solve overflow should consider but are not limited to the following measures.

(1) Increase the system design pressure and/or the set pressure of the pressure relief facility as allowed by the pressure design specifications.
(2) Design a pressure relief system that can safely accommodate the overflow medium, including the impact of operator intervention response (usually consider 10–30 min).
(3) Install safety instrument systems to avoid liquid overflow. The measures adopted still need to evaluate all conditions, especially pay attention to the start-up stage and process conditions (such as flow rate, temperature and density) deviating from normal values, and abnormal operating conditions that are more likely to cause overflow compared to normal operation.

6.3.5 Control Valve Failure

Although API Std 520 Sizing, selection, and installation of pressure-relieving devices stipulates that “process self-control regulating valves are generally installed at the inlet or outlet of the equipment, when the regulating valve installed at the equipment inlet fails and closes, there is no need to consider pressure relief measures when the equipment is overpressured", but in actual engineering, it is often handled conservatively according to "all regulating valves are assumed to be normally open during accidents, regardless of the accident assumptions at the time of design". If a failure occurs, the inlet valve is fully open or partially open, then pressure relief measures may need to be set up to prevent overpressure. If the same accident causes one or several outlet valves to close, while the inlet valve is still open, then the required discharge amount is the difference between the maximum inlet flow and the maximum outflow of the still open outlet valve.

Generally, the failure of a control valve does not affect other control valves. If the opening and closing of a faulty control valve affects the function of other control valves, then the discharge amount of the safety valve needs to be increased. There are two reasons for control valve failure, namely instrument compressed air failure and spring failure.

Sometimes the situation is much more complex. For example, a high-pressure container with liquid level control at the bottom, where the liquid is discharged into a low-pressure system. Under normal operation, high-pressure gas will not enter the low-pressure system, high-pressure liquid is discharged into the low-pressure system, and part of the liquid flashes. However, during design, it is necessary to consider the possibility of high-pressure gas entering the low-pressure system due to the loss of liquid level under high pressure. If the amount of gas entering the low-pressure system is quite large, or if the high-pressure gas source is "infinite", the low-pressure system may quickly overpressure. In this case, the pressure relief measures of the low-pressure system need to accommodate all the gas entering the low-pressure system through the liquid level control valve. When the amount of high-pressure gas is not large and the capacity of the low-pressure system is relatively large, the high-pressure gas entering the low-pressure side will increase the pressure on the low-pressure side, while the pressure on the high-pressure side will decrease accordingly; at this time, consider the discharge amount of high-pressure gas under normal replenishment conditions, plus a certain surplus.

Due to temporary driving, accident handling, or due to emission volume, when partially opening the bypass valve, consider opening the bypass valve 25% under the condition of full opening of the regulating valve. The calculation of the relief volume of the safety valve when the regulating valve fails can be calculated using the formula for calculating the flow coefficient cv value of the regulating valve.

6.3.6 Device Power Failure

When considering the size of the safety valve required for device power failure, a detailed analysis of the scope of the power failure and its impact on production must be made. Because during a power outage, it may affect the operation of electric actuators such as pumps, fans, compressors, and valves, and sometimes it may also affect the operation of instrument compressed air.

6.3.7 Abnormal Process Heat Input

The reasons for abnormal heat input are:

(1) The control valve of the heating medium (steam, heat transfer oil, etc.) abnormally opens.
(2) The calorific value of fuel oil increases or the control valve of the fuel oil heater abnormally opens.
(3) The heat exchanger needs to be cleaned.
(4) The heat transfer of the heat exchanger fails, transferring the high-temperature medium to the next heat exchanger.

When determining the size of the safety valve, the potential ability of process heat input, i.e., design margin, should be considered, not just normal heat input. For example, the design margin of the furnace burner is 25%, and the maximum steam generation amount minus the normal condensation amount or steam outflow amount should be calculated at 125% of the furnace nameplate load to select the safety valve. When designing the system, if future expansion is considered, the safety valve and piping size should meet the needs after expansion, but the size of the safety valve nozzle must be considered based on the current design volume. For reboilers and similar tubular heat exchangers that are heated by steam, when determining the heat exchange conditions during the failure of the regulating valve, it should be assumed that the tube is clean and free of dirt.

6.3.8 Liquid Expansion

General pipelines in the device do not need to consider overpressure caused by liquid thermal expansion. For pipelines operating at temperatures below ambient temperature, when both ends of the valve may be cut off, and the vapor pressure of the medium (such as liquefied petroleum gas, refrigerant, etc.) under ambient temperature may exceed 133% of the maximum allowable working pressure of the pipeline, protective measures should be taken.

For overhead liquid pipelines outside the device, when the diameter is greater than or equal to 200 mm, the length exceeds 30 m, and it is possible to cut off the valve at both ends, a safety valve for liquid expansion pressure relief should be set.

The inlet diameter of the safety valve is DN20mm, and the set pressure is the highest working pressure allowed by the pipeline flange at the working temperature. If a bypass with a check valve is set next to the cut-off valve, when the pressure in the pipeline rises, the check valve can play a pressure relief role, and a safety valve for liquid expansion can be avoided.

Both thermal expansion and fluid vaporization (if the heat source is sufficient to vaporize the fluid on the low-temperature side of the heat exchanger under the pressure set by the pressure reducing device) should be considered.

The analysis should consider the flow change rule. The transition period may include two-phase flow, which can increase the required orifice plate area. The required discharge volume is the maximum vapor generation rate under pressure relief conditions, including any non-condensable substances produced by over-heating, which is less than the normal condensation or vapor outflow rate. The potential possibilities of the system and each of its components should be considered, such as factors that may limit the heat flux of fuel or heat medium control valves or pipes. Consistent with the practice of other overpressure causes, design values should be used for valve size selection; however, the reserved surplus capacity (because in general practice, the burner has 125% of the heater design heat input) must be considered.

When installing a limit switch on the valve, the wide-open flow should be used, not the restricted flow. In shell-and-tube heat exchange equipment, the calculation of heat input should be based on a clean state, not a dirty condition. When there are volatile liquids in the distillation system or storage container, once there is heat input, it is possible to vaporize the liquid.

Vaporization can cause a large increase in the volume of vapor. The condensation of vapor requires the removal of the heat source or cooling.

For distillation systems or storage tanks that lose cooling, this may lead to a large increase in non-condensable vapor. If the normal vapor handling capacity of the process system is overloaded due to a large increase in fluid volume, the pressure relief system must vent excess vapor to prevent overpressure of the equipment.

The formula for calculating the amount of release required for liquid expansion is shown in Eq. (6.1):

$$W = BH/c_{\mathrm{p}} \tag{6.1}$$

where

W—Mass flow rate, kg/h;

B—Volume expansion coefficient, °C^{-1}, see Table 6.3;

H—Total heat transfer, kJ/h, for heat exchangers, the maximum heat load during operation can be taken;

c_{p}—Specific heat capacity at constant pressure, kJ/(kg·°C).

6.3.9 External Fire

The equipment of flammable liquid processing plants may be exposed to external fires, causing liquid evaporation and gas expansion overpressure. To determine the

Table 6.3 Expansion coefficients of hydrocarbon liquids and water at 15.6 °C

Liquid gravity, API	Value, $°C^{-1}$
3–34.9	0.00072
35–50.9	0.00090
51–63.9	0.00108
64–78.9	0.00126
79–88.9	0.00144
89–93.9	0.00153
94 and lighter	0.00162
Water	0.00018

size of the pressure relief system, it is assumed that the process equipment inlet and outlet are cut off and isolated from other flows. If the pressure relief system can ensure that the equipment will not exceed the design pressure, and the cut-off valve has a lock or lead seal, a single pressure relief device can protect multiple equipment.

The storage pressure of liquid hydrocarbons should be greater than or equal to the vapor pressure corresponding to the storage temperature. When the storage tank is exposed to flames, the material stored in the container is heated due to radiation, convective heat transfer, and direct contact with the flame, and the pressure rises until the safety valve opens, so that the pressure in the container does not exceed the maximum allowable pressure. If the discharge capacity of the safety valve is less than the amount of vapor generated, the pressure in the container will rise above the maximum allowable pressure, which is unsafe.

Before the container is exposed to flames, the discharge amount required by the safety valve must be calculated according to the heat input to the container. API Std 520 Sizing, selection, and installation of pressure-relieving devices provides a method for calculating the safety valve of a storage tank during a fire based on test data. The calculation is based on the heat absorption of the liquid-containing surface of the container (called the wet surface) during a fire, and the heat received by the non-liquid container surface is ignored. Only equipment with a flame height of 7.5 m (25 ft) or less needs to be considered. The height of the flame is based on the platform of the ground-style accumulable liquid material device (which can form a large flame). If the platform is a grille and cannot accumulate liquid, it cannot be used as a calculation basis.

6.3.10 Heat Exchanger Tube Rupture

The definition of tube rupture is a sudden complete rupture of a heat exchange tube, allowing high-pressure liquid to flow through the opening from the shell side or tube side, and the opening is twice the cross-sectional area inside the heat exchange tube.

Unless the corrected low-pressure side hydrostatic test pressure exceeds the design pressure of the high-pressure side, all heat exchangers should assume that the heat exchange tube ruptures. In some cases, the design pressure on the low-pressure side has been increased to higher than the maximum possible system pressure on the high-pressure side. In this case, the maximum possible system pressure on the high-pressure side can be used for calculation. For equipment manufactured according to current specifications, this means that if the design pressure on the low-pressure side is 10–13 times the design pressure on the high-pressure side, pressure relief protection is not required. Please note that all components on the low-pressure side that may be subjected to overpressure must be checked to confirm the possible pressure in the event of a heat exchange tube rupture, not just the heat exchanger, and other factors such as static pressure head must also be considered.

If the high-speed flow caused by the rupture leads to a short residence time, the heat transfer inside the tube rupture can be ignored. Therefore, the calculation of tube rupture should be based on adiabatic. If the inflow calculated from the tube rupture exceeds the normal working flow of the high-pressure side, it should be verified that this flow can be maintained on the high-pressure side of the system. If the flow cannot be sustained, use the normal flow. For types of heat exchangers other than shell and tube, including internal coil in the vessel, reasonable engineering judgment must be used to determine whether the seal rupture between the two fluids should be considered.

For very high-pressure situations, some customers have a standard for calculating flow, as if two or more heat exchangers burst at the same time. This is because experience shows that a ruptured heat exchanger may collide with adjacent heat exchangers, damaging them.

When calculating the flow through a ruptured heat exchanger, the customer's method should be followed. In the absence of a customer's method, in addition to the factors discussed in API Std 521 Pressure-relieving and depressuring systems, the following factors should also be considered:

(1) If the tube plate thickness is less than 4 in, and the high-pressure side fluid is two-phase flow or flash steam, the HEM assumption may not be conservative enough. For these situations, assume that the heat exchanger rupture occurs on the tube plate. Then the path length of one branch of the fluid only spans the tube plate, so the unbalanced effect that may lead to a larger flow should be considered.
(2) For the fluid branch along the tube path, the friction along the tube path can be considered to achieve possible load reduction. For U-tube bundles, assume the path length is twice the straight tube length; for single-pass tubes, assume the path length is equal to the straight tube length.
(3) The friction effect of the long tube path is usually more significant in reducing the load than the extra load generated by the unbalanced effect through the entire tube plate path, so the ideal tube nozzle HEM analysis can usually be used for both paths, ignoring the unbalanced and friction effects.

(4) The driving force of the flow is the difference between the maximum possible system pressure on the high-pressure side and the accumulated release pressure on the low-pressure side.
(5) The treatment of the flow into or out of the ruptured heat exchanger is the same, that is, the shell side or tube side can be the high-pressure side.
(6) Determine the effect when the high-pressure fluid pressurizes the low-pressure side. Higher pressure in the equipment or pipeline may hinder the normal flow of the medium. An example is a cooling water pipe network, where high-pressure fluid releases through the cooling water pipe, replacing the normal cooling water flow. If the pressure is high enough, it may prevent or stop the flow of cooling water. This will result in the loss of cooling accidents, which may involve multiple situations.
(7) If the ruptured heat exchanger is treated as a flow-limiting orifice plate, a reasonable orifice coefficient should assume a critical gas flow of 0.84 and a liquid flow of 0.65.

6.3.11 Instrument Air or Electric Instrument Failure

Local failure means that the control valve or switch valve loses instrument air or power, at which time the valve will enter the failure position. Instrument air system failure may lead to the following consequences:

(1) A process unit has no instrument air supply.
(2) Part of the main manifold has no instrument air supply.
(3) Completely no instrument air supply.

To minimize the risk of instrument air system failure, various schemes should be considered in the design, such as increasing the size of the instrument air storage tank, setting multiple instrument air compressors and dryers, adopting a ring-shaped instrument air manifold, setting emergency backup instrument air (factory air or nitrogen), etc.

6.3.12 Mechanical Equipment Failure

Mechanical equipment failure can be regarded as a single point failure, examples are as follows:

(1) Pump failure (power loss or steam turbine air loss).
(2) Compressor failure (power loss or steam turbine air loss).
(3) Heat exchanger failure (such as tube path rupture of a tube and shell heat exchanger, etc.).
(4) Failure of vessel internals.

The consequences of each equipment failure need to be analyzed separately according to the cause of the accident. See Table 6.4 for a summary table of single safety valve release quantities.

6.4 Calculation of Safety Valve

6.4.1 Gas Discharge

In the pressure calculation of the safety relief valve involved in API, follow formula (6.2), that is, adiabatic reversible process:

$$pV^k = \text{constant} \tag{6.2}$$

where

p—pressure;
V—volume;
k—ratio of specific heat of ideal gas (cp/cV).

6.4.1.1 Critical Pressure (Absolute Pressure)

The critical pressure is calculated by formula (6.3)

$$\frac{p_{\mathrm{cf}}}{\mathrm{p}_1} = \left[\frac{2}{\mathrm{k}+1}\right]^{\frac{\mathrm{k}}{\mathrm{k}-1}} \tag{6.3}$$

where

P_{cf}—valve nozzle critical flow pressure (absolute pressure), psi;
p_1—relief pressure (absolute pressure), psi;
k—ratio of specific heat of ideal gas.

The pressure at which the flow speed at the valve nozzle reaches the speed of sound is the critical pressure. During critical flow, even if the downstream pressure of the valve nozzle (back pressure of the valve) is very low, the actual pressure at the valve nozzle will not be lower than the critical flow pressure, and an irreversible expansion process occurs from the valve nozzle to the downstream of the valve nozzle, with energy diffusion to the surrounding fluid due to turbulence.

Table 6.4 Single safety valve release quantity

Serial number	Condition	Liquid release	Gas release
1	Vessel outlet valve closed	Maximum pump liquid flow	Total vapor and vapor feed plus the amount of vapor generated under release conditions
2	Condenser cooling water failure	–	Total vapor entering the condenser under release conditions
3	Tower top reflux failure	–	Total vapor and vapor feed plus the amount of vapor reduced by the side line reflux failure under release conditions
4	Side line reflux failure	–	The difference in vapor quantity entering and exiting the equipment under release conditions
5	Non-condensable gas accumulation	–	The effect on the tower is the same as serial number 2, the effect on other vessels other than the tower is the same as serial number 1
6	Entry of highly volatile substances; water entering hot oil; light hydrocarbons entering hot oil	–	Adopt optional protective measures to avoid this situation. For guidance on heat exchanger rupture, refer to item 15
7	Overflow	Maximum pump liquid flow	–
8	Automatic control system failure (1) Inlet control facilities and bypass (2) Outlet control facilities (3) Valves that maintain status in case of failure (4) Throttle valve	–	Based on each operating condition analysis
9	Abnormal process heat or steam input (1) Input of abnormal process heat (2) Valve opened due to negligence (3) Check valve failure	–	Estimate the maximum steam generation, including non-condensable gases generated by overheating

(continued)

Table 6.4 (continued)

Serial number	Condition	Liquid release	Gas release
10	Thermal expansion (1) Cold source cut off (2) Pipeline outside the process area cut off	See the description in the previous related sections	–
11	External fire	See the description in the previous related sections	–
12	Heat exchanger tube rupture	Liquid flow through the rupture area twice the cross-sectional area of a heat exchanger tube	Steam flow through the rupture area twice the cross-sectional area of a heat exchanger tube
13	Power failure (steam, electricity or other reasons)	–	Study the entire set of devices to determine the impact of a power outage, and determine the size of the safety valve based on the worst possible scenario
14	Air cooler	–	Fan stops rotating, safety valve size determined by the difference between normal heat load and accident heat load
15	Buffer tank	–	Maximum liquid entry

6.4.1.2 Critical Flow

Critical flow refers to the downstream pressure of the valve nozzle being less than or equal to the critical pressure, with formulas (6.4)–(6.7):

$$A = \frac{W}{CK_d p_1 K_b K_c}\sqrt{\frac{TZ}{M}} \tag{6.4}$$

$$A = \frac{2.676\ V\sqrt{TZM}}{CK_d p_1 K_b K_c} \tag{6.5}$$

$$A = \frac{14.41V\sqrt{TZG_v}}{CK_d p_1 K_b K_c} \tag{6.6}$$

$$C = 0.03948\sqrt{k\left(\frac{2}{k+1}\right)^{\frac{(k+1)}{(k-1)}}} \tag{6.7}$$

where

A—Diameter of the pressure relief hole, mm;

W—Pressure relief flow rate, kg/h;

V—Pressure relief flow rate (standard conditions), m^3 /min;

C—The coefficient determined by the specific heat of the ideal gas under the inlet temperature of the relief valve, which can be determined by formula (6.7);

k—The ratio of specific heat (cp/cV) of the ideal gas under the relief temperature;

K_d—Effective discharge coefficient, taken as 0.975 when calculated with the above formula;

K_b—Back pressure correction factor, this factor is 1 for conventional safety valves and pilot safety valves, for balanced bellows safety valves it can be provided by the manufacturer, or found from Fig. 6.30 of API Std 520 Sizing, selection, and installation of pressure-relieving devices;

K_c—Combination factor related to the rupture disc, taken as 1 when there is no rupture disc, and 0.1 when a rupture disc is installed;

p_1—Inlet pressure of the safety relief valve, this pressure should be the set pressure of the safety relief valve plus overpressure plus atmospheric pressure, kPa;

Z—Compression factor under the inlet conditions of the safety relief valve;

M—Relative molecular mass under the inlet conditions of the safety relief valve;

T—Inlet temperature of the safety relief valve, K;

G_v—Relative density of the gas under relief conditions compared to air at 1 bar and 15.6 °C conditions.

6.4.1.3 Subcritical Flow

Subcritical flow refers to the downstream pressure of the valve nozzle being greater than the critical pressure.

(1) (Non-steam) Calculation is given by formulas (6.8)–(6.11):

$$A = \frac{17.9W}{F_2 K_d K_c}\sqrt{\frac{ZT}{Mp_1(p_1 - p_2)}} \tag{6.8}$$

$$A = \frac{47.95V}{F_2 K_d K_c}\sqrt{\frac{ZTM}{p_1(p_1 - p_2)}} \tag{6.9}$$

$$A = \frac{258V}{F_2 K_d K_c}\sqrt{\frac{ZTG_v}{p_1(p_1 - p_2)}} \tag{6.10}$$

$$F_2 = \sqrt{\left(\frac{k}{k-1}\right) r^{\left(\frac{2}{k}\right)} \left[\frac{1 - r^{\left(\frac{k-1}{k}\right)}}{1 - r}\right]} \tag{6.11}$$

where

F_2—Subcritical flow coefficient, calculated by formula (6.11);

r—The ratio of back pressure to upstream relief pressure, p_2/p_1;

p_2—Back pressure of the relief valve, kPa;
k—The ratio of specific heat (cp/cV) of the ideal gas under the relief temperature.

(2) For steam, calculate according to formula (6.12):

$$A = \frac{190.5W}{p_1 K_d K_b K_c K_N K_{SH}} \tag{6.12}$$

In formula (6.12), KSH is the correction factor for superheated steam, and for saturated steam at any pressure, $KSH = 1$. KN is the Naiper equation correction factor, when $p1 < 10339$ kPa, $KN = 1$; when 10339 kPa$<p1 \le 22057$ kPa:

$$K_N = \frac{0.02764p_1 - 1000}{0.03324p_1 - 1061} \tag{6.13}$$

6.4.2 Liquid Discharge

6.4.2.1 Confirm the Liquid Discharge Capacity of the Relief Valve

The method to confirm the liquid discharge capacity includes determining the discharge coefficient of the liquid relief valve designed according to 10% overpressure, according to the discharge capacity required by the ASME standard, the size calculation is given by formulas (6.14)–(6.16):

$$A = \frac{11.78Q}{K_d K_w K_c K_v}\sqrt{\frac{G_1}{p_1 - p_2}} \tag{6.14}$$

$$K_v = \left(0.9935 + \frac{2.878}{Re^{0.5}} + \frac{342.75}{Re^{1.5}}\right)^{-1.0} \tag{6.15}$$

$$Re = \frac{Q(18800G_1)}{m\sqrt{A}} \tag{6.16}$$

where

Q—Flow rate, L/min;

K_d—Effective discharge coefficient, taken as 0.65 when a rupture disc is installed, and 0.62 when there is no rupture disc;

K_v—Viscosity correction factor, can be obtained from API Std 520 Sizing, selection, and installation of pressure-relieving devices, as found in Fig. 37, or calculated by formula (6.15);

K_w—Back pressure correction factor, when the back pressure is atmospheric pressure, $Kw = 1$; For balanced bellows relief valves, refer to API Std 520 Sizing, selection, and installation of pressure-relieving devices, Fig. 6.31;

p_1—Inlet pressure of the safety relief valve (gauge pressure), kPa;

p_2—Back pressure of the relief valve (gauge pressure), kPa;

G_1—Relative density of the liquid in the flowing state relative to water at 21 °C;

R_e—Reynolds number, calculated by formula (6.16);

μ—Viscosity of the liquid at the flowing temperature, cP.

6.4.2.2 No Need to Verify the Liquid Discharge Capacity of the Relief Valve

Before the ASME standard unified the verification of the discharge capacity of the relief valve, the liquid discharge was usually calculated by formula (6.17) to size the relief valve. This method assumes a discharge coefficient $K_d = 0.62$ and 25% overpressure. For overpressure beyond 25%, the additional K_p can be found in API Std 520 Sizing, selection, and installation of pressure-relieving devices Fig. 38.

$$A = \frac{11.78Q}{K_d K_w K_c K_v K_p}\sqrt{\frac{G}{1.25p_s - p_2}} \tag{6.17}$$

where

p_s—Set pressure (gauge pressure), kPa;

K_p—Overpressure correction factor, when overpressure is 25%, $K_p=1$; when overpressure exceeds 25%, it can be found in API Std 520 Sizing, selection, and installation of pressure-relieving devices, Fig. 38.

6.4.3 Two-Phase Discharge

When the safety relief valve discharges both gas and liquid at the same time, the following steps can be taken:

(1) Determine the flow rate of the discharged gas and liquid.
(2) Calculate the valve hole area required for the discharge of gas according to the aforementioned method.
(3) Calculate the valve hole area required for the discharge of liquid according to the aforementioned method.
(4) Add the calculated areas of gas and liquid to get the total valve hole area required.

6.4.4 *Fire Discharge*

According to the requirements of API Std 521 Pressure-relieving and depressuring systems, for the discharge of fire safety relief valves, only one of the gas expansion and liquid evaporation conditions is selected for calculation. The calculation formula is as follows:

6.4.4.1 Calculation of Heat Absorption

The heat absorption is calculated by formula (6.18):

$$Q = C_1 \cdot F \cdot {A_{ws}}^{0.82} \tag{6.18}$$

where

Q—Heat absorption, W;

C_1—When there are fire-fighting facilities and appropriate discharge capacity (from one area to another), $C_1 = 43{,}200$; when there are no fire-fighting facilities and appropriate discharge capacity, $C_1 = 70{,}900$;

F—Environmental factor, see Table 6.5;

A_{ws}—Wetted surface area, m.

Table 6.5 Environmental factor *F*

Equipment type		Environmental factor *F*
Exposed container		1
Insulated container (thermal conductivity of insulation material exposed to fire conditions), W/(m^2·K)	22.71	0.3
	11.36	0.15
	5.68	0.075
	3.80	0.05
	2.84	0.0376
	2.27	0.03
	1.87	0.2026
Exposed container with cooling water facilities		1.0
Pressure reduction and emptying facilities		1.0
Tank covered with soil		0.03
Underground storage		0.00

Note According to API Std 521 Pressure-relieving and depressuring systems

6.4.4.2 Calculation of Discharge Hole Diameter

For the discharge valve hole diameter of supercritical fluid, gas or steam exposed to open fire, it is calculated by formula (6.19):

$$A = \frac{F' \cdot A'}{\sqrt{p_1}} \tag{6.19}$$

where

A—Effective discharge area of the valve, in^2;

A'—Area of the container exposed to fire, ft^2;

p_1— Valve pressure release (absolute pressure), which is the set pressure plus the allowable overpressure plus atmospheric pressure, in psi.

The F' in formula (6.19) can be calculated by formulas (6.20) and (6.21). If the calculated result is less than 182, take 182; if the data is incomplete and cannot be calculated by formula (6.20), roughly take 821.

$$\mathrm{F}' = \frac{0.2772}{\mathrm{C} \cdot \mathrm{K_D}} \cdot \left[\frac{(\mathrm{T_w} - \mathrm{T_1})^{1.25}}{\mathrm{T_1^{0.6506}}}\right] \tag{6.20}$$

$$\mathrm{T_1} = \frac{\mathrm{p_1}}{\mathrm{p_n}} \cdot \mathrm{T_n} \tag{6.21}$$

where

T_w—The maximum temperature that the container wall can withstand, in K. For ordinary carbon steel, the recommended maximum container wall temperature is 866.16 K (593 °C);

T_1—The adiabatic temperature of the gas in the release state, calculated by formula (6.21);

p_n—Normal operating pressure, in psi;

T_n—Normal operating temperature, in K.

6.4.4.3 Calculation of Release Amount

The release amount can be calculated by formula (6.22).

$$\mathrm{q_{m,relief}} = 0.2772\sqrt{\mathrm{M} \cdot \mathrm{p_1}} \cdot \left[\frac{\mathrm{A}'(\mathrm{T_w} - \mathrm{T_1})^{1.25}}{\mathrm{T_1^{1.1506}}}\right] \tag{6.22}$$

6.4.5 Dynamic Calculation

Under fire conditions, as the steam inside the container is released, the composition of the steam and liquid inside the container changes, and the temperature and latent heat value also change. The maximum release amount depends not only on the heat absorption rate but also on the actual composition of various components in the container. Therefore, using conventional methods, it is difficult to determine the maximum release amount of the PSV and the characteristic parameters of the released fluid. For multi-component mixtures with a wide boiling point range, a time-dependent model can be established to calculate the maximum release amount of steam.

Chen Wenfeng and others used HYSYS dynamic simulation to carry out dynamic simulation of a three-phase separator [size: 4000 mm (I.D) × 16000 mm (T/T); design parameters: 1.1 MPa/105 °C; operating parameters: 0.45 MPa/75 °C; operating liquid level height: 2000 mm; pressure safety valve set point: 1.1 MPa] under fire conditions, and obtained the dynamic changes of the release amount and the characteristics of the released substances over time, which are of great significance for the safety protection of the entire production facility and personnel (Figs. 6.5, 6.6, 6.7 and 6.8).

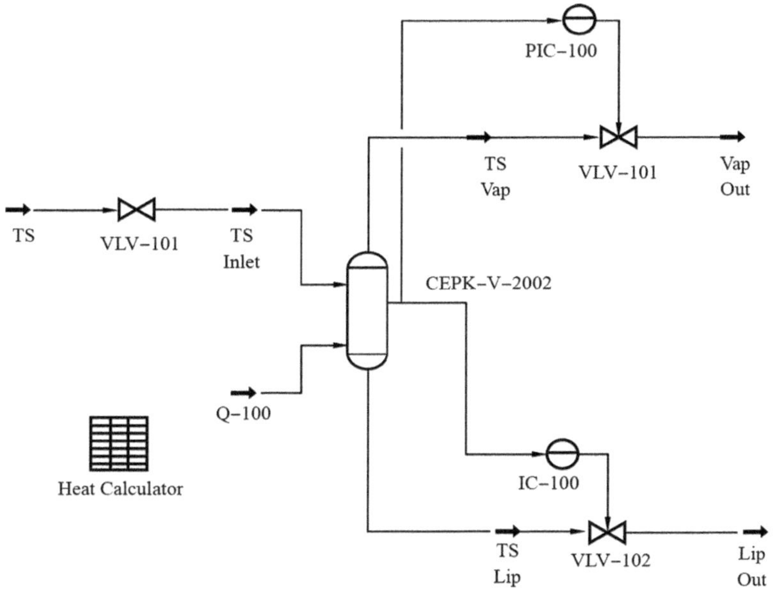

Fig. 6.5 PSV fire condition HYSYS dynamic model of the separator

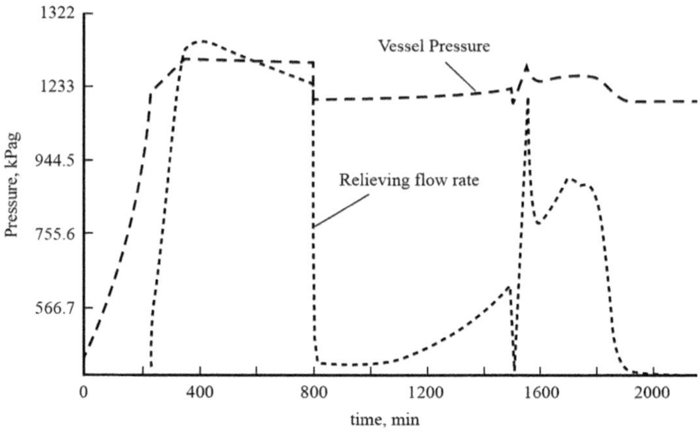

Fig. 6.6 Changes in the pressure inside the container during the release process

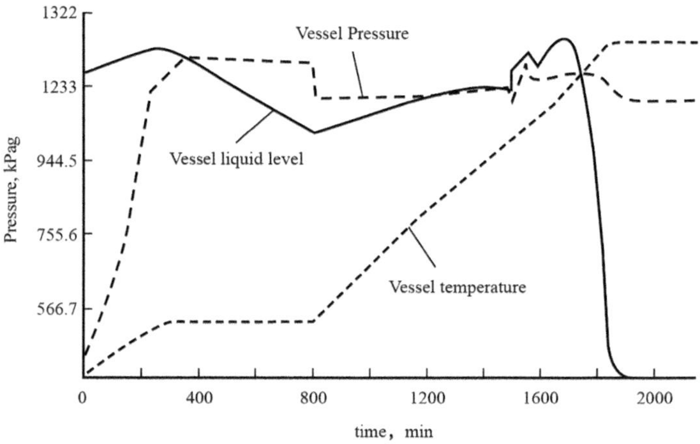

Fig. 6.7 Changes in the liquid level and temperature of the fluid inside the container during the release process

6.4.6 PSV Calculation Example

The inlet three-phase separator needs to be equipped with an appropriate Pressure Safety Valve (PSV) to handle several potential overpressure conditions. Using the HYSYS safety analysis environment, various overpressure situations can be defined in a PSV. The PSV in the following example is used to handle both outlet control valve blockage and fire situations simultaneously.

Select the Safety Analysis button in the lower left corner of the HYSYS window to enter the safety analysis environment (Fig. 6.9).

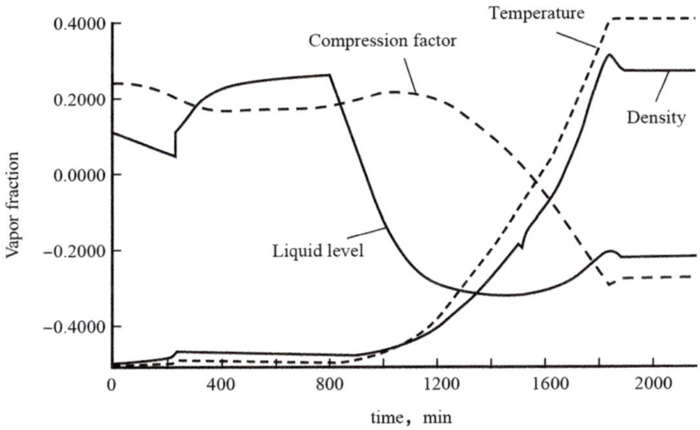

Fig. 6.8 Changes in the characteristics of the released substances

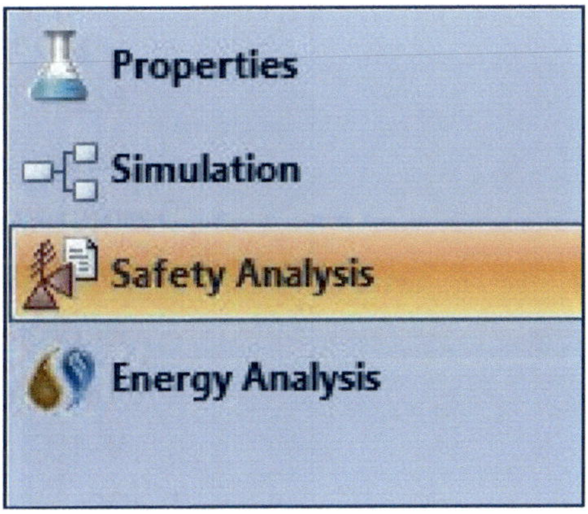

Fig. 6.9 Entering the safety analysis environment

Click on the Home menu bar, there are two ways to add a PSV to the simulation. Select the Add PSV icon on the Home tab in the function area and connect it to the appropriate process diagram object, or right-click on the unit operation in the navigation pane and select the appropriate connection (Fig. 6.10).

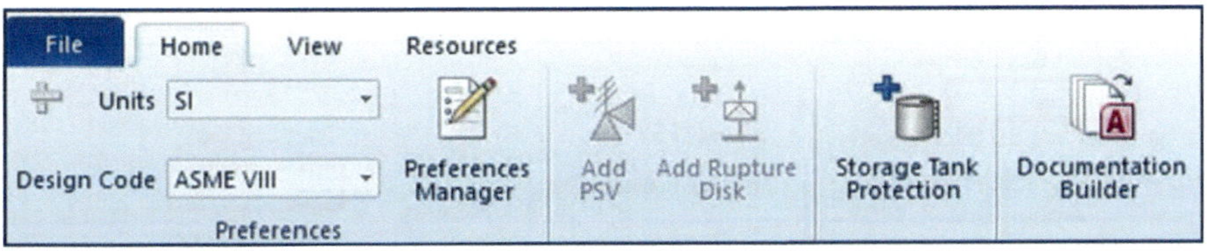

Fig. 6.10 Adding a PSV

Expand the Unit Operations menu item in the navigation pane (Fig. 6.11). Find and right-click on the Inlet Sep unit operation (Fig. 6.12).

Expand the Create PSV menu item and select Inlet Sep Vap. This will add a PSV to the gas outlet side of the inlet separator. In the main process diagram of the safety analysis environment, the PSV will be connected to the Inlet Sep Vap stream (Fig. 6.13). Double-click on the PSV icon on the process diagram to view the Equipment tab of the 100 PSV 001 menu, and enter the design conditions of the

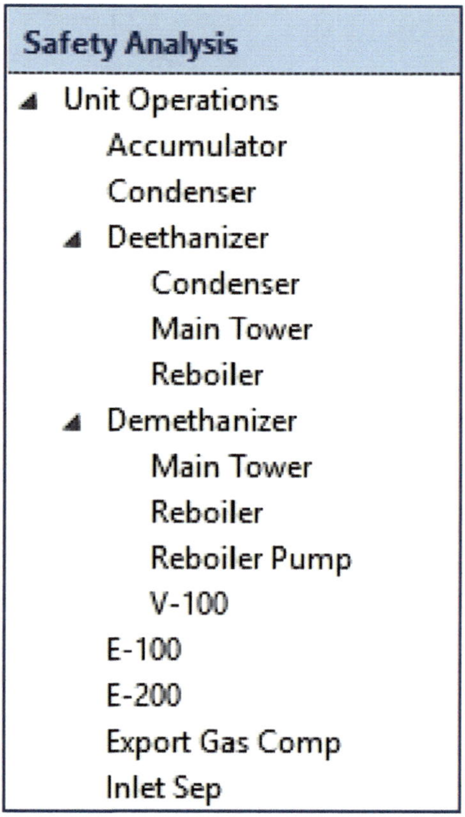

Fig. 6.11 Expand unit operation menu item

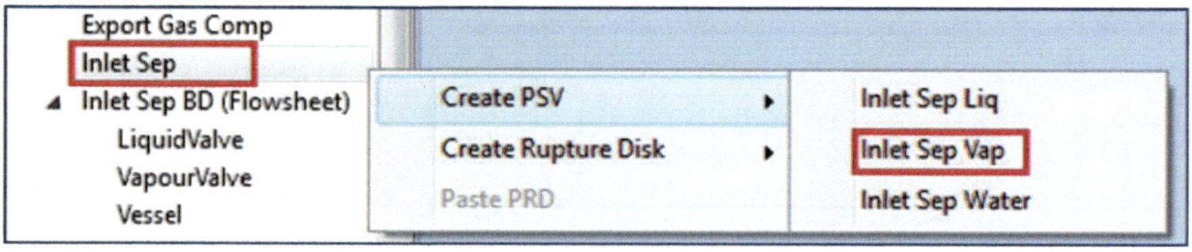

Fig. 6.12 Inlet Sep unit operation

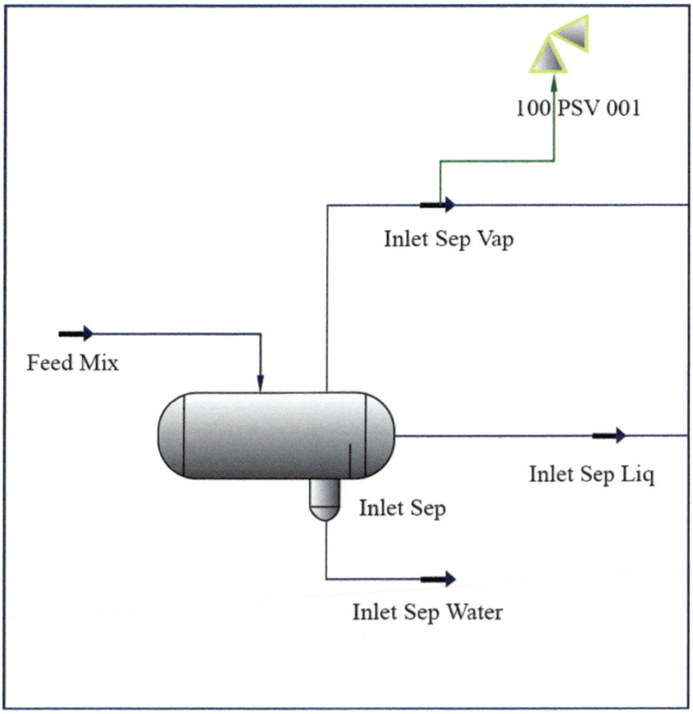

Fig. 6.13 PSV connected to Inlet Sep Vap

protected device. The design temperature is 40 °C (104°F), and the design pressure assumed value (gauge pressure) is 37ba (r536.6psi).

The design temperature and pressure are the conditions of the device you want to protect and should be less than or equal to the maximum allowable operating temperature and pressure of the protected device. By default, the set pressure of the PSV will be equal to the design pressure of the device. If the user wants a different set pressure, they can also customize the set pressure (Fig. 6.14).

At this point, the minimum input conditions for the protected vessel should be completed. To design the size of the safety valve under certain pressure relief conditions, the conditions represent overpressure situations, such as power failure, control valve closure, factory fire, etc. Add any number of conditions to the given PSV, and calculate the PSV size requirements for each condition. Choose a PSV size from these results to handle all conditions.

Select the Scenarios tab, add the pressure relief conditions considered by this PSV. Click Create Scenario, define a pressure relief valve size calculation condition (Fig. 6.15).

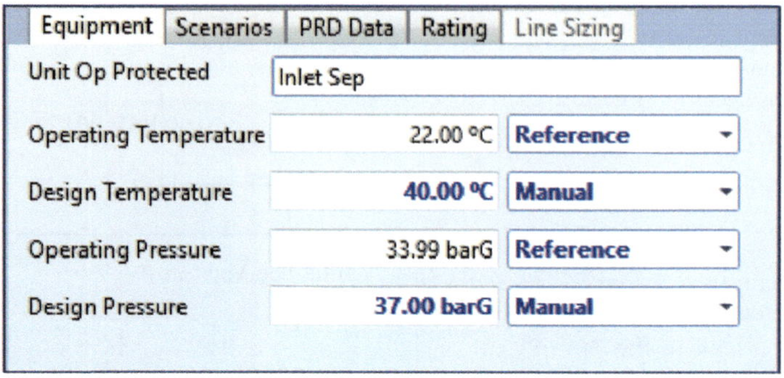

Fig. 6.14 Equipment tab

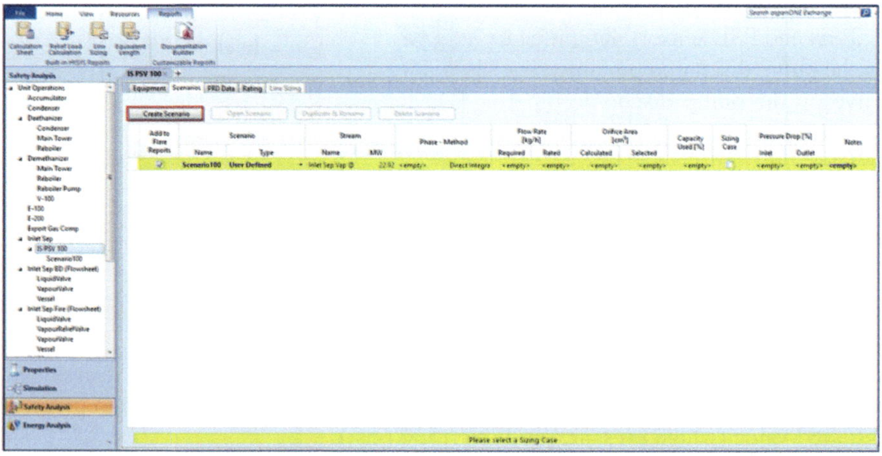

Fig. 6.15 Define pressure relief valve size calculation condition

6.4.6.1 Blockage Condition

Click the drop-down menu in the Scenario Type area, select the Blocked Outlet option (Fig. 6.16). This scenario will evaluate the pressure relief requirements when the outlet control valve on the Inlet Sep Vap stream fails. Assuming that the valve fails to close, it is necessary to release the gas flow of the Inlet Sep stream. To set the flow rate of the released fluid (equivalent to the Inlet Sep Vap stream flow rate), click the Reference button at the bottom of the form. This will call the flow rate through the PSV's Reference Stream (in this case, the Inlet Vapor flow). Select Reference from the Relieving Temperature drop-down menu (Fig. 6.17). Choosing "Reference" as the relieving temperature and flow rate means that steady-state process flow conditions will be passed to the PSV size calculation. These inputs can also be overridden by user-defined values. In addition, for many types of emergency conditions, there is little consensus in the industry on how to determine the required relief flow. Therefore, it is necessary to check and ensure that the program's inputs are reasonable and follow design standards and specifications.

Click the Edit button next to Relieving Pressure. Keep the default Allowable Overpressure at 10% of the set pressure, and click OK (Fig. 6.18).

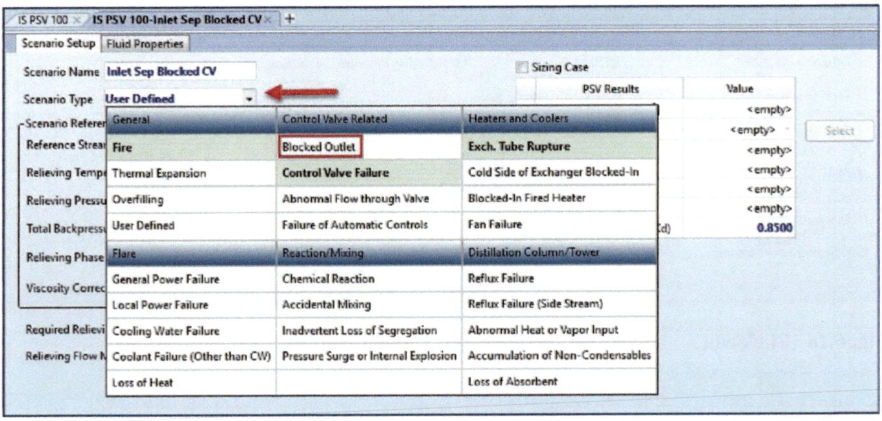

Fig. 6.16 Select blocked outlet option

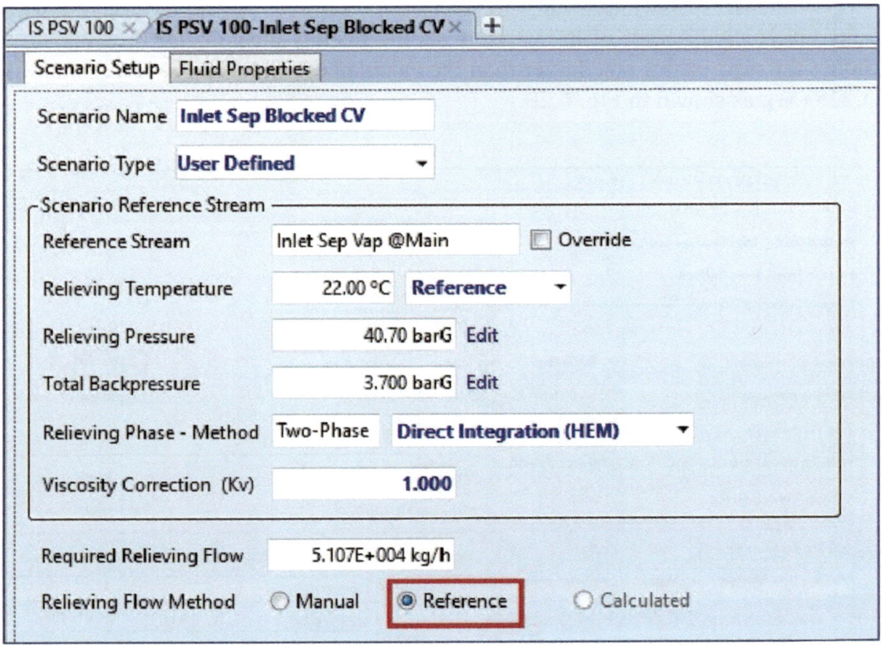

Fig. 6.17 Select reference

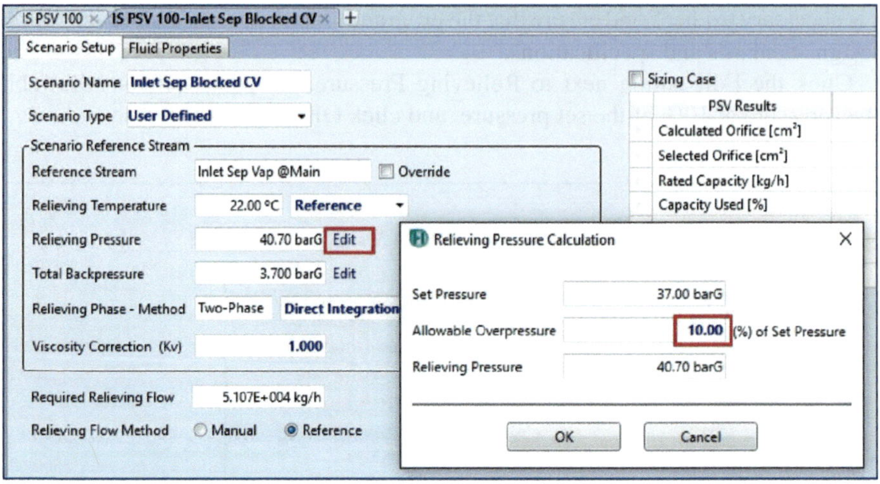

Fig. 6.18 Click edit

Click the Edit button next to Total Backpressure. Define Variable Superimposed BP pressure as 2.2 bar (31.91psi). Assume some backpressure fluctuations caused by other PSVs in the system. After setting, click OK (Fig. 6.19).

HYSYS will display that the calculation is complete. The Valve Results table in the upper right corner of the window should display the calculated orifice area. Choose the next orifice size larger than the calculated orifice area, which should be 23.225 (M), as shown in Fig. 6.20.

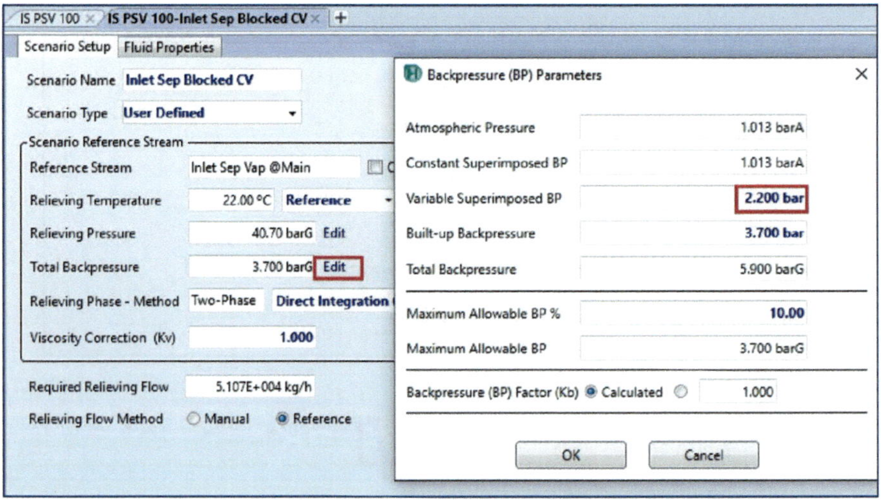

Fig. 6.19 Define variable superimposed BP pressure

PSV Results	Value
Calculated Orifice [cm²]	18.67
Selected Orifice [cm²]	23.225 (M)
Rated Capacity [kg/h]	6.352E+004
Capacity Used [%]	80.40
Orifice Designation	4 M 6
In/Out Flanges	300 x 150
Discharge Coefficient (Kd)	0.8500

Fig. 6.20 Valve results table

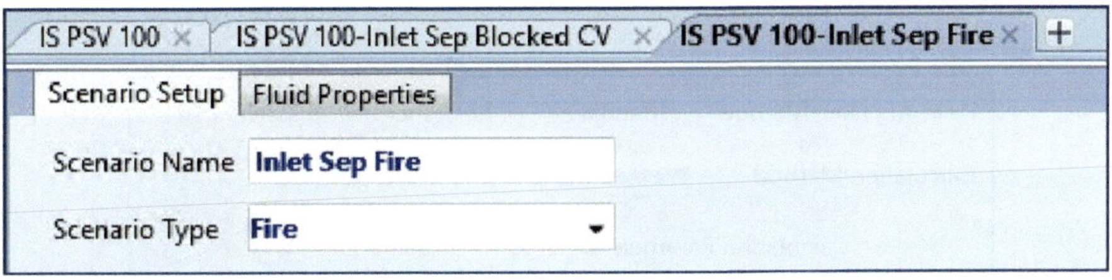

Fig. 6.21 Fire situation analysis

6.4.6.2 Fire Condition

Click the drop-down menu in the Scenario Type area, select the Fire option, and analyze the potential fire situation around the inlet separator (Fig. 6.21).

For fire conditions, HYSYS can use the API Std 521 Pressure-relieving and depressuring systems method, which is suitable for gas, liquid, or multiphase containers. It is important to ensure that the condition references the appropriate flow that best represents the entry into the protected unit operation. HYSYS provides several methods for calculating the required relief flow. Depending on the phase state under initial conditions, the program will use the calculation method with suitable options.

In this scenario, the feed mix entering the Inlet Sep separator should be used as the reference stream. As it is a multiphase mixture, the API Std 521 Pressure-relieving and depressuring systems calculation standard will apply the Wetted API, Semi-Dynamic Flash, or Supercritical models.

The Calculation Method selects Wetted (API). Next to the Reference Stream, click the Override checkbox, select the Feed Mix stream, and click OK (Fig. 6.22).

Please note that the Required Relieving Flow is calculated using the Wetted (API) method. Confirm that the Drainage & Firefighting parameters are set to Absent, and the Calculate Latent Heat parameter is set to Yes (Fig. 6.23).

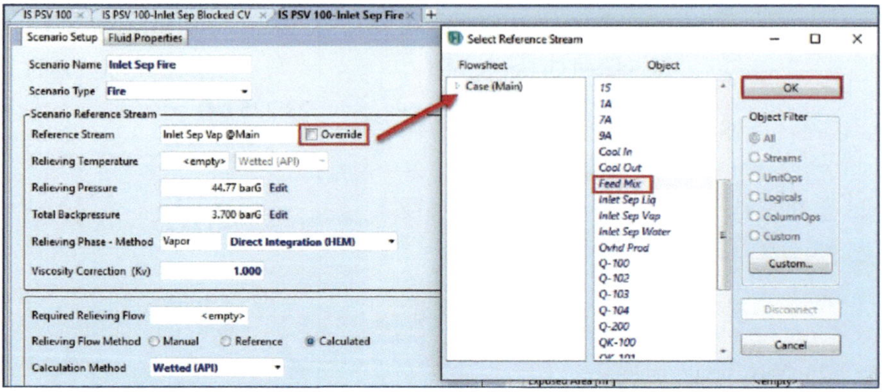

Fig. 6.22 Select feed mix stream

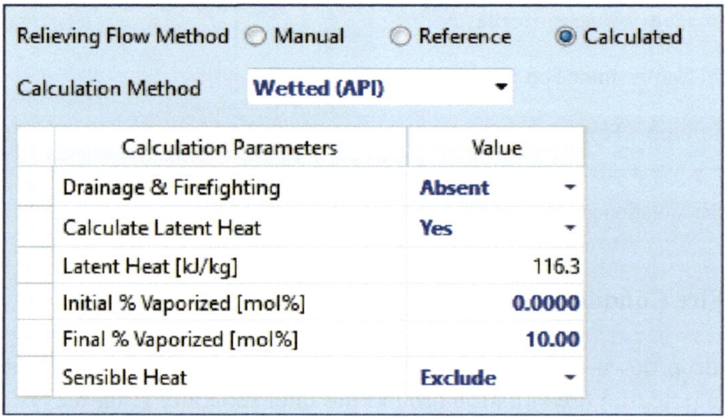

Fig. 6.23 Confirm parameters

Set the vessel dimensions in the Vessel Parameters section. Set the Diameter to 1.524 m (5 ft), Vessel Tan/Tan to 5.486 m (18 ft), and assume the Liquid Level is 0.6096 m (2 ft). Now, this scenario should be solved (Fig. 6.24).

After calculating the PSV size for the fire and outlet blockage scenarios of the Inlet Sep separator, it was found that the outlet blockage scenario requires a larger PSV orifice than the fire scenario. Since the vessel must be protected under both scenarios, a PSV capable of handling the worst-case relief must be selected. This may cause problems with the fire scenario relief due to the oversized PSV, leading to valve chattering and other adverse reactions. The HYSYS safety analysis environment can set multiple safety valves for a given verification scenario. This method allows the relief flow to be diverted to multiple PSVs under overpressure conditions. Therefore, the combined area of the two valves can support a larger relief situation (outlet

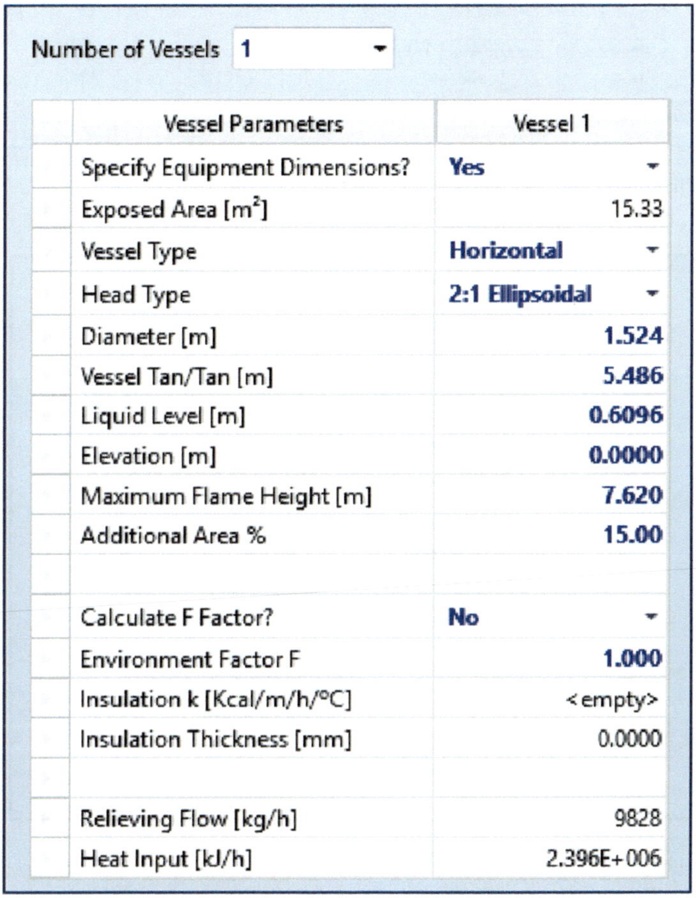

Number of Vessels 1

Vessel Parameters	Vessel 1
Specify Equipment Dimensions?	Yes
Exposed Area [m²]	15.33
Vessel Type	Horizontal
Head Type	2:1 Ellipsoidal
Diameter [m]	1.524
Vessel Tan/Tan [m]	5.486
Liquid Level [m]	0.6096
Elevation [m]	0.0000
Maximum Flame Height [m]	7.620
Additional Area %	15.00
Calculate F Factor?	No
Environment Factor F	1.000
Insulation k [Kcal/m/h/°C]	<empty>
Insulation Thickness [mm]	0.0000
Relieving Flow [kg/h]	9828
Heat Input [kJ/h]	2.396E+006

Fig. 6.24 Set vessel dimensions

blockage), while the smaller of the two valves can support a smaller relief situation (fire scenario).

On the Scenarios tab, click the Sizing Case checkbox for the Blocked Outlet. Set the outlet blockage scenario as the verification scenario for calculating the final valve size. The verification scenario must be the emergency scenario with the maximum relief flow (Fig. 6.25).

Once an appropriate PSV is selected, use the Line Sizing option in the safety analysis environment to design the inlet and outlet pipe sizes for the selected PSV. HYSYS can check for excessive pressure loss in the pipeline. Finally, use the Documentation Builder option in the safety analysis environment to record the results. The safety analysis environment also provides a pipe size calculation function, which can be used to define the PSV inlet and outlet pipe specifications and ensure that their sizes can handle the relief flow.

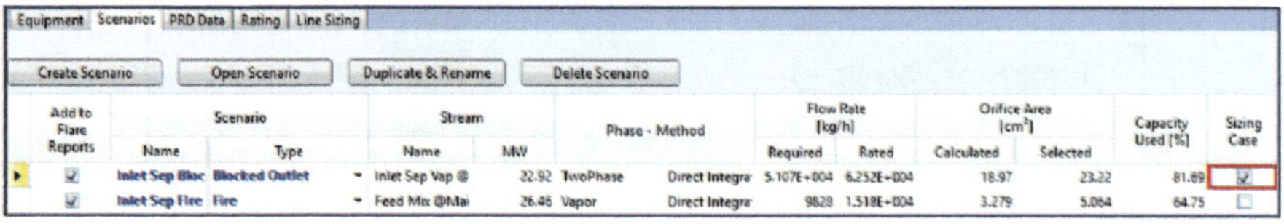

Fig. 6.25 Click the sizing case checkbox

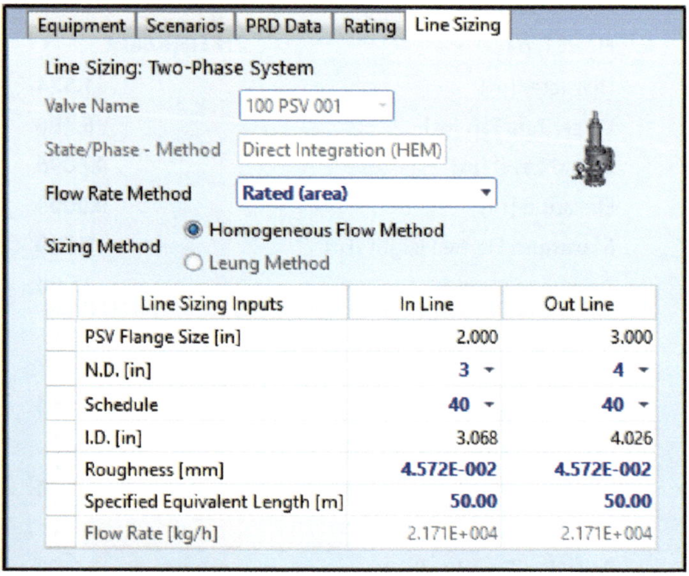

Fig. 6.26 Enter pipe specifications

Click the Line Sizing tab on the PSV window. Keep the connecting pipe the same size as the PSV inlet and outlet flanges. Enter the pipe specifications (Fig. 6.26).

By clicking the Constraint Setting button, you can view the constraint conditions for pipe size calculation (Fig. 6.27), and you can modify the limits of inlet pipe pressure drop, outlet pipe average flow speed, and outlet pipe outlet speed.

The pipe size design function also allows the design of the lengths of the inlet and outlet pipes, as well as any fittings on these pipelines. Check the Calculate Equivalent Length checkbox to view the options for calculating the equivalent lengths of the inlet and outlet pipes. Click the Calculate Equivalent Length checkbox to view the options that appear at the bottom of the table (Figs. 6.28 and 6.29), and enter the PSV inlet and outlet pipe information as shown in Fig. 6.30.

Proper and effective relief system design documentation is essential for compliance with relevant regulations. In the safety analysis environment, you can use the ABE datasheet to prepare design basis documents for all PSVs, rupture disks, and

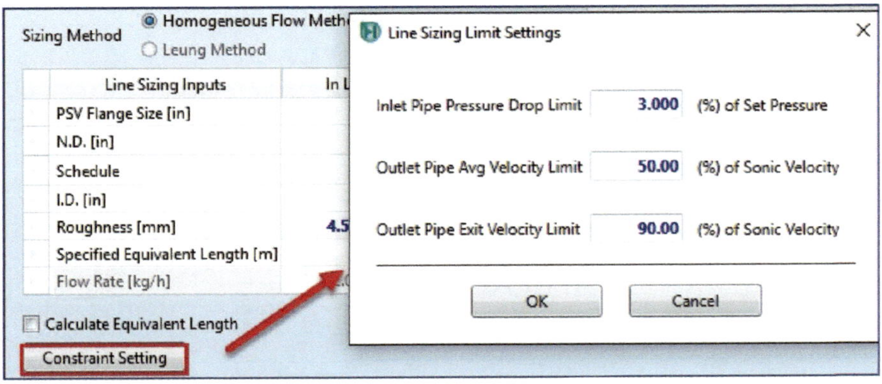

Fig. 6.27 View pipe size calculation constraint conditions

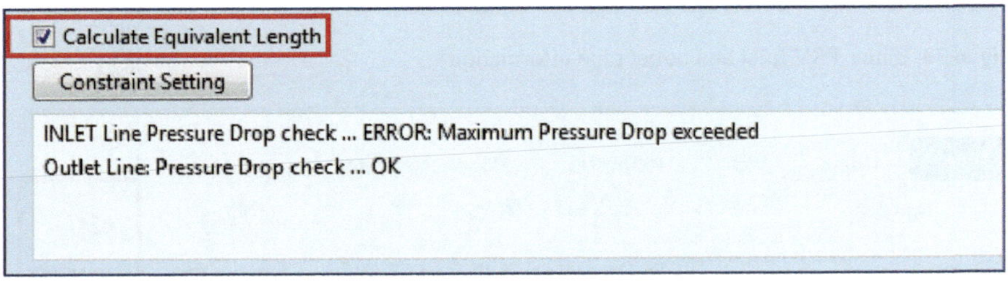

Fig. 6.28 Equivalent length options

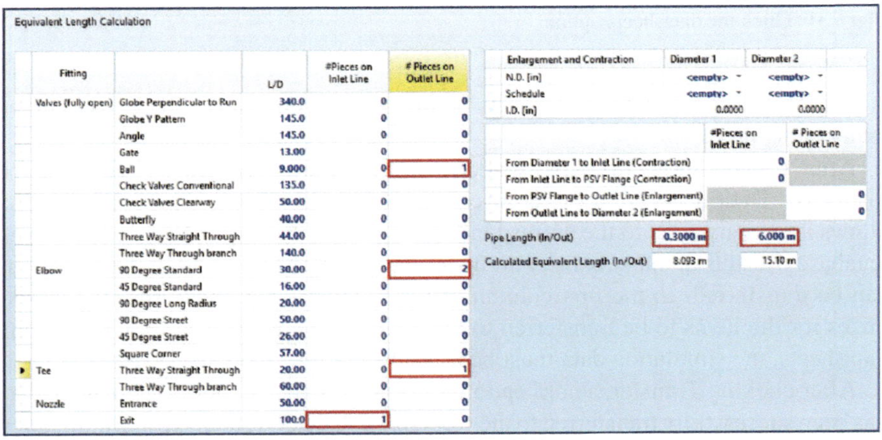

Fig. 6.29 View options at the bottom of the table

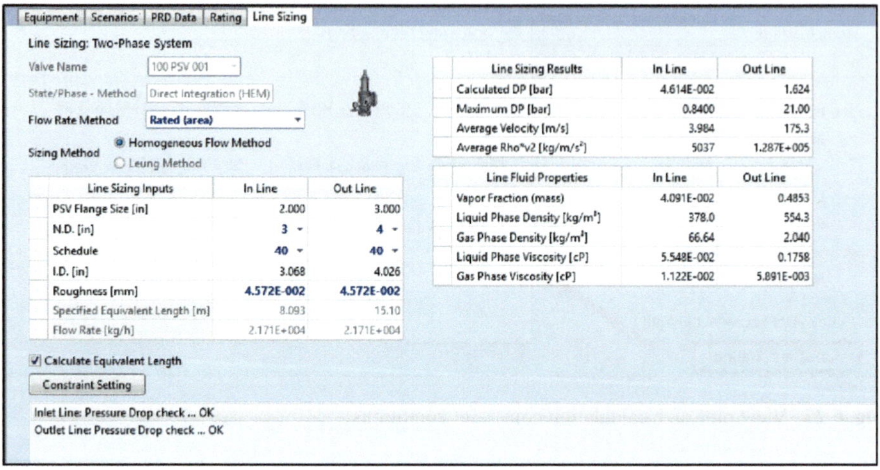

Fig. 6.30 Enter PSV inlet and outlet pipe information

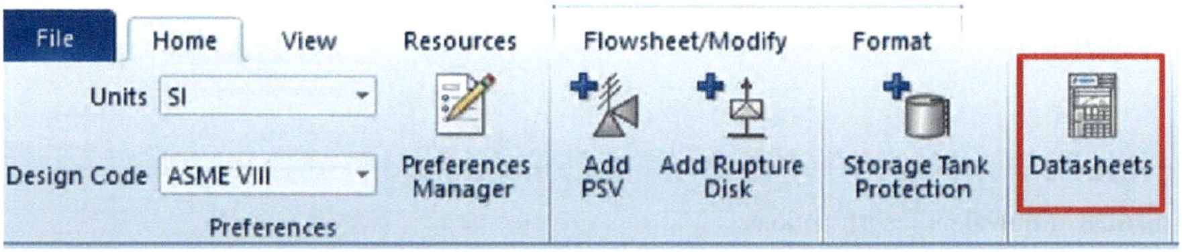

Fig. 6.31 Click the datasheets option

storage tanks [6]. Click the Datasheets option on the Home tab or Reports tab in the menu bar (Fig. 6.31).

After connecting to the ABE workspace, the Datasheets Mapper table will be displayed. Datasheets Mapper retrieves simulation data from Aspen HYSYS and allows it to be mapped to the corresponding objects in the Aspen Basic Engineering database. Results about streams, unit operations, components, and unit connections can be transferred. In the first column of the Mapper table, make sure the check-boxes for the items to be transferred to the workspace are checked. Before creating datasheets, the simulation data must be transferred to the workspace (Fig. 6.32).

After clicking Transfer, a notification window will appear, notifying that the data has been successfully transferred to the workspace. Click OK to make the notification window disappear. On the Datasheets tab, click explorer. In the Workspace Objects table, click IS PSV 100A to create a new datasheet for this valve. After the safety relief valve calculation, the throat diameter and pipe size should be determined according to the manufacturer's specific requirements, or they can be roughly selected according to Table 6.6.

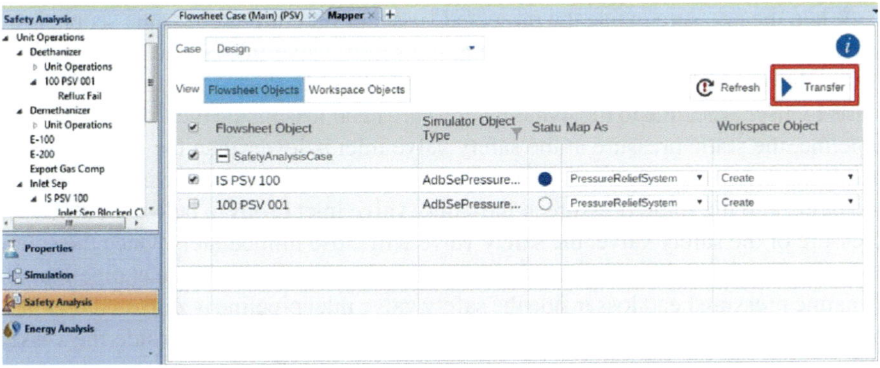

Fig. 6.32 Transfer simulation data to the workspace

Table 6.6 Throat diameter and pipe size

Standard orifice size	Aperture size cm^2	Aperture size in^2	1 × 2 (25 × 50)	1.5 × 2 (38 × 50)	1.5 × 2.5 (38 × 62)	1.5 × 3 (38 × 75)	2 × 3 (50 × 75)	2.5 × 4 (38 × 100)	3 × 4 (75 × 100)	4 × 6 (100 × 150)	6 × 8 (150 × 200)	6 × 10 (150 × 250)	8 × 10 (200 × 250)
D	0.710	0.110	•	•	•								
E	1.265	0.196	•	•	•								
F	1.981	0.307	•	•	•								
G	3.245	0.503			•	•	•						
H	5.065	0.785				•	•						
J	8.303	1.287					•	•	•				
K	11.858	1.838							•				
L	18.406	2.853							•	•			
M	23.226	3.60								•			
N	28.000	4.34								•			
P	41.161	6.38								•			
Q	71.290	11.05									•		
R	103.226	16.0									•	•	
T	167.742	26.0											•

Valve size in (mm) (inside diameter × outside diameter)

6.5 Safety Valve Inlet and Outlet Settings

The installation of the safety valve should carefully consider the inlet pipeline, outlet pipeline, pressure sensing pipeline (if any), and startup method.

6.5.1 Inlet Piping

The design of the inlet piping for the safety valve is particularly important. When installing the valve, the pressure loss of the inlet piping should be considered first, and then the convenience of the operating position. The ideal installation location is directly on the protected equipment to minimize the loss before the valve [7].

When the pressure inside the protected vessel or pipeline exceeds the set pressure of the safety valve, the static pressure at the safety valve inlet is the static pressure inside the vessel before the safety valve starts to discharge. After the safety valve starts to discharge, due to the dynamic pressure head loss inside the safety valve inlet pipeline, the static pressure at the safety valve inlet is lower than the static pressure inside the vessel. At this time, if the pressure drop in the safety valve inlet pipeline is too large, and the static pressure at the safety valve inlet drops to below the reseating pressure of the safety valve, the safety valve will close immediately. Once the safety valve is closed and there is no medium flow inside the safety valve inlet pipeline, the dynamic pressure head loss inside the safety valve inlet pipeline is zero, and the static pressure at the safety valve inlet rises back to the static pressure inside the vessel. When the static pressure at the safety valve inlet exceeds the set pressure of the safety valve again, the safety valve opens again. In this way, the safety valve repeatedly opens and closes, causing vibration or chattering. Therefore, it is necessary to control the pressure drop in the safety valve inlet pipeline to prevent vibration or chattering.

API Std 520, Part II Sizing, Selection, and installation of pressure-relieving devices—Part II—Installation recommends that the maximum pressure drop before the valve is 3% of the set pressure, which is also required by ASME related specifications. The pressure loss before the valve is the sum of the valve inlet pressure loss, the pressure loss in the pipeline before the valve, and the cut-off valve pressure loss. This pressure loss is calculated using the maximum rated flow through the safety valve, and the flow speed should not exceed 40 m/s. If the pressure drop in the safety valve inlet pipeline exceeds 3%, the inlet diameter can be increased, multiple smaller safety valves can be used, or the pipeline layout can be changed to reduce the pressure drop. The inlet diameter should be greater than or equal to the flange diameter of the safety valve inlet; the inlet pipeline should be as short as possible. When using a safety valve with a guide valve, because the safety valve with a guide valve has a separate pressure-taking pipe, there is no need to consider the impact of the inlet pipeline pressure drop on the performance of the safety valve.

When the safety valve inlet and outlet pipelines may be blocked, anti-blocking measures need to be taken, such as using steam or gas back blowing, steam tracing, etc. to prevent blocking. When using these methods, consider the impact of adding steam or gas on the process and how to check for gas supply failures. In these situations, consider whether the safety valve can be installed in a place that will not come into contact with the blocking material, such as upstream or downstream of the vessel, and consider anti-blocking measures only when it is not feasible in practice.

To prevent the safety valve from being corroded by corrosive media in the transport pipeline and equipment, or to prevent media leakage, a rupture disc can be added before the safety valve. When a rupture disc is installed, the ASME BPVC Section VIII specification requires a pressure gauge or other facilities that can check for leakage or rupture of the rupture disc between the rupture disc and the safety valve. GB 150.1 "Pressure Vessel Part 1: General Requirements" B.6.1 requires that the cavity between the rupture safety device and the safety valve should be equipped with a pressure gauge, exhaust port, and alarm indicator.

The safety valve installed on the pipeline should be located in a place where the fluid pressure is relatively stable and a certain distance from the fluctuation source. The safety valve should not be installed at the dead end of the horizontal pipeline to avoid accumulation of liquid or impurities.

The safety valve inlet pipeline should not have a U-shaped belt, the cut-off valve should be full diameter, and a position lock or lead seal should be set.

6.5.2 Outlet Piping

For gases with a relative molecular mass less than 80, if they do not cause toxicity, corrosiveness, unpleasant odors, and other harms to the nearby ground or equipment, they can be considered for direct discharge into the atmosphere. Depending on the nature of the gas, it may be necessary to obtain the approval of the environmental protection department. When the safety valve discharges to the atmosphere, the outlet of the discharge pipe should be vertically upward and cut into a square mouth to avoid obvious directionality of the safety valve's discharge, which is conducive to the diffusion of the safety valve's discharge. According to GB 50,160 "Design Fire Protection Standard for Petrochemical Enterprises", the discharge pipe outlet should be more than 3 m above the operating platform or building within an 8 m radius centered on the discharge outlet. A small hole with a diameter of 6–10 mm needs to be opened at the low point near the elbow of the safety valve outlet to prevent rain, snow, or condensate from accumulating in the discharge pipe. When discharging directly into the atmosphere, the flow speed of the discharge pipe should not be less than 60 m/s.

It is important for conventional safety valves to have appropriate outlet pipeline sizes. If the size is not reasonable, it will cause the valve to fail. Pressure loss can occur at the outlet manifold, causing excessive back pressure on the valve, which may cause the relief valve to close. The opening and setting pressure of the pilot-operated safety valve that vents to the atmosphere is not affected by back pressure; of course, if the outlet pressure exceeds the inlet pressure (such as a storage tank storing low vapor pressure medium), a check valve needs to be set. Compared with conventional safety valves, the setting pressure of balanced safety valves is less affected by back pressure. If the outlet pressure exceeds the critical flow pressure, the discharge capacity of any relief valve is limited by back pressure. The outlet pipeline should be at least the same size as the safety valve outlet, but generally larger than the valve outlet diameter to limit back pressure.

When calculating the pipe diameter of the safety valve inlet and outlet pipelines, because the back pressure of the safety relief valve is very low, the gas speed in the tail pipe (from the safety valve outlet to the branch manifold or main manifold) is very high, it is recommended to set a reducer at the safety valve outlet flange according to the tail pipe Mach number and back pressure, increase the tail pipe diameter, thereby reducing the gas speed in the tail pipe, the Mach number of the tail pipe should not exceed 0.7. The safety valve outlet pipeline should not have U-shaped belts, should have a slope (minimum 1:500, slope towards the torch manifold), the isolation valve should be full diameter, and a position lock or lead seal should be set. If the fluid discharged by the safety valve may freeze or wax or asphalt at ambient temperature, heat tracing can be carried out at the safety valve inlet and outlet tail pipes to prevent blockage when a small amount of medium leaks. For the temperature drop formed by the JT effect when a large amount of wet gas is normally discharged, it is difficult to deposit and form freezing due to the fast flow rate. When the outlet of the wet gas safety valve is connected to the torch pressure relief main pipe, it should be inserted into the main pipe at a 45° angle along the flow direction from the upper part to avoid condensate in the main pipe flowing into the branch pipe, and can reduce the back pressure of the safety valve.

When selecting safety, attention should be paid to the reaction force and vibration when the safety valve is discharged, and it should be provided to the piping professional or the manufacturer is required to provide it in the calculation book, so that the piping professional can set appropriate fixings and supports. When the pressure loss of the valve's front pipeline is too large, the inherent sound frequency of the valve's front pipeline is close to the inherent mechanical frequency of the main moving parts of the valve, the relief valve will resonate. The higher the set pressure, the larger the valve aperture, and the greater the inlet pipeline loss, the easier it is to resonate. The impact force generated by resonance is very large, which may cause the safety valve to destroy itself.

6.5.3 Reaction Force

The discharge of the pressure relief device will impose a reaction force due to the flow of the fluid, this force will be transmitted to the pressure relief device, the installation hole and the adjacent support container shell. The exact size of the load stress and induced stress depends on the reaction force and the configuration of the piping system. The size of the reaction force will vary depending on whether the discharge system is open or closed.

(1) In an open discharge system, for gases and vapors, compressible fluids discharge to the atmosphere in a critical stable state through elbows and vertical discharge pipes. The reaction force (F) includes the impact and static pressure effects, see formula (6.23).

$$F = 129W\sqrt{\frac{kT}{(k+1)M}} + \frac{(Ap)}{1000} \tag{6.23}$$

where

F—Reaction force to the atmospheric discharge point, N;
W—Flow rate, kg/s;
k—Heat capacity ratio;

T—Inlet temperature, K;
M—Relative molecular mass;
A—Outlet area of the discharge point, mm^2;
p—Static pressure (gauge pressure) at the discharge point, kPa.

In an open discharge system, for two-phase discharge, assuming that the two-phase mixture is in a uniform flow state, it is calculated according to formula (6.24).

$$F = \frac{10^6 W^2}{A} \cdot \left(\frac{x}{\rho_g} + \frac{1-x}{\rho_l} \right) + \frac{Ap}{1000} \tag{6.24}$$

where
F—Reaction force to the atmospheric discharge point, N;
W—Flow rate, kg/s;
x—Gas phase mass fraction under outlet conditions;
ρ_g—Gas phase density under outlet conditions, kg/m^3;
ρ_l—Liquid phase density under outlet conditions, kg/m^3;
A—Outlet area of the discharge point, mm^2.
p—Static pressure (gauge pressure) at the emission point, kPa.

(2) For closed emission systems, if the discharge is carried out under steady flow conditions, the pressure relief device generally does not exert a large force and bending moment on the emission system, only at the place where sudden expansion occurs in the emission pipeline, the calculated reaction force will be relatively large. The calculation of this force is complex and requires analysis of the complex time relationship of the piping system.

6.6 Setting of Bursting Discs

In places where safety relief devices must be installed and safety valves are not suitable, bursting discs should be installed or used in series with safety valves. A bursting disc refers to a pressure-sensitive element that can act quickly due to overpressure, used to seal pressure and control the bursting pressure. When the pressure difference on both sides of the bursting disc reaches the predetermined value at the predetermined temperature, the bursting disc immediately acts (ruptures or falls off), releasing the pressure medium. A bursting disc device refers to a pressure relief safety device composed of a bursting disc (or bursting disc assembly) and a holder (or support ring). Bursting discs can mainly be divided into positive arch type and reverse arch type, with further subdivisions shown in Table 6.7.

One of the following conditions should be met to prioritize the use of bursting discs:

(1) The pressure may rise rapidly.
(2) The discharge medium contains particles, is prone to sedimentation, crystallization, polymerization, and has a high viscosity.

Table 6.7 Classification of bursting discs

Type	Name
Positive arch type	Standard type Slotted type Back pressure bracket type Reinforced ring type Soft pad type Grooved type
Reverse arch type	Clamp type Back pressure bracket type Knife holder type Alligator tooth type Grooved type

(3) The discharge medium is highly corrosive, and its value is high when using a safety valve.
(4) The process medium is very expensive or highly toxic, and no leakage is allowed during the work process, it should be used in series with a safety valve.
(5) When the working pressure is very low or very high, it is relatively difficult to manufacture a safety valve.
(6) The use temperature is low and affects the working characteristics of the safety valve.
(7) A larger discharge area is needed.

For one-time use pipeline systems (such as the pipeline venting system for vehicle blowdown), where the rupture of the bursting disc does not affect operation and production, a bursting disc can be set. To reduce the loss of process medium after the bursting disc ruptures, it can be installed in series at the entrance of the safety valve. The calibrated bursting pressure of the bursting disc is the same as the set pressure of the safety valve. The nominal diameter of the bursting disc should not be less than the inlet diameter of the safety valve. The discharge area after the bursting disc ruptures should not be less than the inlet area of the safety valve, and it should be ensured that the fragments of the bursting disc do not affect the normal operation of the safety valve.

If there is a possibility of a corrosive gas environment in the total discharge pipe, the bursting disc should be installed at the outlet of the safety valve to protect the safety valve from corrosion. At this time, the medium inside the container should be clean, without adhesive substances or blocking substances.

The maximum design bursting pressure of the bursting disc does not exceed 10% of the set pressure of the spring safety valve, the nominal diameter of the bursting disc is the same as the outlet diameter of the safety valve, and the discharge area of the bursting disc should not be less than the inlet area of the safety valve. To prevent the pressure inside the pressure vessel from rising rapidly under abnormal conditions, or to increase the discharge area in case of fire, one or several bursting discs can be installed in parallel with the safety valve. The calibrated bursting pressure of the bursting disc is slightly higher than the set pressure of the safety valve and should not exceed the design pressure of the container. The bursting disc should have enough discharge area to meet the requirements of protecting the container.

6.7 Water Hammer Relief

Water hammer, also known as hydraulic shock, occurs when the motion state of the liquid in the pipeline suddenly changes (such as the sudden closing or opening of a valve, the sudden start or stop of a water pump, the sudden change of load in a water turbine or hydraulic cylinder, etc.). Due to the sudden and rapid change in the liquid flow rate, the fluid inertia inevitably causes violent fluctuations in the pressure inside the pipe, that is, the sudden rise and fall of pressure, and propagates throughout the length of the pipe. The sudden change in pressure causes the pipe wall to vibrate and is accompanied by a hammer-like sound, hence this phenomenon is called the water hammer phenomenon or hydraulic shock. When the valve is closed quickly, the flow rate inside the pipe drops sharply, and the pressure rises rapidly, which is called positive water hammer. Positive water hammer can cause the pipeline to burst. When the valve is opened quickly, the flow rate inside the pipe rises sharply, and the pressure drops rapidly, which is called negative water hammer. Negative water hammer can cause the pipeline to produce a vacuum and cavitation, causing the pipeline to deform. When the pressure rise caused by the water hammer phenomenon is slight, it only manifests as noise and vibration; when it is severe, the pressure change can exceed tens or hundreds of times the original normal pressure inside the pipe, or even exceed the allowable stress of the pipe wall material, causing deformation or even rupture of the pipe and pipe fittings. The propagation speed of the water hammer wave in the pipeline oil flow is calculated by formula (6.25).

$$c = \sqrt{\frac{1}{\rho\left(\frac{1}{K} + \frac{d}{\delta \cdot E}\right)}} \tag{6.25}$$

where:

C—the propagation speed of the water hammer wave in the pipeline oil flow, m/s;

d—the inner diameter of the pipeline, m;

δ—the wall thickness of the pipeline, m;

E—the elastic modulus of the pipe material, Pa, for steel pipes $E = 2.06 \times 10^{11}$ Pa;

K—the bulk modulus of the liquid in the pipe, Pa, for crude oil $K = 1.7 \times 10^9$ Pa;

ρ—the density of the liquid in the pipe, kg/m^3.

The pressure increase value Δp caused by water hammer is directly proportional to the change value Δv of the oil flow speed in the pipeline. It can be calculated by formula (6.26):

$$\Delta p = \rho c \Delta v \tag{6.26}$$

The Δp calculated according to formula (6.26) is the water hammer pressure value generated when the flow rate changes instantaneously.

6.8 Acoustic Vibration Analysis

In the pipeline system, especially in the pressure reducing valve, safety valve, and other pressure changing areas, large vibrations will be generated due to sound pulsation, causing pipeline damage. Therefore, acoustic vibration analysis (AIV) should be carried out where there are pressure reducing valves in the natural gas system. Acoustic vibration (AIV) can cause damage and rupture to the operating pipeline or small diameter connections, and generate sound radiation in the frequency domain of 500–2000 Hz. Figure 6.33 on the left shows that the acoustic vibration excites the pipe wall vibration, and the pipe wall vibration causes the small diameter pipe to vibrate. If not controlled, AIV can cause major pipeline failures (Fig. 6.33 on the right). This internal sound energy can also generate external noise, affecting the safety of operators (loss of hearing).

When acoustic vibration occurs, the risk is greatest at the connection of small diameter pipes or at the welding support points of branch pipes. Excitation at different frequencies can cause resonance at these positions, leading to cracking and loss damage. AIV loss damage occurs in a very short time (from a few minutes to a few hours). The sound power level (PWL) can be used to measure the sound energy generated by the fluid passing through the pressure reducing valve or pressure reducing equipment. Acoustic vibration research (AIV design research) calculates PWL and judges whether the pipeline system is sufficient to resist the loss brought by AIV vibration. If necessary, it is necessary to change the PWL or modify the pipeline system. There are usually two methods to evaluate AIV:

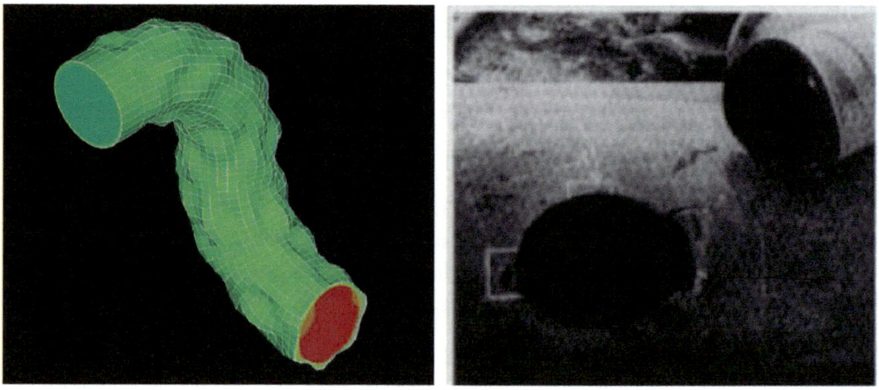

Fig. 6.33 AIV pipeline failure example

(1) Eisinger or "*D*/*t*" method. This method checks the pipeline and corresponding acoustic loss by comparing the actual diameter (*D*)/thickness (*t*) ratio with empirical values. PWL is a parameter to determine the design limit, published by Eisinger, Carucci, and Mueller in the NORSOK (Norwegian Petroleum Standardization Organization) standard.

(2) American Energy Research Institute Standard (2008). This method checks branches and welding points. The calculation of PWL is similar to the *D*/*t* method, but there are some differences in the method of deriving the loss limit.

References

1. API Standard 521 Pressure-relieving and depressuring systems, Seventh Edition, June 2020.
2. Edited by Wang Zizong. Petrochemical Design Manual, Volume Four: Process and System Design. Beijing: Chemical Industry Press, 2015.
3. Compiled by Sinopec Shanghai Engineering Co., Ltd. Chemical Process Design Manual. Beijing: Chemical Industry Press, 2018.
4. HG/T 20570.2 Setting and Selection of Safety Valves.
5. SY/T 10044 Recommended Practices for Determining the Size, Selection, and Installation of Refinery Pressure Relief Devices.
6. HG/T 20570.3 Setting and Selection of Bursting Discs.
7. Chen Wenfeng, Liu Peilin, Guo Zhou, et al. HYSYS Dynamic Simulation of the Discharge Process of Safety Valves in Complex Pressure Vessels [J]. Natural Gas and Petroleum, 2010, 12: 55-57.

Chapter 7
Design and Calculation of Blowdown Valves (BDV)

7.1 Introduction to BDV

Blowdown Valves (BDV) and throttling orifices are used together as the system's emergency pressure relief devices, and their reasonable selection and design have a significant impact on the system setup and back pressure of the piping network under fire or accident conditions.

Currently, the BDV pressure relief medium set in the system is only gas phase. In a fire, to reduce the internal pressure and wall pressure of the container, so as not to add more fuel to the fire source, the container needs to be depressurized. According to API Std 521 Pressure-relieving and depressuring systems, the equipment pressure should be reduced to 690 kPa or below 50% of the equipment design pressure, whichever is lower, within 15 min [1].

BDV and throttling orifices are used together as the system's emergency pressure relief devices, which can automatically and quickly depressurize the target equipment, unit, system, or pipeline. Their reasonable selection and design have a significant impact on the system setup and back pressure of the piping network under fire or accident conditions.

The triggering of BDV can be automatic or can be opened after confirmation by the operator. Automatic trigger signals can be high pressure, shutdown signals, or fire alarm signals. If it is the emergency release of a single unit, it can be automatically triggered, but if it is the emergency release of the system or the entire field, it is recommended to be triggered after confirmation by the operator.

Under normal production conditions, oil and gas processing facilities are filled with high-pressure hydrocarbon fluids. In an emergency situation, such as a fire, these high-pressure hydrocarbon fluids will cause ignition or explosion. Setting up automatic relief devices to timely release high-pressure hydrocarbon fluids to the flare collection system can reduce operational risks while also protecting the safety of production facilities and personnel.

K. Fang, *Onshore Oilfield Surface Facilities: Process Safety Engineering Guide*,
https://doi.org/10.1007/978-981-96-7830-3_7

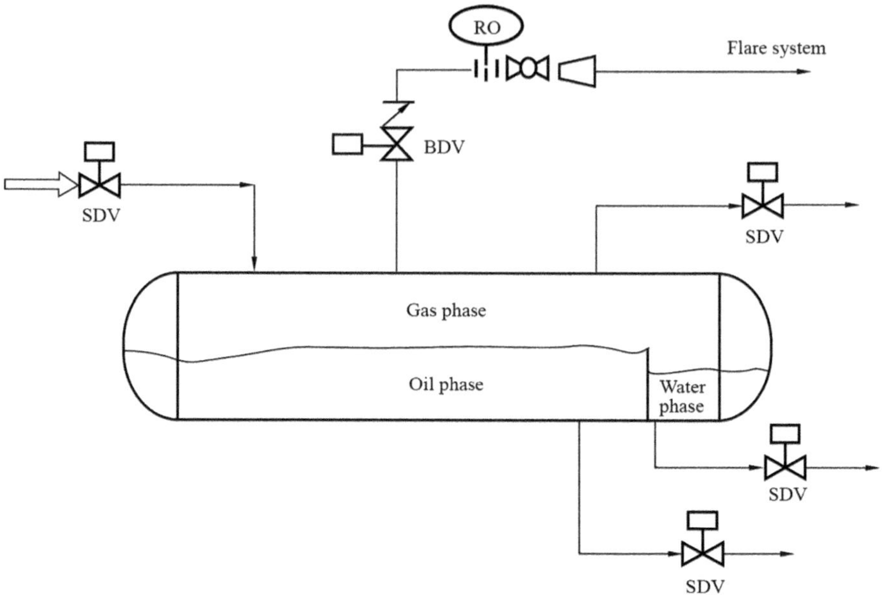

Fig. 7.1 Typical emergency relief device

The main function of the emergency relief device BDV is to timely depressurize the high-pressure system in emergency situations such as fire, leakage, or for the purpose of equipment maintenance. The typical emergency relief device mainly consists of a relief valve, an orifice plate, and the upstream and downstream pipelines. In the event of a fire, the gas detection signal in the control room triggers the inlet and outlet shutdown valves of the separator and automatically closes them, while automatically opening the relief valve, starting to depressurize the high-pressure system. When performing equipment maintenance, depressurization can also be achieved by manually opening the relief valve (Fig. 7.1).

7.2 Setting and Calculation of BDV

7.2.1 Regulations for BDV Setting

According to API Std 521 Pressure-relieving and depressuring systems, for pressure vessels or systems operating at or above 1724 kPa (gauge pressure) and containing a considerable amount of hydrocarbon compounds, BDV needs to be set. Shell industry standards require that all pressure vessels containing 4 m^3 of propane or other hazardous compounds need to set BDV; for high-pressure operating facilities (gauge pressure at or above 35 bar), and for all manned platform high-pressure facilities, BDV needs to be set. The Norwegian Classification Society stipulates that all pressure vessels or pipe sections containing 1 t of untreated hydrocarbon compounds at shutdown, with an operating pressure (gauge pressure) at or above 450 kPa, need to set BDV. ExxonMobil stipulates that operating pressure at or above 1720 kPa, containing 5.6 m^3 of flammable liquid, needs to set BDV. Total company requires that operating pressure (gauge pressure) greater than 7 bar and $p \cdot V$gas > 100 bar·m, or light hydrocarbons (C4 and lighter components) content at or above 20 t, need to set BDV.

In summary, for pressure devices, BDV is generally a standard configuration; for container types, the size of the volume and operating pressure determine whether to configure BDV.

If there is only a pipeline between two SDV/ESD, unless there are special requirements, such as the need for rapid depressurization for maintenance, containing a large amount of light hydrocarbons, or required after safety analysis, BDV is generally not considered. When determining whether to set BDV, other factors to consider include the type of platform, the quantity and composition of the release source material, and the operating conditions (whether high pressure). From the actual design situation of domestic and foreign oil and gas fields, the main reference is the recommended practices in the API standard, and sometimes BDV is also considered for particularly large separators.

During BDV venting, the unit or system is generally in a state of isolation: that is, all material inlets and outlets SDV/ESD are closed (including liquid phase), and the BDV only vents the residual and fire-induced vaporized materials of the unit or system. The materials discharged through the BDV are pure gas phase. The purpose of the BDV remote venting process equipment retaining liquid is to reduce the pressure inside the container, so that the wall stress is still below the fracture yield stress when heated, to minimize the possibility of container rupture during a fire.

Whether a BDV needs to be installed depends on various standards mentioned in API Std 521 Pressure-relieving and depressuring systems. If a loss of tightness could escalate an emergency situation, emergency depressurization facilities need to be installed. The curve shown in Fig. 7.2 can serve as a guide to determine whether a system needs to install a BDV. This curve is based on fire conditions, with a duration of 10 min, an initial leakage rate of 1 kg/s, and a spacing of 4 m between containers.

API Std 521 Pressure-relieving and depressuring systems provide depressurization experience guidelines based on general container insulation capacity data:

(1) When the liquid phase catches fire, reduce the pressure inside the container to 50% of the design pressure within 15 min.
(2) When the gas phase (gas phase generated by system depressurization) catches fire, reduce the pressure inside the container to 690 kPa (gauge pressure) within 15 min.

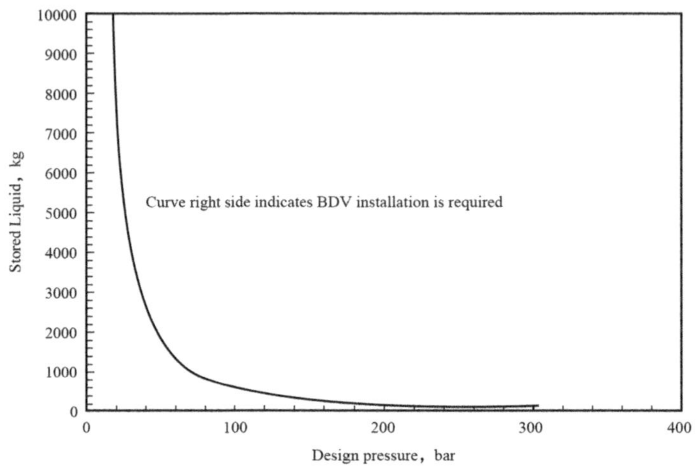

Fig. 7.2 Fire condition limit curve

When a pool fire or jet fire occurs, in some cases it may be necessary to analyze and evaluate the insulation capacity of the container to determine the depressurization rate of the container (according to API Std 521 Pressure-relieving and depressuring systems). It must be recognized early in the design that compared with API's experience guide, the API standard may require a faster depressurization rate (for example, 5 min) and a larger depressurization system for the container's insulation capacity. The insulation capacity of the container is determined by the heat flux outside the container, and the radiation heat of the flame needs to be calculated. API Std 521 Pressure-relieving and depressuring systems provide a method for calculating radiation heat, and professional software such as PHAS can be used for more accurate radiation heat calculations. This software can calculate the radiation heat of the flame after a leak of a certain size aperture is ignited. The size of the aperture usually includes 100, 50, 25, and 10 mm, depending on the nature of the equipment being evaluated. The heat flux generated by the 10 mm aperture is smaller and lasts longer, while the heat flux generated by the 100 mm aperture is larger and lasts shorter. Therefore, it is necessary to analyze the reliability of the aperture size range, clarify the worst conditions (which may not be the maximum aperture situation). The maximum radiation heat received by the container should come from the system with the highest pressure and the largest residual liquid. AP also provides suggestions for mitigation measures that can be used to reduce heat load and thus reduce depressurization requirements (such as passive fire protection, water spray fire extinguishing) [3].

7.2.2 *BDV Calculation*

7.2.2.1 BDV Design Requirements

(1) The initial venting pressure of the BDV is the high-high pressure alarm value of the equipment or the set pressure value of the safety valve.
(2) The depressurization system should have the ability to reduce the initial venting pressure of the BDV to 690 kPa (gauge pressure) or 50% of the container design pressure within 15 min.
(3) The venting capacity of the depressurization system is generally based on the normal operating liquid level of the container and includes the volume of connected equipment and pipelines, while the effective vapor generation surface area is the area of the container wetted by the liquid phase.
(4) For normal venting in fire conditions and adiabatic conditions, the initial venting temperature is the operating temperature of the venting facility. For cold venting in adiabatic conditions, the initial venting temperature is the lowest temperature of the environment where the venting facility is located.
(5) For all gas phase containers, the isentropic efficiency is 87–98%; for gas–liquid two-phase containers, the isentropic efficiency is 70%; for all liquid phase containers, the isentropic efficiency is 40–70%.
(6) Gas phase pipe inventory: All gas phase volumes in the SDV/ESD isolation system, whether or not in the fire zone.
(7) Liquid phase inventory: Only the range of fire inspection is calculated, that is, the vaporization amount produced by the fire within 15 min needs to be included.
(8) BDV valve front and rear pressure: The upstream of the flow restriction orifice plate belongs to the high-pressure section, its pressure is equal to the pressure inside the container, and it gradually decreases with the venting time; the downstream of the orifice plate belongs to the low-pressure section, its pressure is equal to the back pressure of the flare system. The BDV upstream of the orifice plate is generally a full-bore valve, so its front and rear pressure is the same as the container.
(9) Discharge volume: The maximum discharge volume is determined by the size of the throttling orifice plate. The pressure difference before and after the orifice plate is the largest at the moment of starting to discharge, and the discharge volume is also the largest. As the pressure in the container decreases and the back pressure of the vent pipe behind the plate increases, when the pressure difference before and after the orifice plate is less than the critical pressure difference, the discharge volume gradually decreases.

7.2.2.2 BDV Calculation

The core issues of BDV calculation mainly focus on analyzing the peak discharge volume that the discharge system needs to withstand during the discharge process and the low-temperature effect. The working conditions considered are fire conditions and adiabatic conditions. Adiabatic conditions refer to the pressure release required when the equipment is undergoing maintenance, with no heat input from the outside world, and the following two situations are considered separately:

(1) Normal discharge: After production stops, the equipment begins to release pressure, and the initial discharge temperature is the operating temperature of the equipment.
(2) Cold discharge: Due to typhoons or other reasons, the system pressure cannot be released in time, and the pressure release only begins when the operating temperature of the system drops to the ambient temperature.

After BDV, a throttling orifice plate should be set up. The type should be a flow-limiting orifice plate to reduce the impact of the vented gas on the vent pipe network and the torch after BDV suddenly opens. It cannot be designed as a pressure-reducing orifice plate. When calculating the orifice plate, the pressure of the gas at the small hole cannot be lower than the critical pressure, and the flow speed of the gas at the small hole cannot exceed the speed of sound. The downstream pressure of the orifice plate is the back pressure that is pushed back from the torch outlet, calculated by formula (7.1).

$$A = \frac{W}{CK_{\mathrm{d}}p_1K_{\mathrm{b}}K_{\mathrm{c}}}\sqrt{\frac{TZ}{M}} \tag{7.1}$$

where

A—The required effective discharge area of the relief valve, in^2;
W—The flow rate required through the relief valve, lb/h;
C—The coefficient determined by the adiabatic index of gas or steam under standard conditions;
K_{d}—Effective discharge coefficient; $K_{\mathrm{d}} = 0.975$;
K_{c}—Comprehensive correction factor, if no rupture disc is installed, the value is 1.0, if a rupture disc is installed, the value is 0.9;
p_1—Upstream relief pressure (absolute pressure), psi;
K_{b}—Back pressure correction factor;
T—Discharge temperature at the inlet, K;
Z—Compression factor;
M—Relative molecular mass of gas or steam.

When calculating, you can refer to the Nomograph for calculation. Or calculate the throat diameter of the orifice plate according to the venting time, see formula (7.2).

$$t = \frac{BV}{C_{\mathrm{d}}A_{\mathrm{r}}} \cdot \ln\frac{p_1}{p_2} \cdot \sqrt{\frac{R_{\mathrm{d}}}{ZT}} \tag{7.2}$$

where

t—Venting time, s;
B—Coefficient, take 0.09;

V—System volume, m^3;
C_d—Discharge coefficient, take 0.85;
A_r—Discharge area of the relief valve, m;
R_d—Relative density of gas;
Z—Gas compression factor;
T—Operating temperature, K;
p_1—Pressure before discharge, kPa (absolute pressure);
p_2—Pressure after discharge, kPa (absolute pressure).

After all the orifice plate calculations in the station are completed, all the throttling orifice plates are modeled in FLARENET. The venting scheme is that the BDV in the field acts by area during a fire, and the BDV outside the total figure of 230 m^2 must consider simultaneous action. According to the tail pipe Mach number not greater than 0.7, the main pipe Mach number not greater than 0.5 (according to NORSOKP-001) into the BDV diameter check calculation, because the Mach number in the pipeline is large during emergency pressure relief, the BDV outlet back pressure may reach 80% of the pressure before the valve, but there is no limit on the BDV outlet back pressure during emergency pressure relief.

Note: The minimum installation distance between BDV and throttling orifice plate is 600 mm.

7.2.3 Introduction to Throttling Orifice Plate

The throttling orifice plate is a concentric sharp orifice plate used to limit the flow rate of the fluid or reduce the pressure of the fluid. The fluid will generate a pressure drop when passing through the orifice plate, and the flow rate through the orifice plate will increase with the increase of the pressure drop [2]. But when the pressure drop exceeds a certain value, that is, exceeds the critical pressure drop, no matter how much the outlet pressure decreases, the flow rate will maintain a certain value and will not increase. The throttling orifice plate is used to limit the flow rate of the fluid or reduce the pressure of the fluid based on this principle (Fig. 7.3).

The main applications in the process system are throttling orifice plates and throttle orifice plates. The comparison between the throttling orifice plate and the throttle orifice plate is shown in Table 7.1.

7.2.3.1 Aperture

To limit the venting volume, reduce the load of the flare system, and meet the pressure relief demand for 15 min.

Fig. 7.3 Throttling orifice plate

Table 7.1 Comparison between throttling orifice plate and throttle orifice plate

Project	Flow restriction orifice plate	Throttle orifice plate
Purpose	Limit flow	Reduce pressure
Applicable situations	Only for flow restriction, such as emergency venting systems	Flow metering, pressure reduction
Expectation of forming flow obstruction	Yes	No
Flow and pressure difference	Flow has only a maximum limit, reaching maximum flow under critical pressure difference	There is a range of flow requirements, hoping to achieve a specified pressure drop at a specified flow
Stages	Single stage is sufficient	Single stage, multi-stage
Single plate pressure drop	No limit, depends on the strength of the plate	Generally does not exceed 50% of the pressure before the plate
$p(d/D)$	No limit	Generally ≥ 0.2
Erosion	Fast flow, severe erosion, recommended for low frequency use	Relatively light, can be used for high frequency applications

7.2.3.2 Back Pressure After the Plate

The flow restriction orifice plate does not have the function of pressure reduction, and the back pressure after the plate depends on the back pressure of the entire flare system and the cumulative back pressure of the tail pipe after the plate. Although the orifice plate is not like a safety valve, there is no specific requirement for the back pressure after the plate, but it should not exceed the design pressure of the flare system. You can first estimate the aperture of an orifice plate for verification calculation, look at the pressure after 15 min, if the pressure is far lower than required, you can reduce the aperture, otherwise increase the aperture, until the pressure just meets 50% of the design pressure or 7 bar (gauge pressure) after 15 min. Note that the time step of the calculation should not be too short, adjust after convergence.

7.2.3.3 Tail Pipe Downstream of the Orifice Plate

The diameter should be changed as soon as possible downstream of the orifice plate to avoid high-speed erosion and high cumulative back pressure. The Mach number of the downstream tail pipe should not exceed 0.5.

7.2.3.4 Isolation Valve Downstream of the Orifice Plate

It should be set after the diameter change and be a full bore valve.

7.2.3.5 Low Temperature Zone Before and After the Orifice Plate

Under high pressure difference, the gas passing through the orifice plate will produce a throttling refrigeration effect, the temperature will decrease, and it may exceed the allowable low temperature of ordinary carbon steel (−29 °C), at this time low temperature type carbon steel can be used. The distance between the orifice plate and the upstream and downstream valves should be at least 600 mm to avoid the impact of severe low temperature.

7.2.4 BDV Inlet and Outlet Settings

The size of the BDV is recommended to be no less than 2 in, and the size of the BDV and the upstream pipeline of the orifice plate meets:

(1) When the operating pressure (gauge pressure) $p \leq 50$ bar, $V \leq 30{,}000$ kg/(m/s^2).
(2) When the operating pressure (gauge pressure) $p > 50$ bar, $V \leq 50{,}000$ kg/(m/s^2).

7.3 BDV Calculation Example

Ensuring the required safety standards while reducing the cost of the pressure reduction system is a major challenge. Therefore, simulation has a wide range of applications in the design and modification of pressure reduction systems, for which accurate models are needed. Usually, for safety reasons, conservative design methods are often adopted, and uncertainties and inaccuracies in the process may lead to overall overdesign.

BLOWDOWN™ was first used in Aspen HYSYS V9.0 to provide accurate low temperature determination, which is one of the keys in the design and operation of each process equipment, because it can not only reduce costs, but more importantly, it can improve the safety of equipment. This example uses a single vessel template to design an orifice plate for a pool fire depressurization scenario.

The Inlet Sep three-phase separator in the main process diagram will be used for some additional research. First, design a suitable discharge valve for the pool fire depressurization condition.

The BLOWDOWN analysis process function is similar to the HYSYS Sub-Flowsheet. However, unlike the traditional HYSYS sub-process diagram, flow connections cannot be added between the BLOWDOWN Sub-Flowsheet's PFD and the main PFD. In addition, the BLOWDOWN Sub-Flowsheet environment does not use the traditional HYSYS solver.

Add a BLOWDOWN unit from the Home menu bar. Under the Safety tab, click on BLOWDOWN and Depressuring, on the Depressuring page, click Add|BLOWDOWN Utility, BLOWDOWN Analysis will appear (Fig. 7.4).

In BLOWDOWN Analysis, in the Template drop-down list, select SingleVessel.blo-Single Vessel BLOWDOWN template (Fig. 7.5).

This template is a pre-configured template on the BLOWDOWN process diagram, used for single vessel structure and its pipeline connections.

Click Start Analysis. On the Design page of the BLOWDOWN Analysis page, you can set the layout of unit operations in the simulation to match the layout of process equipment (Fig. 7.6).

On this table, select the appropriate Inlet Line, Blowdown Line, and Vapor Outlet Line configurations (Fig. 7.7).

For the Inlet Line, select the Without pocket option. This means that all liquids in the inlet pipe section will flow to the main container. For the Blowdown Line, select

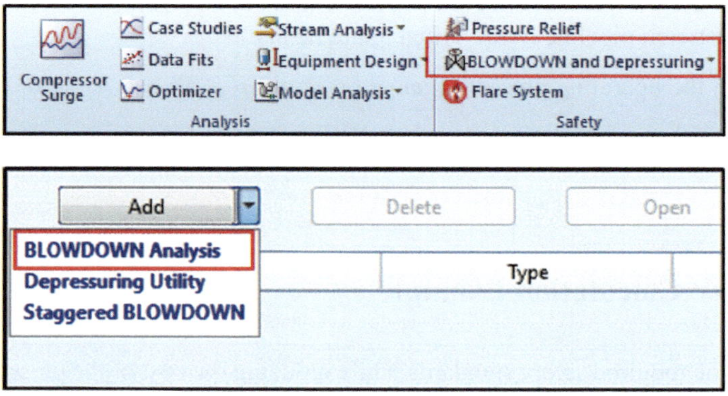

Fig. 7.4 BLOWDOWN analysis interface

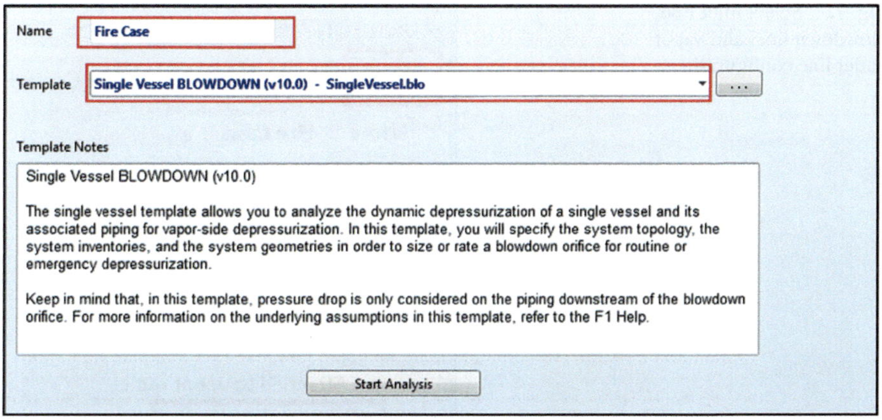

Fig. 7.5 Select SingleVessel.blo-Single Vessel BLOWDOWN template

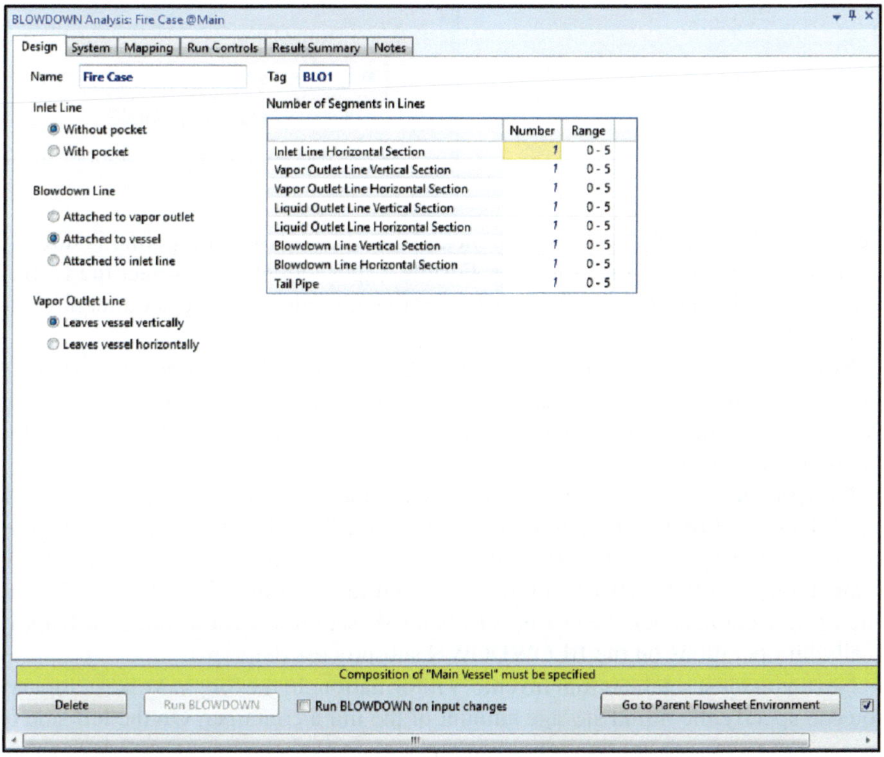

Fig. 7.6 Design page

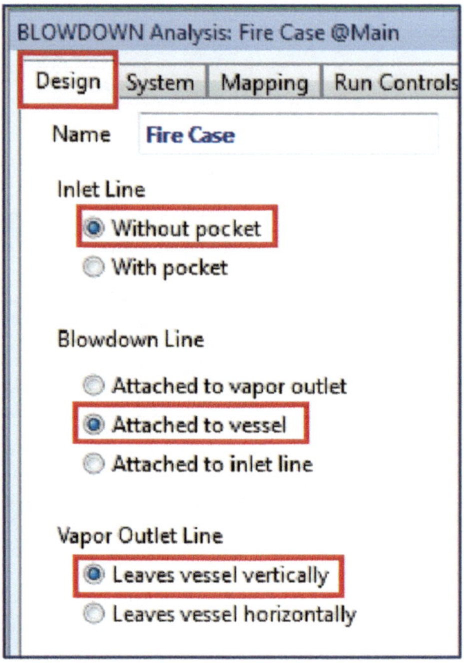

Fig. 7.7 Select inlet line, blowdown line, and vapor outlet line configurations

the Attached to vessel option, which means that the blowdown pipeline is directly connected to the top of the container. For the Vapor Outlet Line, select the Leaves vessel vertically option, indicating that the initial direction of the gas outlet pipeline is vertical.

Next, set the Number of Segments in Lines for each pipeline, according to the configurations of the Inlet Line, Blowdown Line, and Vapor Outlet Line. This example will simulate a standalone container, so no pipelines are needed, and the line segments are set to 0 (Fig. 7.8).

Next, on the System page, set the equipment, fluid, and environmental information for all BLOWDOWN unit operations. First, set the Template Parameters that apply to the entire template (Fig. 7.9). In the Ambient Air Properties section, set the Air Temperature to 21 °C (70°F) and the Air Speed to 2.438 m/s (8 ft/s). In the Define Fire Zone section, check the box next to Main Vessel (Note: These values will apply to all unit operations on the BLOWDOWN sub-process diagram).

Next, you must set the Initial Inventory information. In the Single Vessel template, you can specify the liquid storage amount of the main container. On the left side of the System page, expand the Initial Inventory|Main Vessel menu (Fig. 7.10).

The composition used in BLOWDOWN Analysis can be linked to the process flow on the flowchart, or manually copied/pasted from the process flow on the flowchart. Select Manual, then click Copy from Stream. In the Select Process Stream window, select the Feed_Mix_Fire_RP stream, then click OK. HYSYS will automatically copy the composition of the selected stream (Fig. 7.11).

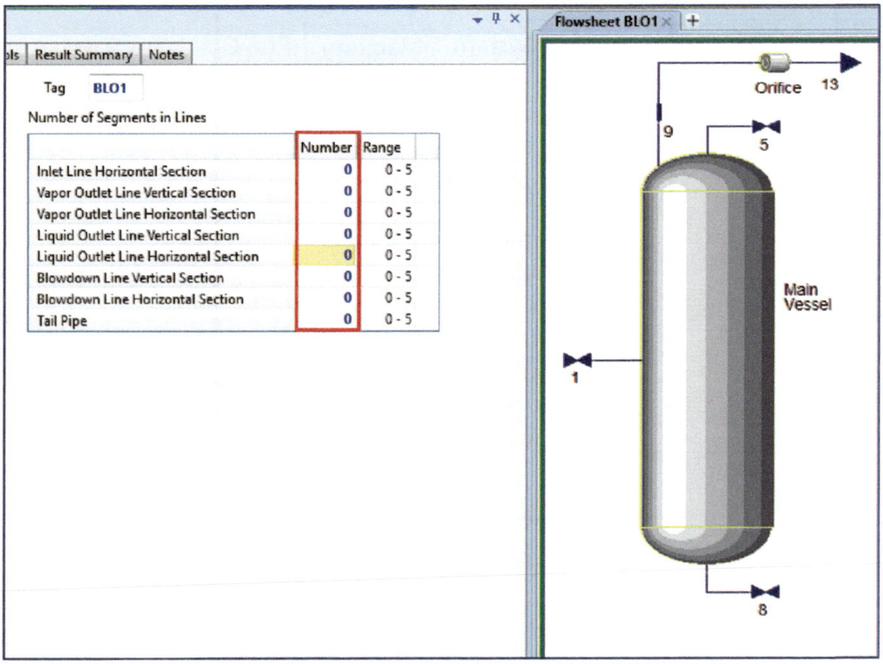

Fig. 7.8 Line segments settings

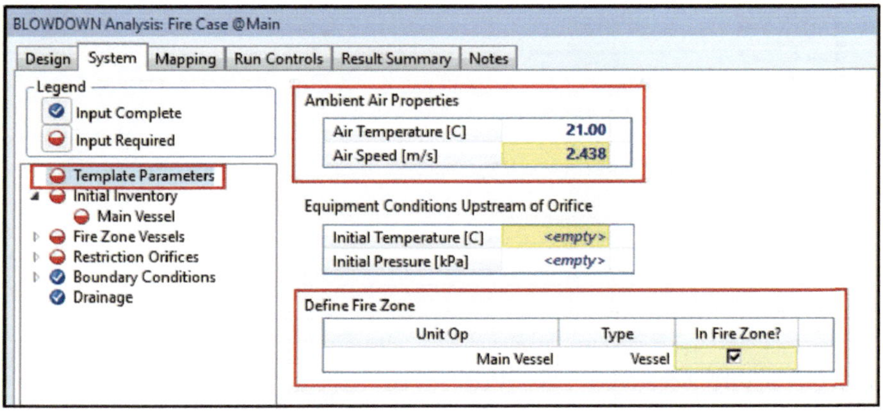

Fig. 7.9 Template parameters settings

Click Copy to Initial Conditions to copy the data to the Initial Temperature and Initial Pressure in the Template Parameters page (Fig. 7.12).

Next, you must set the vessel-related parameters. In the left pane, select Fire Zone Vessels, and set the vessel configuration on the Geometry, Heat Transfer, and Initial Condition tabs (Fig. 7.13).

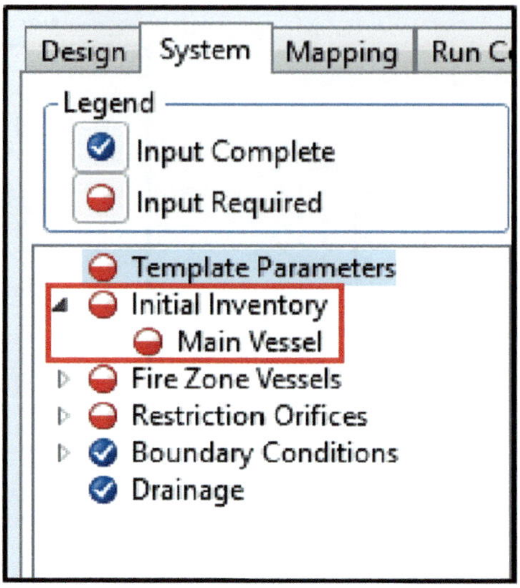

Fig. 7.10 Set initial inventory information

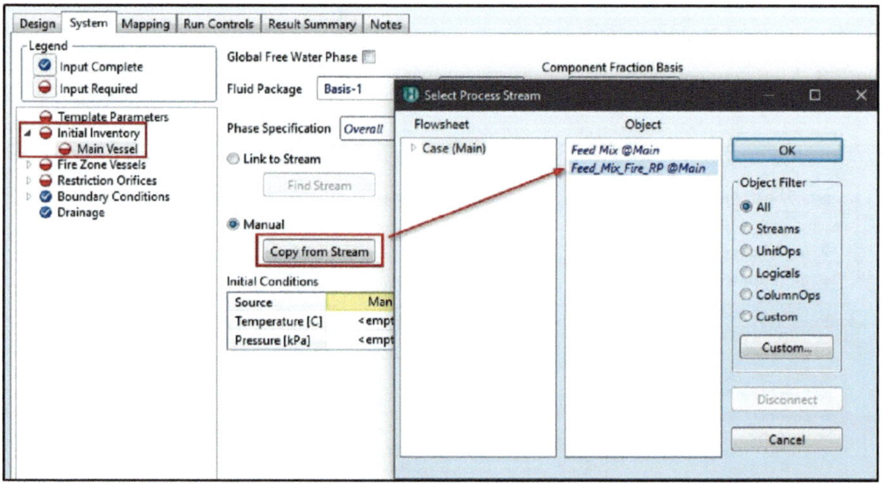

Fig. 7.11 Copy the composition of the selected stream

On the Geometry tab, enter the following geometric dimension specifications (Fig. 7.14).

Go to the Simple Fire tab and confirm that the Heat Flux Method is set to Apply to liquid (Fig. 7.15).

On the Initial Conditions tab, set the Liquid Volume Percent (%) to 50%, which is the total volume percentage filled with liquid at the start of the discharge (Fig. 7.16).

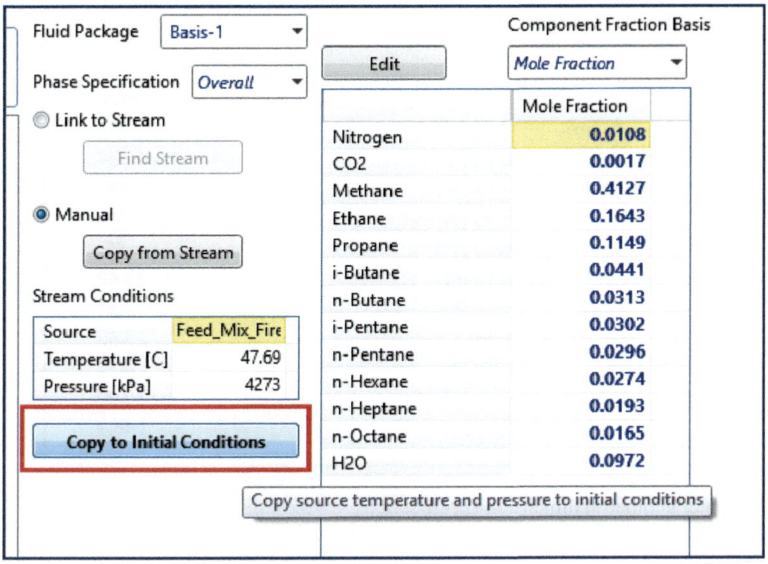

Fig. 7.12 Click copy to initial conditions to copy data

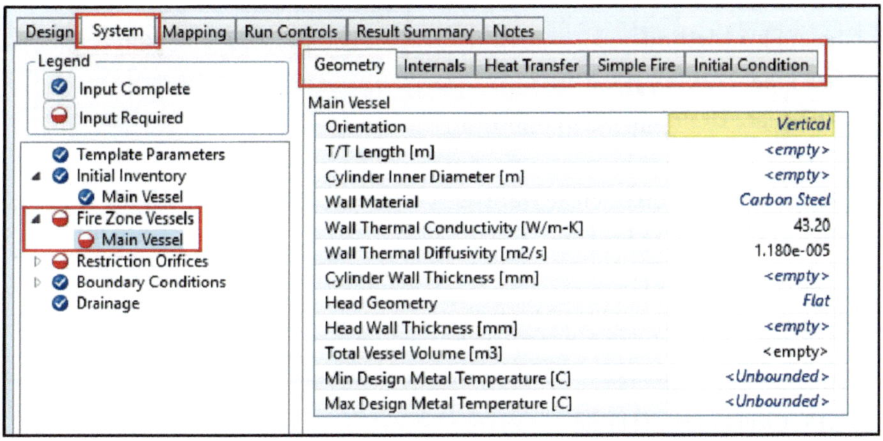

Fig. 7.13 Set vessel configuration

Next, you must set the information for the restriction orifice. For the Restriction Orifice size information (Fig. 7.17), set the orifice Diameter to 12.70 mm (0.50 in) and the Discharge Coefficient to 1.0.

View the component list (the left pane shows the HYSYS simulation components, the right pane shows the BLOWDOWN components), and note that water is not automatically mapped (Fig. 7.18).

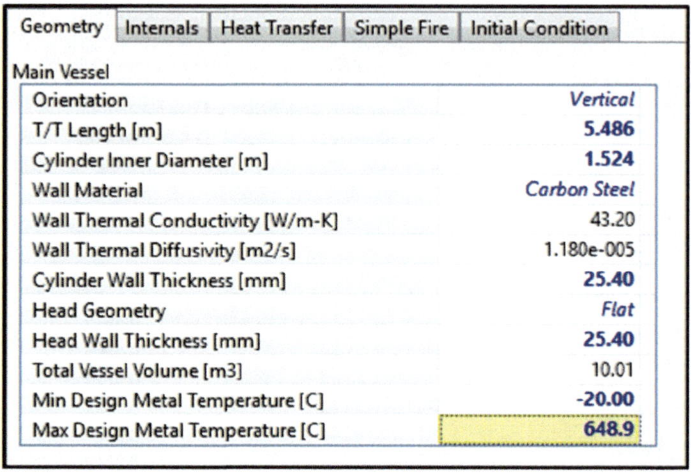

Fig. 7.14 Enter geometric dimension specifications

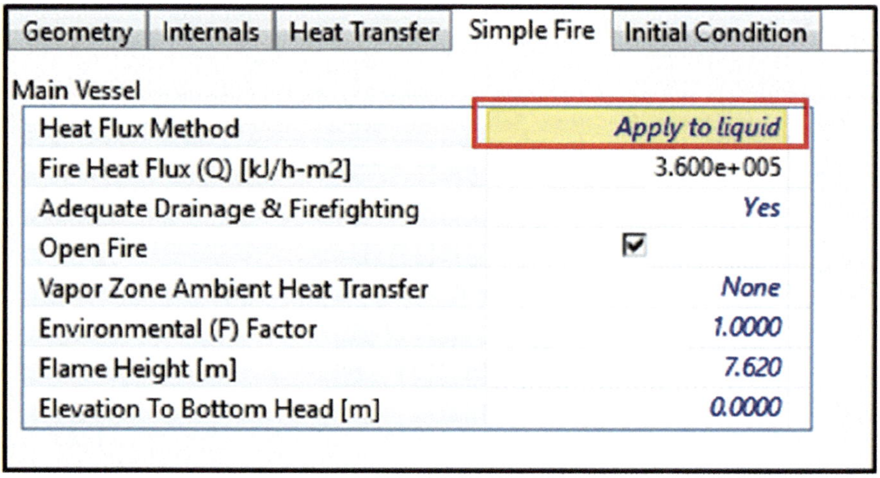

Fig. 7.15 Confirm apply to liquid

In BLOWDOWN, water can be simulated as dissolved in the hydrocarbon phase or as a separate free water phase. The component mapping information provides how to choose the water treatment option. To simulate water as a free water phase, you must return to the System tab and check the "Global Free Water Phase" option. To simulate it as dissolved in the hydrocarbon phase, you must manually add water to the BLOWDOWN component list and perform component mapping. HYSYS V10.0 requires manual selection of water treatment options to ensure special consideration for this input.

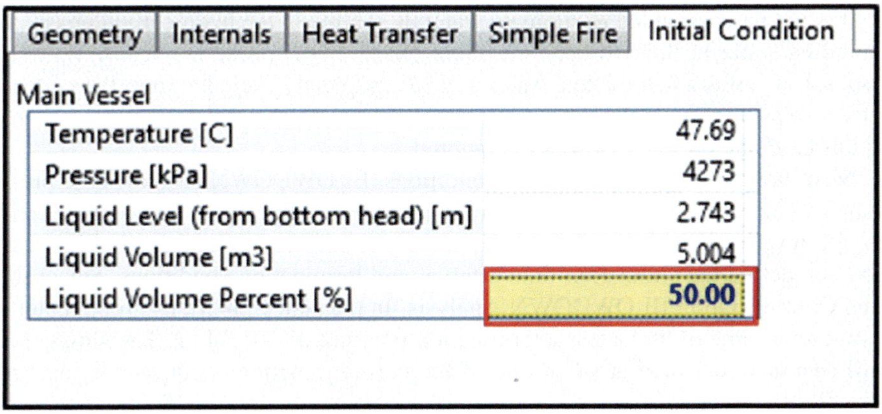

Fig. 7.16 Set liquid volume percent (%) to 50%

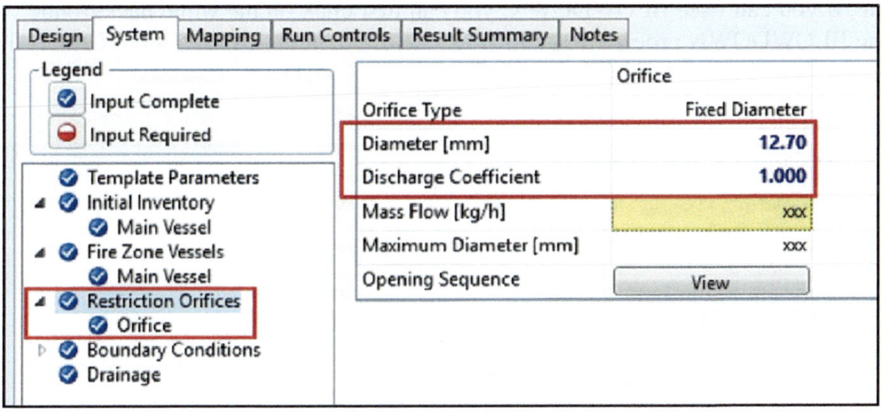

Fig. 7.17 Set restriction orifice size information

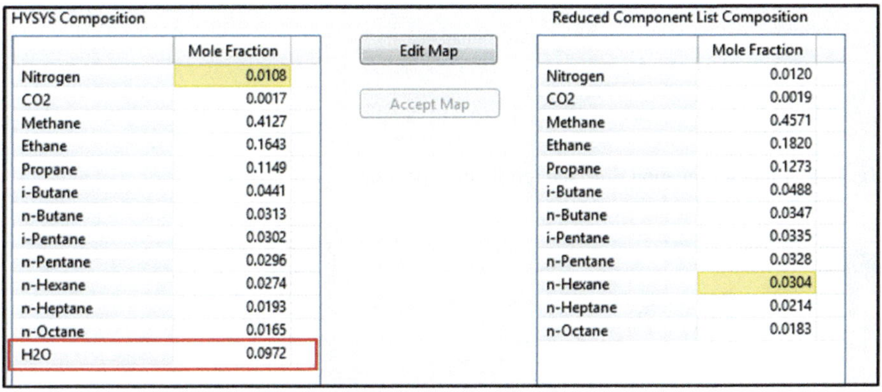

Fig. 7.18 View component list

It is not recommended to simulate water as soluble in the hydrocarbon phase, as including water in the hydrocarbon liquid phase or gas phase may cause flashing issues. If no choice is provided, water will be completely excluded from the system, which helps to provide the most conservative temperature prediction in the system. In this case, the presence of water is ignored.

Next, you must specify the Halt Conditions. BLOWDOWN can calculate problems over any time range and display the results, while reporting events second by second. Users can control the behavior of the solver. On the Run Controls tab, you can define the simulation time interval and termination conditions. Select the Run Controls tab in BLOWDOWN Analysis. In the Halt Conditions group, enter a termination time of 900 s and a termination pressure of 101.3 kPa. The simulation will terminate calculation when one of the two termination conditions is reached (Fig. 7.19).

Now, all the inputs required for the BLOWDOWN case run have been completed. To run the BLOWDOWN simulation, you can click Run on the top BLOWDOWN tab, if you can't see BLOWDOWN, you can first click on the white background of the BLOWDOWN process diagram (Fig. 7.20). Or, in the BLOWDOWN Analysis form, click Run BLOWDOWN at the bottom (Fig. 7.21).

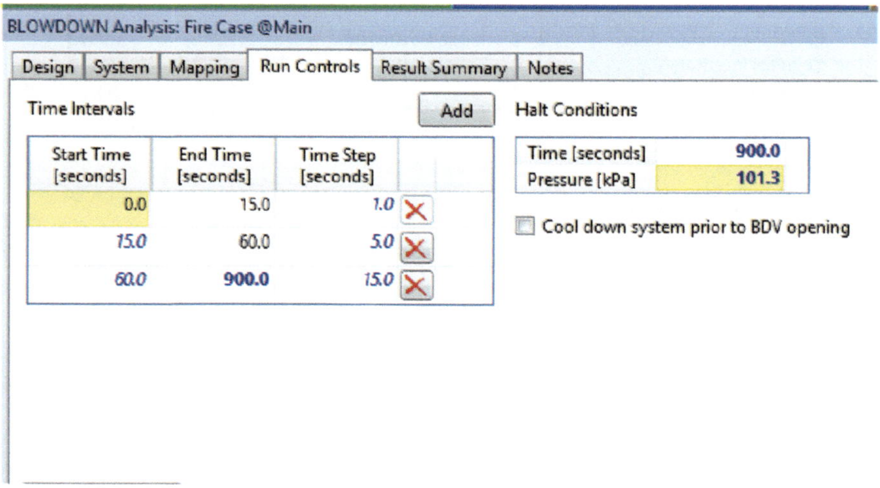

Fig. 7.19 Enter termination time and termination pressure

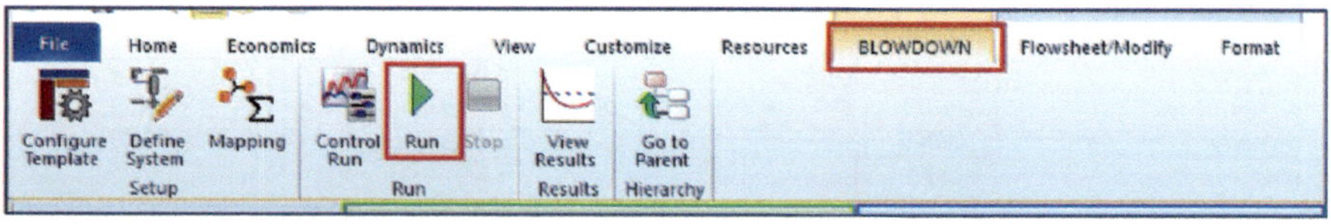

Fig. 7.20 Run BLOWDOWN simulation (one)

Fig. 7.21 Run BLOWDOWN simulation (two)

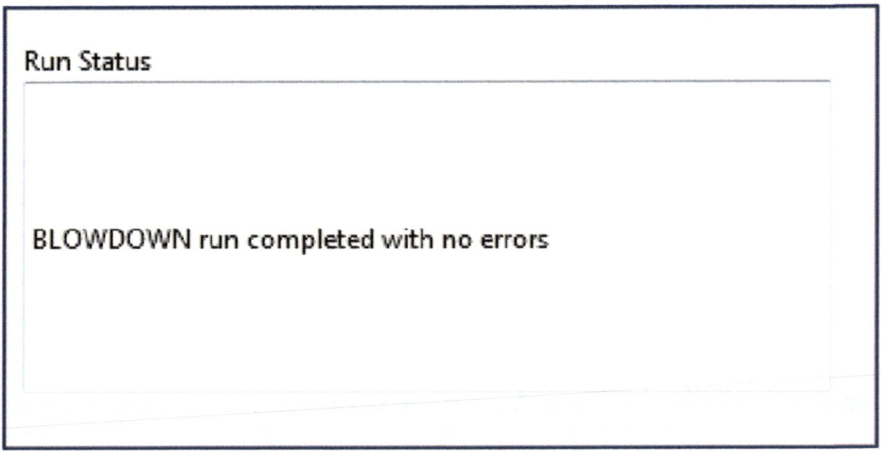

Fig. 7.22 Simulation run completed

After the BLOWDOWN simulation is completed, a message will appear in the status bar indicating that the simulation run has been completed (Fig. 7.22).

On the Results Summary page of BLOWDOWN Analysis, you can view the entire system result diagram, key result information, and warning information from the simulation. Select the Major Findings page to view the main results of the BLOWDOWN simulation. Select Plots under BLOWDOWN Analysis to view the graphical results (Fig. 7.23).

Next, you can use the Adjust unit operation on the main process diagram to design the BLOWDOWN orifice plate. Go to Halt Conditions and set the final pressure to 0. Ensure that the simulation always ends at the termination time, not under the termination pressure condition. Check the Run BLOWDOWN on input changes checkbox at the bottom (Fig. 7.24).

Click the BLOWDOWN tab or the Go to Parent button on the Flowsheet/Modify tab to return to the main process diagram (Fig. 7.25).

From the main process object panel, add an Adjust module. On the Connections tab|Connections page, specify Orifice Diameter as the Adjusted Variable (Fig. 7.26). Specify Final Pressure as the Target Variable (Fig. 7.27).

In the Target Value group, select the User Specified option and specify 50% of the design pressure (gauge pressure) 1825.015 kPa (250 psi) as the target value (Fig. 7.28).

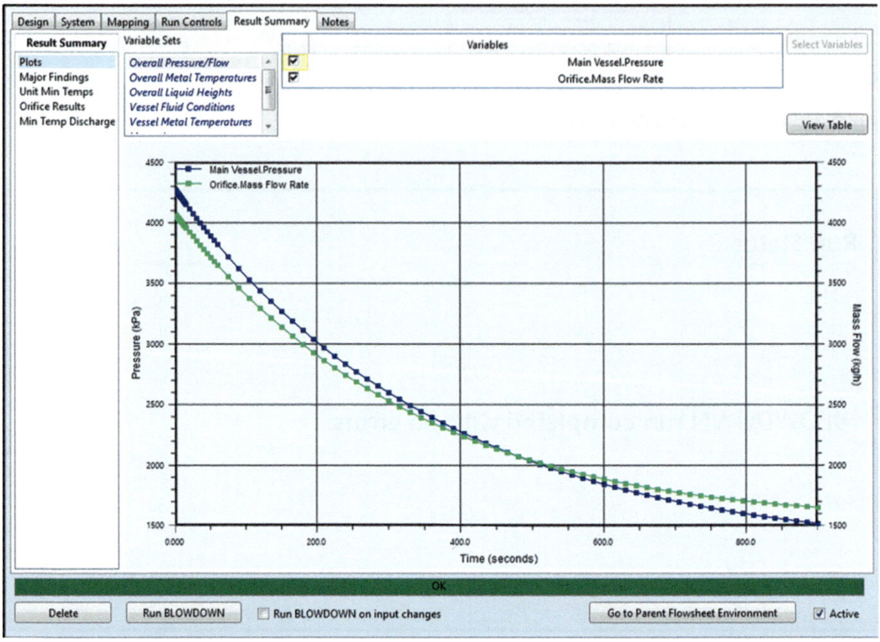

Fig. 7.23 View graphical results

Fig. 7.24 Check Run BLOWDOWN on Input Changes

Fig. 7.25 Return to main process

On the Parameters tab|Parameters page, under the Tolerance item, specify 1–10% of the required final pressure as the tolerance, i.e., specify 27.6 kPa (4 psi), Step Size is 2.54 mm (0.1 in).

Start the convergence calculation of the Adjust module.

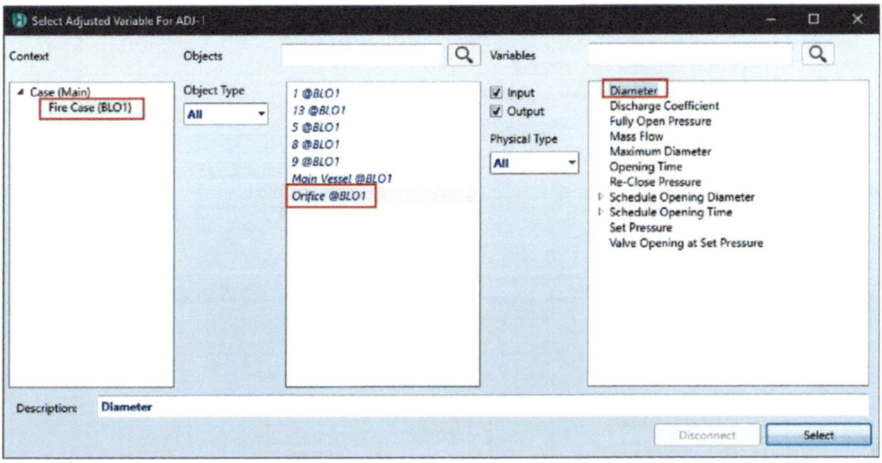

Fig. 7.26 Specify orifice diameter

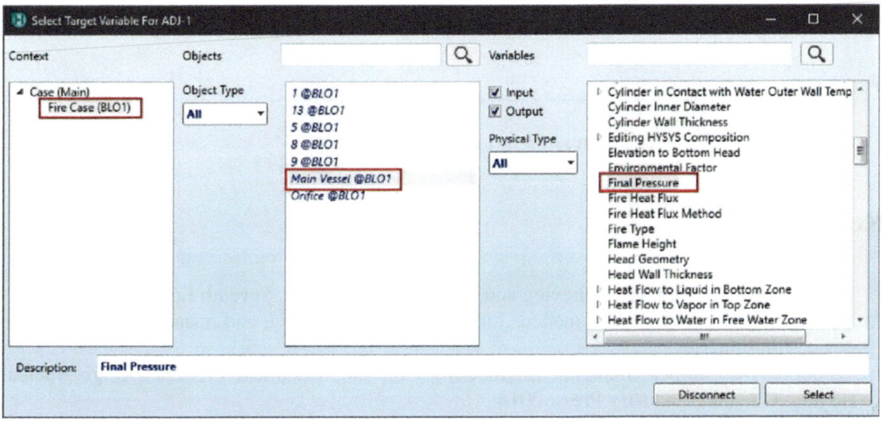

Fig. 7.27 Specify final pressure

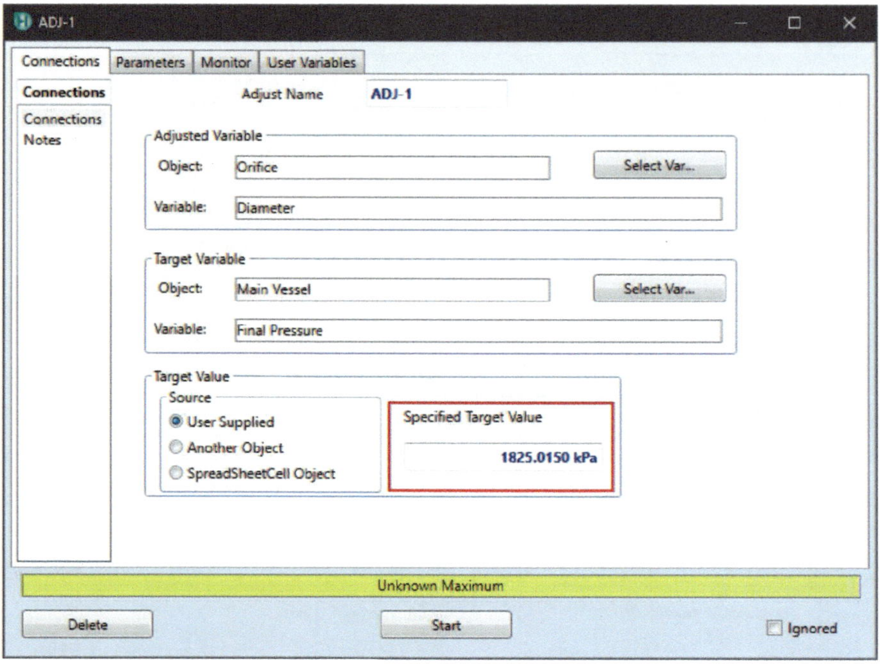

Fig. 7.28 Specify 50% design pressure target value

References

1. API Standard 521 Pressure-relieving and depressuring systems, Seventh Edition, June 2020.
2. SY/T 10044 Recommended practices for determining, selecting, and installing pressure relief devices in refineries.
3. Compiled by Sinopec Shanghai Engineering Co., Ltd. Chemical Process Design Manual. Beijing: Chemical Industry Press, 2018.

Chapter 8
Analysis and Calculation of Venting Systems

With the rapid development of the oil and gas industry, the capacity of production facilities tends to be large-scale, the construction and production of facilities are integrated, and public utilities are centrally supplied. Due to the scale of the facilities and the relationship of process production, the facilities will emit a large amount of waste gas under accident conditions such as start-up and shutdown, equipment failure, water stoppage, gas stoppage, power outage, fire, and misoperation, and these waste gases are mostly flammable, explosive, toxic and harmful gases. In order to prevent accidents and ensure equipment and personal safety, venting and flare systems need to be set up to handle the emitted flammable and explosive substances. The general method is to collect the venting branch pipes into the venting main pipe, send the gas to the flare through the venting main pipe, and discharge it into the atmosphere after combustion treatment, minimizing its impact on the environment and production safety.

The design of the venting system is very complex and needs to solve a large number of problems. If the design is not appropriate, it may seriously affect the normal production of the facility and even pose a huge risk to the operators. The venting and flare system is designed based on start-up and shutdown, daily operation and various sudden conditions. If the flare system cannot operate normally, it may cause damage to assets. Judging from the venting and flare systems that have been built, there are significant differences in the design methods and strictness requirements of this system among various companies. This book aims to clarify the minimum requirements for designing venting and flare systems and provide good and optimal design cases.

K. Fang, *Onshore Oilfield Surface Facilities: Process Safety Engineering Guide*,
https://doi.org/10.1007/978-981-96-7830-3_8

8.1 Principles of Venting System Design

8.1.1 Design Principles

(1) The venting system should be designed with safety and environmental protection as the principle in engineering design, and should be planned, economical, reasonable, and convenient for operation and maintenance according to the direction of the process flow, so as to ensure that the exhaust gas can be discharged safely and promptly during normal and abnormal production of the oil and gas equipment.
(2) Focus on the economic, environmental and social benefits of project construction, and pay attention to occupational health, safety, and environmental protection.
(3) The venting system should be centrally controlled and arranged as much as possible to reduce land occupation and facilitate management.
(4) Determine the setting and design scale of the venting system based on the characteristics and pressure settings of the emission sources of each device to ensure that it complies with local laws and regulations.

8.1.2 Design Standards

The design standards for venting and flare systems are as follows:

(1) API Std 520, Part I Sizing, Selection, and Installation of Pressure-Relieving Devices—Part I—Sizing and Selection.
(2) API Std 520, Part II Sizing, Selection, and Installation of Pressure-Relieving Devices—Part II—Installation.
(3) API Std 521 Pressure-Relieving and Depressuring Systems [1].
(4) API Std 537 "General Refinery and Petrochemical Facility Flare Specifications".
(5) ASME B31.3 "Process Piping".
(6) ASME BPVC Section VIII "ASME Boiler and Pressure Vessel Code".
(7) ISO 23251 (equivalent to API Std 521).
(8) ISO 25457 (equivalent to API Std 537).
(9) EN 13,445 "Unfired Pressure Vessels".
(10) SH 3009 "Design Specification for Combustible Gas Emission Systems in Petrochemical Industry".
(11) HG/T 20,570.12 "Flare System Installation".
(12) SH/T 3029 "Design Specification for Exhaust Stacks and Flare Towers in Petrochemical Industry".

8.2 Composition of Venting System

The venting and flare system consists of four main components, which as part of the plant safety system need to be quantified and designed.

(1) The source of the venting system, which is the connection between the process and utility facilities and the venting system.
(2) The venting collection system, which collects fluids from various sources and transports them to a centralized disposal device.
(3) The separation system, which can remove liquids from the flow stock for separate treatment or recovery.
(4) The flare system, which can safely release flare gas into the environment.

The following will discuss the first three parts of the venting system, and the flare system will be detailed in Chap. 9.

8.2.1 Venting System Source

The venting system source includes:

(1) Pressure safety (relief) valve.
(2) Bursting disc.
(3) Automatic relief valve.
(4) Manual relief valve.
(5) Maintenance venting.
(6) Control valve.

8.2.2 Venting Collection System

The venting collection system consists of main headers and branch pipes, which collect gas flows from various sources and merge them to the flare separation system and flare system.

8.2.3 Flare Separation System

The role of the flare separation system is to remove the liquid discharged into the flare collection system, or to remove the liquid condensed in the flare collection system before discharging it into the environment. Without a flare separation system, liquid-containing venting gas will be directly discharged into the flare system or vent pipe, and the flare system and vent pipe cannot handle gas–liquid mixed fluid, which will cause pollution around the venting point, and the flare system may cause a fire rain, burning liquid may fall to the ground near the venting point, causing casualties and local fires.

API Std 521 Pressure-relieving and depressuring systems provide clear requirements for the design of flare separation systems such as flare liquid tanks.

8.3 System Design

8.3.1 Design Process

Table 8.1 lists the design process for a new station field venting system.

8.3.2 System Analysis

The reasons for the pressure relief and venting of oil and gas field stations (plants) vary, such as the venting of gas during production, the venting of facilities for maintenance, and the venting of pressure regulating systems, as well as the pressure relief and venting in case of accidents. Different venting purposes have a large difference in venting volume. The venting volume during facility maintenance can be artificially controlled to a smaller scale, while the pressure relief and venting in case of accidents must meet the pressure relief requirements within the specified time.

The design of the venting system of the oil and gas field station (plant) is related to the safety of the station (plant), and how to determine the design scale of the station (plant) venting system is not yet regulated. The design concept of full venting is not scientific and cannot guarantee the safe operation of the station (plant) venting system. The maximum venting rate of the station (plant) should be calculated based on the setting of the self-control system under accident conditions, and the venting system scale should be determined accordingly. Adopting the design principle of "shutdown first, then vent", simulate the venting process of the station (plant) under steady-state and dynamic conditions, study the changing trend of gas venting pressure, temperature, and rate under different venting conditions, take reliable measures to control the initial venting rate, ensure safe and effective venting, and optimize the venting system scale.

For emergencies such as earthquakes, floods, and fires that can easily cause major disasters, the ESD system should be set to the highest level, and the oil and gas field station (plant) must stop production and urgently vent; for emergencies such as the failure of public systems like instrument air systems, cooling water systems, and other important equipment shutdowns, the whole plant must stop production, and whether urgent venting is needed can be determined after manual judgment; for small accidents in some processes, public devices, and some equipment device accidents,

Table 8.1 Venting system design process

Step	Content
1. Establish venting and flare system design guidelines	Refer to international standards, national standards, and industry standards to establish system design guidelines, and reach consensus with the owner
2. Screen for the lowest temperature	Screen for low-temperature venting sources, judge the impact on venting system materials and piping design
3. Determine all possible venting conditions for a single venting source	Refer to the project design guidelines, combined with the specific process flow, analyze the venting load of a single venting source under each venting condition, for specific analysis methods and calculations, see Chap. 6
4. Determine the maximum venting load for a single accident	List all venting sources and all conditions for each venting source, find the maximum venting load for a single venting source under a single accident condition
5. Determine the number of system isolation sections and emergency relief valves	In order to ensure the safety of the station field, determine the number of system isolation sections and corresponding emergency relief valves based on the process system flow and configuration
6. Determine the venting load of a single emergency relief valve	Based on the system liquid volume, gas volume, and venting rate requirements under venting conditions for a single system isolation section, determine the venting load of a single emergency relief valve, for specific analysis methods and calculations, see Chap. 7
7. Determine the maximum load of multiple venting sources venting simultaneously in the same accident	Use the matrix method to arrange and combine the possibilities and maximum venting load of multiple venting sources venting cumulatively under various conditions
8. Determine the maximum load of different emergency relief valves venting simultaneously	Use the matrix method to arrange and combine the possibilities of multiple different emergency relief valves venting cumulatively under various conditions of the possibility and maximum discharge load, while the discharge may be due to a combination of automatic, remote operation, and manual pressure relief

(continued)

Table 8.1 (continued)

Step	Content
9. Evaluate and determine whether to adopt delayed or staged venting	Under the same accident conditions, if the safety valve and emergency pressure relief valve cannot be vented simultaneously or there is danger in simultaneous venting, it is necessary to consider whether to classify and grade the system, such as • Maximum allowable back pressure • Gas–liquid mixture • Venting temperature range. The venting system should ensure that all operations or emergency conditions such as start-stop, maintenance, etc., can be smoothly vented
10. Determine the control conditions of the flare system	Combine all the analyzed conditions to determine the control conditions of the flare system design
11. Flare radiation spacing	Calculate the flare radiation heat value under control conditions, mainly determining the following indicators • Flare equipment height/location • Design acceptable radiation heat intensity. The flare system should also meet the requirements of gas cold diffusion in the case of extinguishing (cold venting), and account for the diffusion of combustible gas caused by site leakage (the flare may ignite)
12. Determine whether to upgrade the device configuration to reduce the scale of the flare system design	Upgrade device configuration options include • Use HIPPS or HIPS to reduce load • Staged or delayed venting • High-performance such as high back pressure flare head
13. Determine the size of the main header and branch pipes of the venting system	The size of the venting system header is mainly restricted by the following factors • Vibration caused by supersonic speed • Vibration caused by flow • The allowable back pressure of a single venting source • The design pressure of the venting and flare system • The form of the safety valve
14. Determine the size of the liquid separator tank and related equipment	

(continued)

Table 8.1 (continued)

Step	Content
15. Determine temperature/material	Evaluate and determine the material of the venting header and related equipment • The highest temperature of the venting source • The lowest temperature inside the flare header • Corrosive medium
16. Determine the flare and venting system	Based on the above content, re-check the design basis and system requirements of the flare system, and finally combine it with the Emergency Shutdown System (ESD)
17. Compile system design documents	Compile flare system design documents

since the accident range can be controlled, there is no need to stop production and vent throughout the plant.

Currently, the ESD of the oil and gas processing station (plant) The design is usually based on a 4-level or 5-level shutdown, as detailed in Chap. 2. In a 4-level or 5-level shutdown, the zero-level and first-level shutdowns are safety measures taken for special conditions, where the Emergency Shutdown Valve (ESD) shuts down and the parallel lock opens the automatic release valve for venting; whereas the second, third, and fourth level shutdowns are safety measures taken for abnormal conditions, where the ESD only shuts down without interlocking the BDV for venting, and when the system is over-pressurized, it can still be relieved through the Pressure Safety Valve (PSV).

Combining the ESD shutdown levels, the venting system corresponding to the accident conditions under different levels is analyzed as shown in Fig. 8.1.

8.3.3 Fire Zones

Typical oil and natural gas process facilities experience a large number of fire incidents, ranging from small electrical fires to large process fires. Fire Safety Assessment (SA) mainly focuses on calculating risk results based on hazards such as fire, explosion, and toxic gas diffusion, and measures whether the fire zones in the project design process are reasonable, whether the active fire protection design meets requirements, whether the fire protection of equipment/structures meets requirements, and whether the escape routes and assembly points are reasonable, etc.

A comprehensive assessment of the materials of each isolation unit can be made according to the PFD, P&ID, general diagram, and material balance table of the station (plant) process flow, forming a list of hazard sources and corresponding consequences. Information on most hazard sources can also be obtained from QRA studies, HAZID and HAZOP studies, and other risk assessment and management

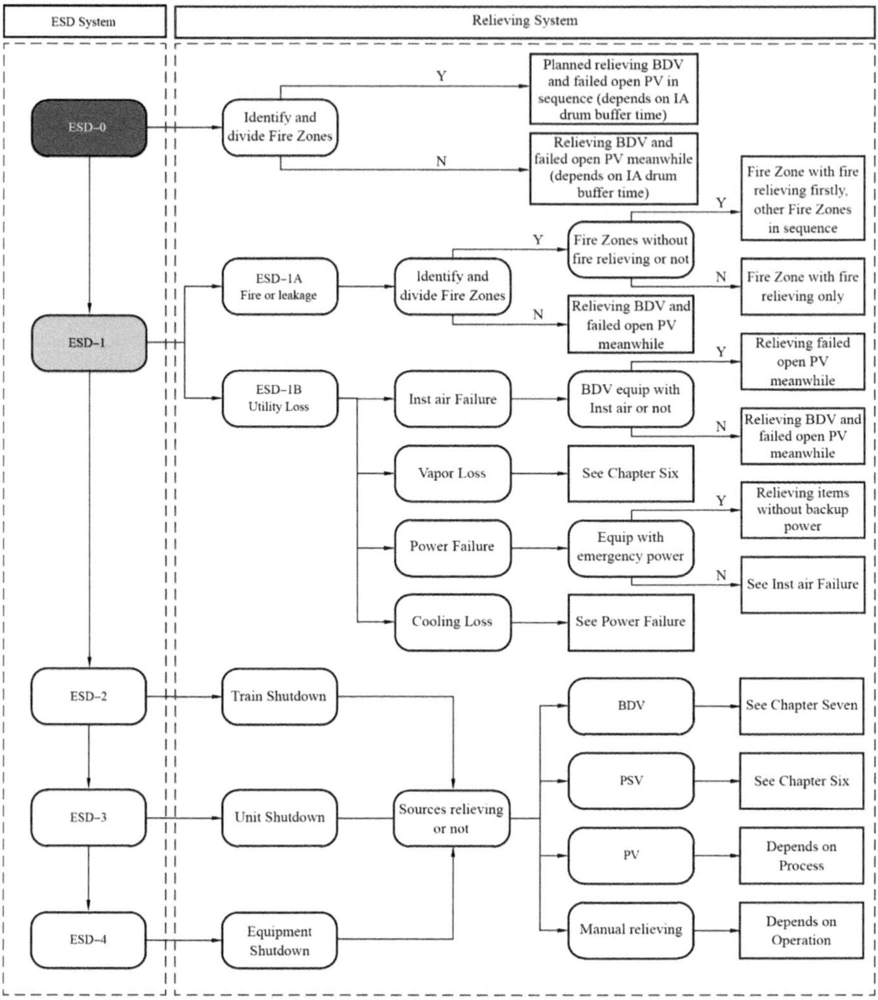

Fig. 8.1 Analysis of venting system corresponding to accident conditions under different levels

studies during the project execution process. The FSA is carried out based on the conclusions of HAZID and HAZOP project studies, the hazard and impact registration table, and existing QRA reports, to determine credible fire, explosion, and toxic gas scenarios and possible consequences related to these scenarios, and by identifying hazard sources (fire, explosion, and toxic gas hazards) in the process flow, determine the location and facilities for control and mitigation based on the isolatable facilities in the process flow diagram.

FSA defines the fire zones of the facility according to the most severe credible fire and escalation scenarios, and defines the fire water requirements for each fire zone. It divides the fire zones of the plant based on the assessment of fire hazards of process unit materials, response means, and the reliability of protective equipment.

The purpose of fire zone division is to limit the consequences of credible events, to ensure its own integrity in the event of fire, combustible gas leakage or explosion scenarios, and not to affect other areas, but not to avoid the occurrence of credible events. From the perspective of fire prevention, the smaller the fire zone division, the more conducive it is to ensure the fire safety of the building; but if the division is too small, it will inevitably affect the use function of the protected area. The determination of the size of the fire zone area should consider factors such as the nature of use, importance, fire hazard, building height, fire fighting capability, and the speed of fire spread of the protected area. Fire zones can be physically isolated from other areas through space, fire barriers, dikes, special drainage systems (e.g., water storage foundations), or a combination thereof.

The maximum size of the fire zone should be determined by the capacity of the available fire extinguishing devices, such as:

(1) At a standard flow rate of 120 m^3/h, the maximum protection radius of a fire monitor is about 45 m.
(2) The maximum pool fire area that a foam system can extinguish is about 7000 m^2.
(3) The maximum fire zone area should be less than 1000 m^2.

Fire zones should be convex, preferably rectangular. The layout should be able to retain or discharge fire water to prevent any danger of liquid ignition. Fire zones should not overlap within equipment units. Overlapping is allowed if no equipment is installed in the overlapping area and only fire-resistant pipes and pipe racks are installed. Fire zones are physically limited by the installation edges, extending at least 1.5 m beyond the edge of the farthest equipment, except for ESDV and passive protection devices (such as firewalls and blast walls, dikes, and other physical restrictions). The principles for specific fire zone and ESDV positioning can refer to TOTAL 321.

8.3.4 Design Guidelines

8.3.4.1 Separate Discharge of Material Flow

In the design, all fluids can flow to a single general vent and flare system. However, the design needs to consider the incompatibility between different discharges and emergency pressure relief, and it may be necessary to segregate discharge fluids. Discharge gases are divided into the following four situations according to the medium state:

(1) Hot gas (discharge temperature above 0 °C, with or without water).
(2) Cold gas (discharge temperature below 0 °C). Here "cold" is defined as a fluid temperature below 0 °C (32°F), which can cause water in the separator, free water in the fluid, or dissolved water to freeze; this situation may occur when

the process equipment contains devices for handling liquefied gases or high-pressure gas flows.

(3) Both cold and hot gases are present, but no water is included.

(4) Liquid discharge system.

For any of the four states of the discharged gas medium, a main pipe should be set up; if it is a combination of the above situations, special attention should generally be paid to the following points:

(1) The mixture of cold (low temperature) gas and hot gas containing water. Low-temperature fluids, such as those released by propane refrigeration, may cause the water-containing gas in the vent and torch system to freeze and block the discharge pipe.

(2) The mixture of high-temperature gas and volatile gas. Extremely hot fluids from process or utility heating systems may cause volatile fluids in the vent and torch system to boil rapidly (explosively).

(3) The mixture of low design pressure sources (such as atmospheric storage tanks) and high design pressure sources. The high back pressure generated by the pressure relief system may exceed the design pressure of the atmospheric storage tank.

(4) The mixture of liquid phase discharge fluid and gas phase discharge fluid. The entry of gas phase into the liquid phase can cause the liquid flow rate to increase and damage the pipes in the vent and torch system.

(5) The mixture of incompatible chemicals. Adverse reactions may cause excessive temperature or pressure, and may damage or contaminate the vent and torch system.

If it is a combination of the above situations, consider:

(1) Separately set up dry torch venting systems and wet torch venting systems.

(2) When the mixing of two streams may produce solids or dangerous physical or chemical changes, the two materials should be separated.

(3) If the mixing of two materials causes the pipe size to increase significantly or the pipe material to be upgraded, the two materials should also be separated.

(4) Generally separate the discharged liquid from the discharged gas. For flows with liquid, a separation facility and a separate liquid phase system should be set up, and attention should be paid to whether there are other materials that freeze or have high viscosity at temperatures above 0 °C (32°F). If the fluid will freeze or form hydrates at temperatures above 0 °C (32°F), consider transporting this part of the medium to a separate torch system.

Under normal circumstances, a separate torch system should be set up for the cold flow, and it should be separated from other torch systems. It can only be merged when the two systems are compatible, and the liquid separation tank of the cold flow should consider the liquid evaporation facility. The discharge of liquid propane and butane often leads to a "cold" two-phase flow. If the torch system is not designed to handle low-temperature liquids, a heating system should be provided to vaporize the liquid, and the special sealing requirements of the cold medium should be considered.

8.3.4.2 Design Documents

The API Std 521 Pressure-relieving and depressuring systems section fully describes excellent torch system design documents. The design documents should record the basic information of the pressure relief load and the combined system load in the vent and torch system.

The best practice is to record all design conditions of the discharge scheme, including special schemes that the design team considers unreliable or irrelevant after evaluation [for example, API Std 521 Pressure-relieving and depressuring systems stipulates: "There is no need to assume that two or more unrelated accidental incidents (also known as double or multiple conditions) occur at the same time"].

8.3.4.3 Use of High Integrity Pressure Protection Systems (HIPPS)

The latest version of API Std 521 Pressure-relieving and depressuring systems includes High Integrity Protection Systems (HIPPS) in process design. For an explanation of the application of HIPPS and the advantages of the HIPPS system, please refer to Sect. 4.1 of Chap. 4.

HIPPS uses the application of high-integrity instruments to alleviate or eliminate discharge conditions and has been used in a series of methods to reduce discharge load conditions, including:

(1) Traditional High Integrity Pressure Protection Systems (HIPPS) isolate high-pressure sources from low-pressure systems.
(2) The HIPPS shutdown set at the separator inlet can prevent high liquid levels from reaching the PSV inlet pipeline. This system prevents two-phase flow from entering the pressure relief device and reduces the liquid load flowing to the torch system.

It is worth noting that although HIPPS reduces the design load, it sometimes cannot completely eliminate it: if the HIPPS system has a bypass, it is still necessary to consider the high pressure release from the upstream equipment to the low-pressure system, but the downstream safety valve only considers the discharge load generated by the failure of this bypass.

8.3.4.4 Stepwise Venting

The characteristic of the discharge load of the emergency relief valve in the venting system discharge source is that the initial flow rate is high, and the flow rate gradually decreases as the system pressure decreases. Stepwise venting can effectively reduce the peak discharge, reduce the design scale of the vent and torch system. For new stations, this method can effectively reduce the size and distance of the vent and torch system; for existing stations, this method is usually used to merge other new

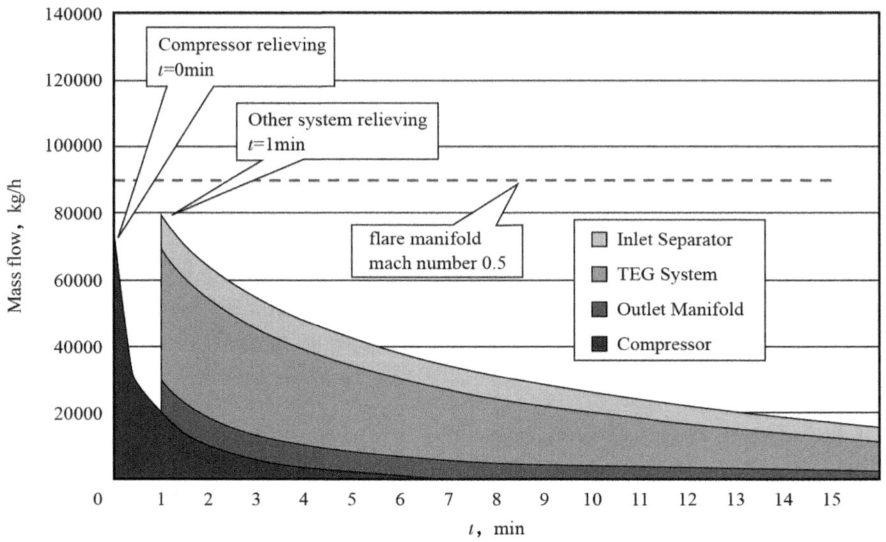

Fig. 8.2 Example of stepwise venting

emergency discharges within the capacity of the existing vent and torch system (Fig. 8.2).

The order of stepwise venting may vary depending on the equipment involved in the accident. For example, in the event of a confirmed fire, the emergency relief valve in the fire area will be activated first, followed by the emergency relief valve far from the accident location. Emergency pressure relief valves in the same fire protection zone are opened simultaneously, and stepwise venting is not allowed.

8.4 Operational Analysis

The main purpose of the venting system is to safely handle flammable and hazardous substances in the plant. The reasons for setting up a venting system are as follows:

(1) Operational reasons. Such as when the flow exceeds the capacity of downstream equipment and excess fluid spills into the flare system.
(2) Maintenance reasons. It allows safe access to equipment and pipelines.
(3) Emergency reasons. Safely handle hydrocarbons to reduce the risk of equipment failure and escalation of emergencies.

Among them, reason (3) is the main reason for installing a general integrated flare or venting system on production facilities.

The operational analysis of the venting system includes the following aspects:

(1) Determine the load of each device's venting and the simultaneous venting load of all possible venting devices under each operating condition.
(2) Operating conditions of simultaneous venting of different devices.
(3) Oil and gas processing stations generally have "operational venting" set up for maintenance and operation process control. Therefore, in addition to the above conditions, operational venting should also be considered.
(4) A flare system usually serves multiple process units and some off-site public facilities, so the possibility of simultaneous venting of these different units needs to be considered. This consideration will increase the venting rate of any part of the flare (flare branches of different units, manifold, off-site flare manifold, total flare manifold).
(5) For each simultaneous venting condition, not only should the pressure safety valve be considered, but also other possible simultaneous venting, including automatic venting of control valves, and the venting rate of automatic or manually initiated emergency pressure relief conditions.
(6) When determining the size of the pressure safety valve, the control of the instrument is not considered. The control of the instrument may reduce the venting volume. However, when determining the size of the flare system, the impact of instrument control on the venting volume can be partially considered, that is, it can be assumed that some instruments are operating normally.
(7) When calculating the size of the flare manifold, it should be based on the venting volume of simultaneous venting. The venting volume of simultaneous venting should be considered as the sum of the calculated venting load and other flare loads, not the maximum venting volume and the sum of other flare loads when all pressure relief valves are fully open. That is to say, it cannot be assumed that all valves will reach the maximum venting volume when fully open at the same time. In fact, these valves will open and close in intervals, and the total venting volume at any moment will not exceed the calculated venting volume.
(8) If the owner agrees, a venting time/duration analysis can also be conducted, instead of simply adding all venting loads together, which will result in a more realistic flare load. In reality, although the same reason will cause venting of different devices, the peak of these venting processes may occur at different moments. For example:

    1. Some device venting starts immediately when an accident occurs, but due to the small amount of hydrocarbons, the venting duration is very short.
    2. Some venting has a long delay, which is because the surge of liquid and pressure accumulation takes time.
    3. For the depressurization process, the venting load will decrease over time, because the pressure inside the device will gradually decrease over time.

In addition, it is necessary to consult the owner about the margin considered for future modifications and expansions.

8.4.1 Instrument Control System

8.4.1.1 Difference Between a Single Venting Valve and an Individual Venting Valve

Chapter 6 mainly discusses how to determine the venting volume and size of a single venting valve. This chapter considers the contribution of a single venting valve to the venting system. The main differences between the two are as follows:

(1) The latter allows the venting volume to be reduced through normal operation of the control system.
(2) The latter allows consideration of some unconventional instruments (such as automatic shutdown and automatic start) to reduce the venting volume of the venting system.

8.4.2 Assumptions about Automatic Control Instruments

When considering the venting volume of the venting system, all control instruments (whether they increase the venting volume or not) need to be considered. The controls that will reduce emissions are operating normally. For various automatic control instruments, the following assumptions can be made:

(1) For a single-loop controller that may increase emissions, it should be assumed to operate normally.
(2) For a single-loop controller that may reduce emissions, it should be assumed to operate normally. However, for the flow of shares, it cannot be reduced to less than 25% of the design flow, and the heat input cannot be corrected.
(3) For cascade control, the following assumptions can be made:

    1. If the main controller increases the emission load, and the auxiliary controller also increases the emission load (without considering the cascade), then use the maximum emission under the condition of the largest increase in emission load (cascade or non-cascade).
    2. If the main controller increases the emission load, and the auxiliary controller reduces the emission load (without considering the cascade), then consider the increased emission load when the cascade control operates normally.
    3. If the main controller reduces the emission load, and the auxiliary controller increases the emission load (without considering the cascade), then consider the increased emission load under the operation of the auxiliary controller without the participation of the main controller.
    4. If the main controller reduces the emission load, and the auxiliary controller also reduces the emission load (without considering the cascade), then

use the emission of the least reduction in emission load (cascade or non-cascade). In any case, it should not be assumed that the flow rate is reduced to less than 25% of the design flow, and the heat input cannot be corrected.

(4) Assume that the part of the automatic instrument that is working abnormally fails to reduce the system venting volume, for example:

1. In the case of abnormally high tower pressure, start the automatic shutdown facility of the steam at the bottom of the tower to cut off the steam heat source, thereby avoiding higher tower pressure due to heat input.
2. Automatically start the backup reflux pump driven by steam turbine during power failure to reduce the tower's emission volume. Such non-continuous operating equipment should be tested regularly.

8.4.2.1 Assumption of the Number of Failed Instruments

For each venting condition, the total number of independent devices (automatic shutdown plus automatic start) that may reduce the venting load should be determined. This number may vary depending on different venting conditions. Each device (automatic instrument system) must be completely independent of all other devices. If two devices have a common component (such as a transmitter or shut-off valve), they must be regarded as one device, because they may both fail due to the damage of the common component. For automatic instruments that are working abnormally, the following standards should be used to determine how many instruments should be assumed to fail:

(1) The number of failures is related to the number of devices that reduce the emission load. In any process unit, it can be assumed that no more than two automatic instruments are working abnormally.
(2) The automatic instrument that should be selected for abnormal operation should be based on providing the maximum venting load.

8.4.3 Fire

The fire venting load of each separate system's safety valve should be determined according to the method in Chap. 6. For protected systems equipped with emergency depressurization systems, the rate of venting to the flare system during emergency depressurization should be used. This emergency depressurization rate should generally be determined according to API Std 521 Pressure-relieving and depressuring systems or customer-related requirements, and should be greater than the venting rate of the fire condition, specifically determined according to the method in Chap. 7.

When determining the occurrence of a fire accident, the layout, trend, and wastewater collection plan of the on-site equipment should be used to estimate the fire area that may produce the maximum venting volume from multiple protection systems.

After confirming the wastewater collection plan and route, the maximum venting volume can be determined by the fire circle according to the following points:

(1) Assume that the maximum fire range that causes multiple protected systems to vent at the same time is within the fire circle diameter range given in Chap. 6.
(2) Divide the fire circle position, these fire circles will completely or partially cover the maximum number of protected systems.
(3) For each fire circle, the total venting load is equal to the sum of the venting loads of all protected systems that are completely or partially covered by the fire circle (emergency venting flow or safety valve venting volume in the fire condition, whichever is larger).
(4) The maximum fire venting condition of each process unit is the venting condition of the fire area with the largest venting volume in all fire areas of the total diagram of this unit, considering only one area having a fire at the same time.

8.4.4 Instrument Air Failure

The release load of each individual system during instrument air failure should be determined according to the method in Chap. 6, and the release load that any failure-open control valve may cause should also be determined, as well as the release load of automatic or manual pressure relief devices, using the larger load. The total venting volume generated by the instrument air failure of the entire plant is the sum of the instrument air failure release loads of all protected systems connected to the flare plus the total load of all possible control valve releases.

Instrument air failure usually does not produce a large release load and should not cause the maximum release load of the flare system. If a large load is generated, the failure position of the device with the largest release load should be questioned, or its release load should be changed in other ways. To avoid multiple emergency pressure relief valves failing open at the same time and generating a load exceeding the venting capacity, consider local instrument air storage tanks or gas cylinders.

8.4.5 Power Failure

The venting release volume of a single protected system during a power failure should be determined according to the provisions of Chap. 6. Then, these loads should be corrected according to the previous description of the impact of instrument control on venting loads.

Due to the failure of the cooling water system caused by power failure, the venting load during power failure should be determined according to the description in Chap. 6. Due to the failure of the instrument air system caused by power failure, the venting load during power failure should be determined according to the above description. The venting load that may be caused by control valves opening due to power failure and discharging to the flare should also be determined. For example, automatic pressure control release caused by shutting down electric compressors.

The total venting load generated by a power failure in the entire plant is the sum of the power failure release loads of all protected systems connected to the flare plus the total load of all possible control valve releases. The impact of automatic shutdown and automatic startup systems on venting loads can be considered according to the description earlier in this chapter.

8.4.6 Steam System Failure

The venting release volume of a single protected system during a steam failure should be determined according to the provisions of Chap. 6. Then, these loads should be corrected according to the previous description of the impact of instrument control on venting loads. The flare load that may be caused by control valves opening due to steam system failure and discharging to the flare should also be determined. For example, automatic pressure control release caused by shutting down steam turbine-driven compressors. The venting load of automatic or manual pressure relief devices should also be determined, using the larger load. The total venting load generated by a steam failure in the entire plant is the sum of the steam failure release loads of all protected systems connected to the flare plus the total load of all possible control valve releases. The impact of automatic shutdown and automatic startup systems on venting loads can be considered according to the description earlier in this chapter.

8.4.7 Cooling Water System Failure

The maximum venting load caused by cooling water failure is usually caused by power failure, steam failure, or simultaneous power and steam failure, and is already included in the analysis of these failures.

8.4.8 Combination of Different Causes

The venting load caused by two or more related pressure relief causes occurring simultaneously should be considered, that is, one cause leading to or increasing the possibility of another cause, or they are all caused by a common cause. For example, the total flare load during power failure and steam supply failure occurring simultaneously, as well as the resulting cooling water and instrument air failure (control valve in the instrument air failure position); the flare load caused by the simultaneous occurrence of plant-wide power failure and instrument UPS backup system failure.

For various combination conditions, the flare release load of a single protected system during failure should be determined according to the provisions of Chap. 6. Then, these loads should be corrected according to the previous description of the impact of instrument control on flare loads. The flare load that may be caused by control valves opening due to combination condition failure and discharging to the flare should be determined. The flare load of automatic or manual pressure relief devices should be determined, using the larger load combination condition. The total venting load generated by the failure is the sum of the combination condition failure release loads of all protected systems connected to the venting system plus the total load of all possible control valve releases. The impact of automatic shutdown and automatic startup systems on flare loads can be considered according to the description earlier in this chapter.

8.5 Venting Pipe Network Hydraulic Calculation

8.5.1 Design Input

In the hydraulic calculation of each venting condition, the following data should be included for the release valve:

(1) Composition of the released gas.
(2) Temperature of the released fluid.
(3) Maximum allowable back pressure of the release source.
(4) Release gas flow rate.
(5) Flow rate-time curve.

8.5.2 Design Criteria

SH The "3009 Petrochemical Combustible Gas Emission System Design Specification" stipulates: The Mach number of the general venting system network should not be greater than 0.7; the Mach number at the end of the combustible gas emission pipeline where condensate may appear should not be greater than 0.5; the pressure of the entire plant's combustible gas emission system network should be maintained at no less than 1 kPa (gauge pressure). The size of the vent manifold, branch pipe, and pressure relief valve (PRV) should be able to meet the maximum back pressure and speed limits under different venting conditions. Different types of pressure relief valves have different back pressure limits, and the back pressure limit of the safety valve should be noted. Many pilot-operated safety valves have a back pressure limit equivalent to the flange rating. However, the back pressure of most conventional safety valves and balanced bellows safety valves is much less than the flange rating. API Std 526 Flanged steel pressure-relief valves provide a variety of valve types, sizes, structural materials, and temperature ranges. In the early stages of design, it is necessary to check the size of the required pressure relief pipeline, and calculate whether increasing the vessel design pressure to reduce the required torch manifold size can save investment.

The total allowable pressure loss of the venting and torch system is usually determined by the back pressure limit of the key pressure relief valve, and the pressure loss comes from the torch body, liquid seal, separator tank, and pipeline, etc. The following factors should be considered when evaluating the back pressure of the venting and torch system:

(1) All potential pressure relief, pressure reduction, and process venting adjustments.
(2) The roughness of the pipeline material, generally using a roughness of 0.45 mm, this roughness is a general empirical value considering some corrosion/dirt/deposits.
(3) The layout of the pipeline, including the pressure loss of fittings and inlets, etc.

Consider the equipment and systems connected to the torch system and their pressure limits. For example, the exhaust ports of analyzers, pump seal gas, centrifuge seal gas, agitator seal gas, other seal exhaust ports, dryers, vessels, and floating roof tanks, etc. This is because back pressure may cause backflow.

Normally, the pressure drop calculation of the venting and torch system starts from the torch head, by calculating the pressure drop of each torch system pipe section with a specific diameter and specific torch load, gradually returning upstream to each pressure relief device, appropriate calculation software (such as Aspen Flare System Analyzer) should be used to calculate the pressure drop of the entire system:

(1) The pressure drop of the torch head is a key parameter, this value should be obtained from the torch supplier.
(2) For the pressure that restricts the size of different parts of the torch system, it should be considered as follows: Preliminary calculations should be made using the pressure relief situation with the highest total venting load determined in this section, which will set size requirements for the torch, torch separator tank, main vent manifold, and most of the torch manifolds of the process unit; then the size of the torch separator tank needs to be checked to ensure that it can accommodate the maximum liquid discharge, so as to determine the position/spacing requirements of the torch and torch separator tank to ensure the thermal radiation safety of the process equipment and personnel. This preliminary calculation can also obtain the back pressure at each pressure relief valve and verify whether it exceeds the back pressure and speed limits.
(3) The pressure drop calculation must consider the pressure loss at the pipe connection. For example, in Aspen Flare System Analyzer, it is recommended to use Miller Tees with Gardel extrapolation, other methods overestimate the pressure loss at the pipe connection at low flow or no flow. In some key system (especially built devices) torch system analysis, sometimes more detailed analysis software (such as Korf Hydraulics) is used, which is more suitable than Aspen

Flare System Analyzer because the former has more comprehensive pipe fitting pressure drop data and is more flexible.

(4) If the off-site torch manifold, process unit torch manifold, torch branch pipe, etc., will bear a torch load higher than the torch load used in the previous step's preliminary screening under some venting conditions, additional pressure drop calculations should be performed. Consider the situation of individual system pressure relief (using the full flow of the pressure relief valve, or the flow of different venting sources venting at the same time), in these situations, although the total load of the torch system is smaller, the branch load is larger, so this situation may become the decisive condition for the size of the branch pipe. These pressure drop calculations can start from the torch head. Or you can simply modify the upstream torch pipeline with a higher rate to simplify the calculation. These simplified calculations give a pressure higher than the complete calculation, but if they prove to meet the PRV back pressure limit, no further calculation is needed.

(5) The same method as above should be used to calculate the pressure drop of the pressure relief discharge branch pipe. The rated flow of the safety valve should be used in the calculation. If a protected system has multiple safety valves that vent to the same discharge branch pipe, use the sum of the rated flows of all safety valves that may open at the same time.

When analyzing plant-wide failure events, it can be considered to reduce the total load of the venting system by staged venting, such analysis can follow ISO 23251 Petroleum, petrochemical and natural gas industries—Pressure-relieving and depressuring systems or API Std 521 Pressure-relieving and depressuring systems.

8.5.3 Network Calculation

The basic principle for determining the size of the discharge pipeline and the relief pipe manifold is: the back pressure that exists or is generated at any point in the system should not cause the release volume of the pressure relief device to be lower than the release volume required to prevent overpressure of the corresponding container. Therefore, the impact of superimposed or accumulated back pressure on valve operation characteristics must be carefully checked. The discharge pipeline system should be designed so that the accumulated back pressure generated by the fluid flowing through the considered valve does not drop below the capacity required by any pressure relief valve that may be released at the same time.

After the size of the discharge pipe manifold and the relief main pipe is determined, the relief accident that generates the maximum back pressure should be clearly defined. Any single relief accident may involve several pressure relief valves. The design of pipeline fixation and support on the discharge pipe manifold should be given special consideration. Sudden changes in flow and temperature can generate large reaction forces. If there is liquid in the relief system, the impact force may be large. This is discussed in more detail in API Std 520, Part II Sizing, Selection, and installation of pressure-relieving devices—Part II—Installation. In addition to the back pressure criterion, flow measurement should also be considered, which forms the basis for calculating the discharge pipeline. The size of the branch and tail pipe of a single device is usually calculated based on the flow rate of the device, and is used to calculate the size of the inlet pipeline connected to the valve (see API Std 520, Part II Sizing, Selection, and installation of pressure-relieving devices—Part II—Installation). The common main pipe system and pipe manifold of multiple devices are generally calculated based on the worst case, that is, the cumulative processing amount required for all devices to possibly discharge simultaneously in a single overpressure accident is reasonably assumed. The inlet size of a single safety valve is determined according to the method in Chap. 6, and the inlet size of a single emergency relief valve is determined according to the method in Chap. 7.

The determination of the size of the vent pipe network generally starts from the system outlet. The pressure at the outlet is known, and then the allowable back pressure of each pressure relief valve in the system is checked in reverse. The discharge back pressure at each node should be lower than the allowable back pressure at that point, which can simplify the calculation and the calculation is carried out in segments according to the pipeline diameter.

After the maximum relief load required and the maximum allowable back pressure of the pipe manifold (according to the type of valve in the system and the applicable specification requirements) are determined, the work of selecting the pipe size is left to the calculation of fluid flow. When the fluid conditions are known, the size of the discharge pipeline can be calculated using several methods. These methods range from treating the fluid in an isothermal process with an appropriate margin of kinetic energy effect to more accurate solutions calculated close to adiabatic, for specific methods see API Std 521 Pressure-relieving and depressuring systems Sect. 6.6, users can choose the method that best suits their needs. Pipe friction loss is generally recommended to use the isothermal flow equation of API Std 521 Pressure-relieving and depressuring systems and SH 3009 "Design Specification for Combustible Gas Discharge Systems in Petrochemical Industry", see formula (8.1). Other methods such as Lapple, Fanno method, etc. see API Std 521 Pressure-relieving and depressuring systems Sects. 5.5.6 and 5.5.7.

$$\frac{fL}{d} = \frac{1}{Ma^2}\left(\frac{p_1}{p_2}\right)^2\left[1 - \left(\frac{p_2}{p_1}\right)^2\right] - \ln\left(\frac{p_1}{p_2}\right)^2 \tag{8.1}$$

where

f—Hydraulic friction coefficient;
L—Equivalent length of the pipeline, m;
d—Inner diameter of the pipeline, m;
Ma—Mach number at the pipeline outlet;
$p1$—Inlet pressure of the pipeline (absolute pressure), kPa;
$p2$—Outlet pressure of the pipeline (absolute pressure), kPa.

The hydraulic friction coefficient f is calculated by formula (8.2).

$$f = 0.0055\left[1 + \left(20000\frac{e}{d} + \frac{10^6}{Re}\right)^{\frac{1}{3}}\right] \quad (8.2)$$

where

e—Absolute roughness of the pipeline, m;

Re—Reynolds number.

The Mach number at the pipeline outlet is calculated by formula (8.3).

$$Ma = 3.23 \times 10^{-5}\frac{q_{\text{m}}}{p_2 d^2}\left(\frac{ZT}{kM}\right)^{0.5} \quad (8.3)$$

where

qm—Gas mass flow rate, kg/h;

Z—Gas compression factor, take the average value of relative segmented calculation;

T—Temperature of the emitted gas, K;

k—Adiabatic index of the emitted gas;

M—Average relative molecular mass of the emitted gas.

The characteristics of the gas flow released from the pipeline are that its density and speed change rapidly, so it is treated as a compressible fluid. Several methods have been studied to calculate the size of the discharge pipe using isothermal or adiabatic flow equations. The actual flow state in the discharge system is generally between isothermal and adiabatic flow states. In most cases, it is recommended to use the more conservative isothermal equation; however, for some uncommon situations, the adiabatic flow equation may be more applicable (such as deep cold situations).

8.5.4 Network Simulation

The main software for calculating the venting network system are Flarenet, Visual Flow, and Inplant.

Flarenet can complete the steady-state design, calculation, and bottleneck elimination of single or multiple flare systems, and can also calculate new flare systems or eliminate bottlenecks in existing pressure relief networks. When designing phases or logistics operations, Flarenet can also be used to identify potential pressure relief hazards. Flarenet can adjust the entire flare and venting system to control pressure and noise.

Flarenet has an intuitive process flow diagram operating environment, which can present the entire flare network, such as pressure relief valves, control valves, pipes, connectors (including expansions, reductions, tees, etc.), separators, and flare heads.

8.5.4.1 Visual Flow

Visual Flow can perform strict steady-state simulation calculations on factory safety systems and pressure relief systems, including fluid reaching critical flow conditions. Visual Flow calculates pressure drop using the industrial standard Beggs and Brill Moddy or Lockhart and Martinelli multiphase methods. For high-speed flow calculations, the specially improved Beggs and Brill methods are combined to ensure accuracy in critical flow applications. In the back pressure calculation of the flare system, the software has high calculation accuracy for safety valve accounting and sizing.

8.5.4.2 Inplant

Inplant is a strict, steady-state fluid mechanics calculation software used for the design, accounting, and analysis of factory piping systems. With Inplant, engineers can quickly evaluate and analyze the factory's piping system. Inplant can be used to design new networks and also to modify existing systems. Its applications range from pipeline size design to accounting for multiphase flow network systems in large, complex loop networks.

Inplant has the same physical property database as PRO/II and a user-friendly interface, and can handle mixtures, single-phase gases, single-phase liquids, vapors, water, and other types of fluids.

8.6 Pipeline Design

The mechanical design of the venting system should be given the same attention as the design of the piping system for processing process fluids. The problems encountered in the design of the discharge pipeline of the pressure relief valve or pressure reducing valve are usually more complex than those encountered in the design of the process system, because the discharge pipeline may be affected by a wide range of operating temperatures, pressures, and the shocks caused by these wide range of operating conditions. In addition, at one time the processing system may contain a certain substance to be processed in the process system; while at another time, it may contain another substance. The main stresses on the discharge pipeline of the venting system are caused by the thermal expansion or contraction caused by cold or hot substances entering the pipeline and the impact caused by the discharge fluid. The flexibility design of the venting system is much more complex than the design of the process pipeline system, because it is necessary to control the impact and thermal expansion.

In most cases, it is possible to keep the stress of the venting system within the allowable range throughout the temperature range by providing guides, fixators, and suitable pipeline configurations.

8.6.1 *Pipeline Design*

8.6.1.1 Design Installation

1. Laying of Venting Network Pipes

The laying of venting network pipes should meet the following requirements:

(1) The pipeline should be laid overhead.
(2) Branch pipes from each venting device should usually enter the manifold from above the manifold, which will prevent the liquid flowing or forming in the manifold from flowing back to each valve through the branch pipe.
(3) New project pipelines should adopt natural compensation methods, expansion and renovation project pipelines should preferably adopt natural compensation methods, and compensators should be installed horizontally.
(4) Flare gas is different from general gas materials, the liquid storage in the network directly affects the safe operation of the network, the pipeline should slope towards the separator, water seal tank, the pipeline slope should not be less than 0.2%; if there is a low point along the pipeline, a separator or liquid collection tank should be set up.
(5) When the pipeline is not sloped, and there is a lowest point in the pipe network, a condensate collection and transfer facility must be set up at the lowest point of the pipe network, such as a small discharge trough or drip pipe, to eliminate the condensate that may accumulate at the lowest point and prevent the pipeline from being blocked and damaging the safe discharge of the flare system. Since there are no dedicated operators for the condensate collection and transfer facilities on the plant's pipelines, a control system for automatic start and stop of the condensate pump can be considered. To promptly grasp the failure of the condensate pump, a high liquid level alarm should be set up for the collection tank, and the alarm signal should be sent to the control room or the corresponding management post. The management of condensate collection and transfer facilities is generally taken care of by the operators of the flare device or other nearby posts, with 1–2 rounds of inspections per shift.
(6) To avoid the flare gas discharged by each production unit carrying hydrocarbon condensate into the main pipe, the pipeline branch should be connected to the main pipe from above, and the branch pipe and the main pipe should be connected at a 45° angle to reduce pressure drop and reaction force.
(7) For maintenance and safety (such as individual units starting, stopping, or having accidents, while other units maintain normal production or undergo maintenance), it should be considered to divide the pipe manifold system into zones using valves. These valves should be equipped with locking or sealing devices. In situations where it is not suitable to install valves, the installation of blind plates should be considered. When setting up zone valves or blind plates, special attention should be paid to their usage to ensure that the operating equipment will not be isolated from the release system. If the pipe manifold

system uses valves, they should be installed in a closed position that will not fail, such as gate valves that can fail in the closed position.

(8) If different manifolds use different materials, considering the possible back-flow, higher quality materials should be used at least 10 m (33 ft) upstream of the process condition changes.

(9) If the plant's pipe network is large, spanning several boundary areas, and each boundary area is composed of several units, when there are partition requirements in the boundary area, cut-off valves should also be set up. All valves should have markers indicating the position of the valve (i.e., open, closed, or degree of openness).

(10) After the construction and maintenance of the pipe network is completed and before it is put into operation, the air in the pipe network should be driven out with purge gas. Before the flare gas discharge pipe network stops operating and prepares for maintenance, the flare gas should also be purged to the flare until it is burned out, and only when it meets the requirements for hot work or maintenance can the purging be stopped for maintenance.

(11) The flare gas pipeline should preferably use elbows with a large radius of curvature to reduce local resistance loss.

(12) The pipeline should be equipped with pipe supports or pads; when the nominal diameter of the pipeline is greater than or equal to DN800mm, measures to reduce the friction coefficient should be taken for sliding pipe supports or pads.

(13) When there is a possibility of vibration and jumping in the pipeline, radial limit measures should be taken at the appropriate position.

(14) The allowable span between active pipe supports depends on the strength and stiffness of the pipe itself and the required laying slope of the pipe. In the design, it should be noted that because the diameter of the flare gas pipeline is generally large, when the diameter is greater than 600 mm, the influence of wind load must be considered in the selection of formulas for calculation; at the same time, radial stability calculations must also be carried out.

2. Purging Facilities

To ensure the safe operation of the vent pipe network, a fixed purging facility is set up at the farthest upstream end of the vent main pipe. This purging facility includes a flow meter, a check valve, and a manual regulating valve. All vent main pipes should be equipped with a hose interface for purging. The purging medium should preferably be nitrogen, but steam can also be used when nitrogen is not available. However, for low-temperature pipelines, the purging gas should not partially or completely condense at the lowest temperature. For this, nitrogen purging is generally used, and the purging gas speed in the largest flare pipe is 0.03 m/s. If the flare system has a water (liquid) seal, the upstream purging gas speed is 0.01 m/s. If there is no water seal, a low flow alarm and a low pressure alarm indicating vacuum should be installed to prevent air from flowing back into the flare system.

3. Insulation

The vent pipe network is generally not insulated, but when the freezing point of the condensate in the discharge pipe is higher than or equal to the coldest monthly average temperature of the area within 10 °C, the pipe should be insulated; when the freezing point of the condensate is higher than the coldest monthly average temperature of the area by more than 10 °C, the pipe should be insulated and equipped with heating measures. When setting up heating insulation, attention should be paid to the fact that heating may cause the temperature of the flare gas to rise, and precautions should be taken to prevent the chemical reaction of the flare gas caused by the temperature rise.

4. Reserved Laying Positions and Interfaces

For the phased production of combustible gas discharge pipes, the laying positions and related interfaces of the later pipes should be reserved in the early design.

5. Anti-External Pressure Design

When the temperature of the combustible gas discharge is greater than 60 °C, the combustible gas discharge pipe before the water seal tank should be designed for anti-external pressure according to GB 150 "Pressure Vessel", and the maximum external pressure should be greater than or equal to 30 kPa.

6. Design Pressure

The design pressure of the pipeline before the water seal tank should not be lower than the design pressure of the separator tank, and the design pressure of the pipeline after the water seal tank should not be lower than the design pressure of the water seal tank.

8.6.1.2 Pipeline Wall Thickness

The choice of wall thickness design for the venting system pipeline network must consider various operating conditions:

(1) Pressure loss at maximum discharge load.
(2) Acoustic vibration generated by pressure relief devices. The downstream of pressure relief devices (such as PSV and BDV) will generate high acoustic energy, which will excite the natural vibration of the pipeline. This kind of acoustic vibration can cause rapid fatigue failure at small diameter connections near the process pipeline or pipeline. Venting pipelines are susceptible to acoustic vibration, high-pressure sources will enter the venting system, and the API standard has currently solved this problem in a rough way.
(3) Flow-induced vibration in the venting pipeline, which is related to high-speed flow through the pipeline.

8.6.1.3 Pressure Fluctuation

ASME B31.3 "Process Piping" stipulates short-term pressure fluctuations, see the allowed pressure and temperature changes in 302.2.4. Occasional situations higher than the design conditions should comply with one of the following pressure design restrictions:

(1) With the approval of the owner, the operating pressure can exceed the pressure rating or design pressure in the case of temperature increase but not exceeding the design temperature.
(2) Exceeding 33% of the design pressure, the time should not exceed 10 h, and not exceed 100 h/a.
(3) Exceeding 20% of the design pressure, the time should not exceed 50 h, and not exceed 500 h/a.

Many design companies do not allow this provision to be used for the design of venting systems, so it should be evaluated in the early stage of the flare system design, such as a reasonably designed step-by-step emergency venting system can use pressure fluctuation specifications to evaluate the results of simultaneous emergency venting.

8.6.2 Structural Materials

The design of the venting system should be combined with the structural material selection of the entire system, including the discharge source, inlet pipeline, emergency venting or pressure relief device, outlet tailpipe, venting manifold, separation system, and flare. When evaluating the required materials, high and low temperatures, high pressure, vibration, and corrosion should be considered. Material selection may affect the design of the flare system, and material restrictions may affect the separation of the discharge source. If different manifolds use different materials, considering possible backflow, higher quality materials should be used at least 10 m (33 ft) upstream of the process condition changes.

8.6.2.1 Extreme Temperatures

Material selection should consider high-temperature discharges of the flare system:

(1) The reduction of material yield stress at high temperatures may affect the requirements for the ANSI Class 300 flange system.
(2) Some materials are prone to stress corrosion cracking at high temperatures (for example, stainless steel is prone to chloride stress corrosion cracking in high salt environments).

Material selection should consider low-temperature discharges of the flare system:

(1) Low-temperature process streams.
(2) Refrigerant streams cool under pressure reduction.
(3) Process streams cooled by Joule–Thomson in pressure reduction devices. The design should also consider the highest and lowest temperatures and the impact of solar radiation that the facility may be exposed to.

8.6.2.2 Corrosion Protection

The corrosion protection design of the flare system needs to consider individual discharges and simultaneous discharges. For example, acidic dry gas is relatively mild, but when mixed with water at high temperatures, it may have strong corrosiveness to carbon steel pipelines. The corrosion protection of the venting pipeline network is the same as that of general pipelines, with no special requirements. For corrosive venting air, consider increasing the wall thickness or applying anti-corrosion coating inside the pipe for corrosion protection.

8.6.3 Pipeline Stress and Support

The flare system has a small load under normal conditions, unless a major emergency venting condition occurs. In this emergency condition, due to the combination of the following factors, the pipeline may generate large stress:

(1) Extreme temperatures.
(2) Liquid phase slug flow.
(3) Load increase, extremely high flow speed.
(4) Thin pipe wall, large size pipeline.

For the design of discharge pipelines, it is required to carefully analyze the possibility of thermal stress and mechanical stress acting on the relevant pressure relief valve. Discharge pipelines that are not separately supported but only supported by the outlet of the pressure relief valve will generate stress on the pressure relief valve and its inlet pipe. The forced alignment of the discharge pipeline will also generate similar stress. Discharge pipelines (including short tailpipes) should be inspected, supported, and carefully corrected according to regulations. Deformations sufficient to cause mechanism failure usually first occur on the inlet pipeline, Moreover, even a small torque can cause serious failures and leaks in the pressure relief valve. When the pressure relief valve discharges, it may also generate stress in the processing pipeline due to the impact of the reaction force it produces. Analysis shows that in necessary situations, measures should be taken to fix or limit the processing pipeline connected to the pressure relief valve. The API Std 520, Part II Sizing, Selection, and installation of pressure-relieving devices—Part II—Installation provides the formula for calculating the reaction force load generated when the pressure relief valve operates.

In the discharge pipeline, impact loads should also be considered. Impact loads may be caused by the sudden inflow of compressible fluid into a multi-directional pipeline system or by the impact of liquid plugs at turning points. In the pipeline, each change of direction may generate a reaction force.

Designing a venting network system with appropriate, sufficient fixation, guidance, and support is very complex. There are various methods for calculating pipeline flexibility, and ASME B31.3 "Process Piping" should be referred to for background discussion. Pipeline stress, anchoring, and support design should consider thermal expansion or contraction, two-phase flow, liquid plugs, vibration fatigue caused by high-speed flow, icing or auto-refrigeration of cold media, and fire protection. To avoid costly overdesign, the mechanical design of the flare manifold should be based on a practical assessment of the highest temperature and duration for each release situation, rather than simply using the highest relief temperature as the design temperature. Once the range of release conditions to be handled is established, apart from considering the impact force, the calculation is no different from most pipeline systems.

8.7 Flare Knockout Drums

8.7.1 Classification Introduction

The flare knockout drum is an important part of the flare system. Each flare discharge manifold should have a knockout drum to separate the liquid droplets carried by the gas or the liquid phase that may occur in the two-phase flow. Normally, a knockout drum is installed in the device to reduce the condensate amount in the flare gas manifold. When the flare is located at a certain distance from the device, the flare gas will generate condensate during transportation, so a knockout drum should also be set up before the flare gas enters the flare drum to separate the condensate again, to prevent liquid droplets from being carried to the flare tip, causing a fire rain phenomenon.

8.7.1.1 Classification of Flare Knockout Drums

Flare knockout drums mainly have horizontal and vertical types. The horizontal knockout drum is divided into the following two types.

(1) The gas enters from the top of one end of the knockout drum and exits from the top of the other end (no internal baffle), which is called single-flow.
(2) The gas enters from both ends of the horizontal axis and has an outlet in the middle; or the gas enters from the middle and exits from both ends of the horizontal axis, which is called double-flow. When the diameter of the knockout drum is larger than 3.6 m, the double-flow type is usually used.

For the vertical knockout drum, the gas inlet is set in the diameter direction of the middle of the vessel, and the outlet is set in the vertical direction of the top of the vessel, and a baffle should be added at the inlet to make the gas flow downward.

When choosing between horizontal and vertical knockout drums, the gas flow rate, required liquid storage volume, and slope requirements of the flare manifold should be considered, and an economic comparison should be made.

8.7.1.2 Types of Gas Inlet and Outlet Channels of the Knockout Drum

The types of gas inlet and outlet channels of the knockout drum can be one of the following:

(1) Horizontal drum: The gas enters from one end of the vertical upper part of the drum axis and exits from the other end. The gas inlet and outlet should face the nearby drum head end.
(2) Horizontal drum: The gas enters from both ends of the vertical upper part of the drum axis and exits from the middle. The gas inlet should face the nearby drum head end.
(3) Vertical drum: The gas enters radially from the drum body and exits from the top of the vertical axis of the drum body. A baffle is used to ensure the direction of the gas flow is downward.
(4) Vertical drum: The gas enters tangentially from the radial direction of the drum body and exits from the top of the vertical axis of the drum body.

8.7.1.3 Liquid Collection Bag

A liquid collection bag should be set up in the horizontal knockout drum. The structural dimensions of the liquid collection bag are as follows:

(1) The diameter of the liquid collection bag should be 500–800 mm, should not be larger than 1/3 of the diameter of the knockout drum, but should not be less than 300 mm.
(2) The height of the liquid collection bag (the distance from the cut line of the liquid collection bag head to the drum wall) should not be less than 500 mm.

8.7.2 Design Principles

The main purpose of the knockout drum is to remove most of the liquid in the gas flow and prevent the liquid from being carried into the flare. The design of the knockout drum should meet the following conditions:

(1) The knockout drum should be as close to the flare as possible, and the possibility of personnel operation and whether to set up a knockout drum downstream should be considered.

(2) The knockout drum should meet the minimum design requirements of ISO 23251 Petroleum, petrochemical and natural gas industries—Pressure- relieving and depressuring systems or API Std 521 Pressure-relieving and depressuring systems.

1. The design of the liquid separator should allow for full vacuum, with a maximum allowable working pressure (gauge pressure) of at least 3.5 bar (50 psi) and an external pressure not less than 30 kPa (gauge pressure).
2. The minimum corrosion allowance for the liquid tank should be 3 mm (1/ 8 in).
3. The liquid separator should be able to separate all free liquid droplets, and the size of the liquid separator should be calculated using the calculation methods in ISO 23251 Petroleum, petrochemical and natural gas industries—Pressure-relieving and depressuring systems, and API Std 521 Pressure-relieving and depressuring systems.

(3) The liquid capacity of the liquid separator should allow for a liquid residence time of 20 to 30 min at the maximum liquid flow rate into the liquid separator, or the liquid separator should have enough space to accommodate the maximum possible amount of liquid. This capacity should be provided between the maximum normal liquid level (i.e., the pump start liquid level) and the maximum allowable liquid level of the liquid separator, taking into account the following situations:

① Requirements for gas/liquid separation.
② Flashing of the liquid under the pressure of the liquid separator.

(4) The size of the liquid separator should be able to remove liquid droplets larger than 300 to 600 μm (0.024 in) at the maximum emergency venting flow rate, and remove liquid droplets larger than 150 μm (0.006 in) at the maximum smokeless flow rate. In special cases, these requirements can be waived for flares that can burn larger size droplets.

(5) To maintain the separation between "cold" flow and "wet" flow, separate separation tanks may be needed. Evaporation facilities can be set up to handle the liquid in the liquid separator. However, if the evaporation facilities are relied upon to prevent the metal temperature from dropping below the minimum design temperature, the long-term reliability of the evaporation facilities should be ensured. Otherwise, low-temperature resistant materials should be used.

(6) The design of the liquid separator should also meet the following conditions:

① All condensate in the flare manifold should be discharged to the liquid separator or receiver, so all pipes should allow the liquid to flow to the liquid separator by gravity.
② The split configuration of the separator inlet pipe should be avoided unless the flow distribution can be ensured. The separator inlet should be equipped with a splitter plate, internal elbow, or baffle to guide the liquid away from

the separator outlet. For more details, see ISO 23251 Petroleum, petrochemical and natural gas industries—Pressure-relieving and depressuring systems or API Std 521 Pressure-relieving and depressuring systems.

③ If the liquid from the closed vent tank is to be transported to the liquid separator, then the size of the liquid separator should be able to accommodate both the maximum liquid volume from the slope pipe and the maximum liquid volume of the closed vent tank. Since this may increase the size of the closed vent tank, it can also be considered to transport the liquid to a low-pressure container.

④ Due to the risk of waxing or scaling, the size of the closed vent tank should not be limited by using a demister.

⑤ The liquid separator should be equipped with an automatic liquid discharge device. The bottom pump of the liquid separator should be able to drain the liquid separator from the highest liquid level to the normal working liquid level or low liquid level within a maximum of 2 h.

⑥ Because the liquid in the liquid separator may be toxic or flammable, or contain dissolved toxic or flammable substances, special care should be taken in design and operation. If there is a risk of releasing toxic substances, the drain pipe should be led to the closed vent system. If there is a risk of material freezing, at least a second series valve is needed.

⑦ Separate facilities for clearing water or heavy hydrocarbons should also be provided. These facilities can be manual or automatic. Special attention should be paid to the danger of releasing flammable or toxic substances into the atmosphere from the drain point.

⑧ Anti-freezing measures should be provided for the liquid separator, and when the metal temperature exceeds 65 °C (150°F), anti-scalding should be set up.

⑨ Facilities for isolating, venting, purging, inspecting, maintaining, and cleaning the liquid separator should be provided.

⑩ Liquids with a temperature exceeding 93 °C (200°F) should not be introduced into a liquid separator containing water or light hydrocarbons to prevent vapor explosion.

⑪ The liquid separator should be equipped with a liquid level gauge, liquid phase thermometer, pressure gauge, high and low pressure and high and low liquid level alarms. The condensate pump should be manually started and a low liquid level interlock stop pump should be set up.

8.7.3 Size Calculation

Determining the size of the liquid separator is usually a trial process, first determining the tank size needed to separate the carried liquid.

(1) The residence time of the gas is greater than or equal to the time required for the droplet to fall through the effective vertical height at the falling speed.

(2) The vertical upward speed of the gas is low enough to allow the droplet to separate when it falls.
(3) The effective height here is generally taken from the distance from the liquid surface. The vertical speed of the gas should be low enough to prevent large liquid slugs from entering the flare.

Since the torch can handle small droplets, the vertical speed allowed in the tank can be determined according to the speed required to separate 300–600 μm diameter droplets. Refer to API Std 521 Pressure-relieving and depressuring systems 5.7.8 to determine the falling speed of droplets in the airflow, calculated by formula (8.4):

$$U_c = 1.15 \times \sqrt{\frac{gD(\rho_l - \rho_v)}{\rho_v C}} \tag{8.4}$$

where

G—gravitational acceleration, taken as 9.81, m/s^2;
D—droplet diameter, m;
Uc—droplet settling velocity, m/s;
ρl—density of droplets under operating conditions, kg/m;
ρv—density of gas under operating conditions, kg/m;
C—drag coefficient of droplets in gas, determined by Fig. 8.3.

The second step in selecting the size of the separator is to consider the impact of the liquid in the tank on reducing the effective volume of gas–liquid separation. The possible sources of this liquid are:

(1) Condensate separated when vapor is released.
(2) Liquid carried out with the release of vapor.

The volume occupied by the liquid should be based on a release duration of 20–30 min. The liquid retained before release (from the pressure relief valve or

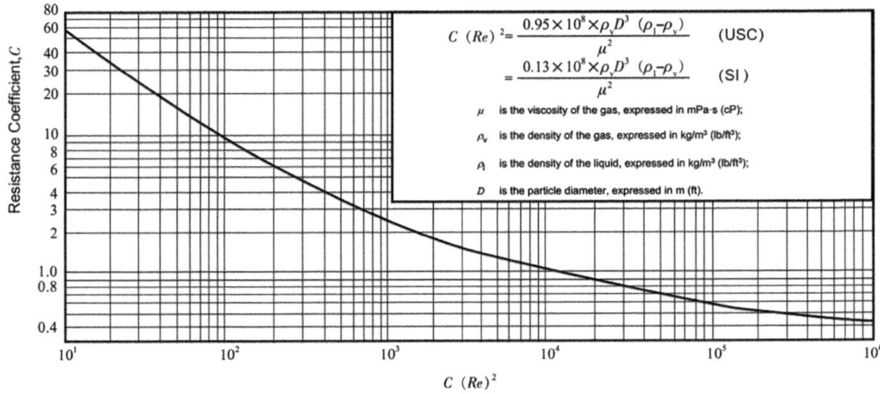

Fig. 8.3 Determination of drag coefficient

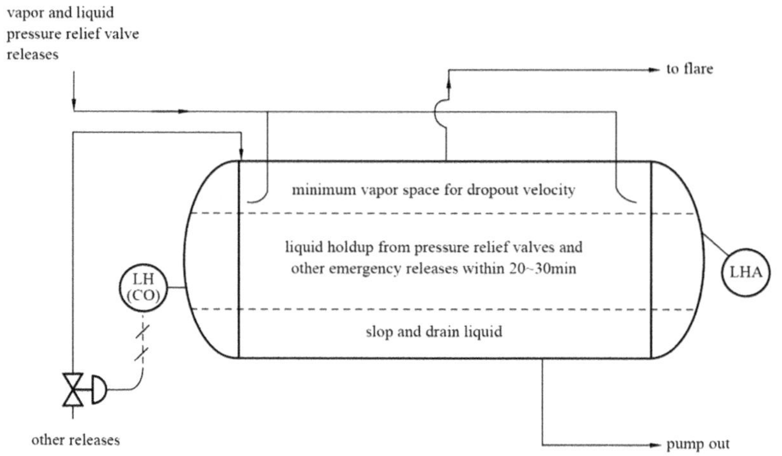

Fig. 8.4 Torch separator

other sources) should be added to the above (1) and (2) liquids to determine the effective gas separation space. However, it is usually not necessary to consider the relationship between volume and gas separation in the following situations, i.e., when the separator is used to accommodate a large amount of liquid released from the pressure relief valve of other systems, there is no significant flash evaporation in the tank, and the liquid can be quickly removed (Fig. 8.4).

In many designs, both horizontal and vertical separators are applicable; when choosing the size of the tank, the economy of the container design should be considered. The economy of the container design may affect the choice between horizontal and vertical tanks. When a large amount of liquid needs to be stored and the gas flow rate is high, the horizontal tank is usually more economical.

8.7.3.1 Trial Calculation of Horizontal Separator Diameter

SH3009 "Design Specification for Combustible Gas Emission System in Petrochemical Industry" provides the calculation steps for the horizontal separator, and the diameter of the separator should be determined by trial calculation according to formula (8.5). When $D\text{sk} \leq D\text{k}$, the assumed Dk is the diameter of the horizontal separator.

$$D_{\text{sk}} = 0.0115 \times \sqrt{\frac{(a-1)q_{\text{v}}T}{(b-1)p\varphi U_{\text{c}}}} \tag{8.5}$$

where

*D*sk—trial calculated diameter of horizontal separator, m;

a—ratio of liquid level height to tank diameter in the tank;

qv—inlet gas flow rate (standard state), m^3/h;

T—temperature of discharged gas, K;

b—ratio of liquid cross-sectional area to total cross-sectional area in the tank;

p—gas pressure under operating conditions (absolute pressure), kPa;

φ—coefficient, should be taken as 2.5–3.0;

Uc—Droplet settling velocity, m/s.

The distance between the inlet and outlet of the horizontal separator is calculated by formula (8.6).

$$L_k = \varphi D_k \tag{8.6}$$

where

Lk—Distance from gas inlet to outlet, m;

Dk—Assumed diameter of the separator, m.

The droplet settling velocity is calculated by formula (8.4).

The ratio of the liquid cross-sectional area inside the tank to the total cross-sectional area of the tank, b, is calculated by formula (8.7).

$$b = 1.273 \times \frac{q_l}{\varphi D_k^3} \tag{8.7}$$

where

q—Amount of condensate stored in the separator, m^3.

The ratio of the liquid level height inside the tank to the tank diameter, a, is calculated by formula (8.8).

$$a = 1.8506b^5 - 4.6265b^4 + 4.7628b^3 - 2.5177b^2 + 1.4714b + 0.0297 \tag{8.8}$$

The gas density under operating conditions is calculated by formula (8.9).

$$\rho_v = \frac{1000Mp}{RT} \tag{8.9}$$

where

R—Gas constant, 8314 N·m/(kg·K);

M—Relative molecular mass.

The drag coefficient C of the droplet in the gas is determined by $C(Re)^2$ from Fig. 8.3, and $C(Re)^2$ is calculated by formula (8.10).

$$C(Re)^2 = \frac{1.307 \times 10^7 D^3 \rho_v (\rho_l - \rho_v)}{\mu^2} \tag{8.10}$$

where

μ—Gas viscosity, mPa·s.

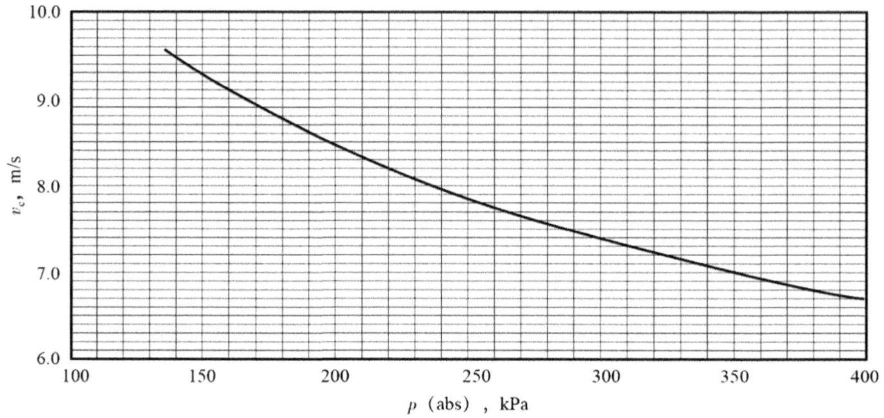

Fig. 8.5 Critical flow speed of horizontal gas flow inside the separator

8.7.3.2 Diameter Calculation of Horizontal Separator

After calculating the diameter of the horizontal separator by formula (8.5), it should be checked by formula (8.11), and the diameter of the separator should meet the check result of formula (8.11).

$$\text{Diameter of horizontal separator} \geq 1.13 \times \sqrt{\frac{q}{v_e} + \frac{q_1}{\varphi D_k}} \tag{8.11}$$

where

q—Volume flow rate of inlet gas under operating conditions, m^3/s;

vc—Critical flow speed of horizontal gas flow inside the separator, m/s, which can be found from Fig. 8.5.

8.7.3.3 Diameter Calculation of Vertical Separator

SH3009 "Design Specification for Emission System of Flammable Gas in Petrochemical Industry" provides the trial calculation formula for the diameter of the vertical separator, which should be determined by trial calculation according to formula (8.12).

$$D_k = 0.0128 \times \sqrt{\frac{q_v T}{p U_c}} \tag{8.12}$$

Generally, the cross-sectional area of gas flow above the highest liquid level inside the horizontal separator (along the radial direction of the tank) should be greater than or equal to 3 times the cross-sectional area of the inlet pipe; the height of the gas phase space inside the vertical separator should be greater than or equal to the diameter of the separator and not less than 1 m; the highest liquid level should be greater than or equal to the diameter of the inlet pipe and not less than 0.3 m. The type of separator should be selected based on the economy of the vessel and flare gas emission system design. When using a horizontal separator, the ratio of its length to diameter should be between 2.5 and 6.0 [2].

References

1. API Standard 521 Pressure-relieving and depressuring systems, Seventh Edition, June 2020.
2. Sinopec Shanghai Engineering Co., Ltd. Chemical Process Design Manual. Beijing: Chemical Industry Press, 2018.

Chapter 9
Flare System

To ensure the stability and safety of the production device system, and to make the exhaust operation affect the normal production operation of the device itself and its surroundings as little as possible, and to protect the surrounding ecological environment, it is necessary to set up a flare system and flare gas recovery system in large oil and gas field station devices to handle the flare gas emitted from various production devices.

Emission operations are mainly divided into atmospheric emissions (cold venting) and flare venting systems (hot venting). This book mainly describes the flare venting system. In many cases, if local environmental laws and regulations allow, pressure relief sources can be safely discharged directly into the atmosphere. As long as it is feasible, atmospheric emission (cold venting) is the simplest, most reliable, and economical method among all treatment methods. When deciding whether to emit hydrocarbons or other flammable or dangerous vapors into the atmosphere, the following issues must be paid special attention to:

(1) Ensure that treatment is achieved without forming a flammable mixture on the ground or elevated structures.
(2) Personnel exposure to toxic gases or corrosive chemicals.
(3) Potential dangers or other problems such as ignition of the vented gas flow at the discharge point, excessive noise, and air pollution. The purpose of the flare system is to convert flammable, toxic, or corrosive gases into less harmful products through combustion. The endpoint of the flare system is the release point of hydrocarbon gases in the environment. The flare system takes into account the three major functions of stable production, safety, and environmental protection. Although people regret that the flare burns a large amount of flammable gas and hope to use these gases and remove the flare system, because the flare gas emission volume varies greatly, from almost zero to hundreds of tons per hour, and the gas composition also changes greatly, it is difficult to recover and use all these gases, so at this stage, the flare should be regarded as an organic part of the production process (Fig. 9.1).

K. Fang, *Onshore Oilfield Surface Facilities: Process Safety Engineering Guide*,
https://doi.org/10.1007/978-981-96-7830-3_9

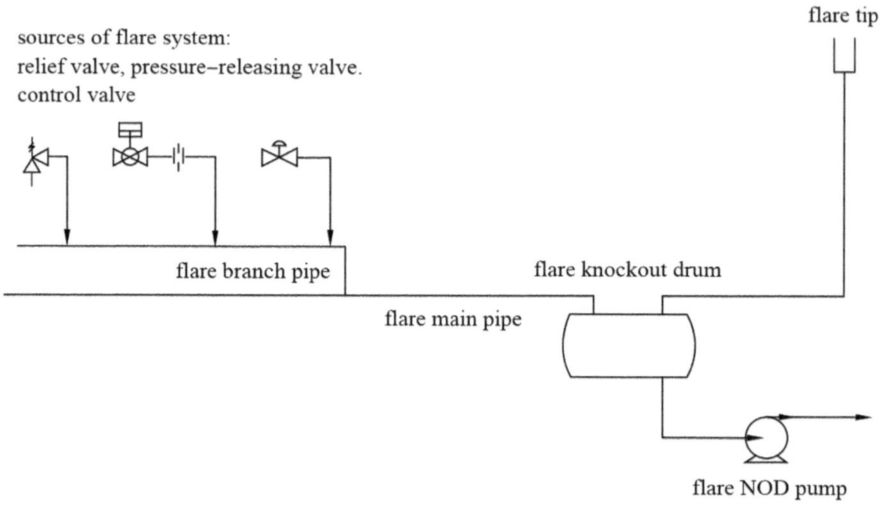

Fig. 9.1 Composition of the flare system

9.1 Classification, Composition, and Selection of Flare Systems

9.1.1 Classification of Flares

There are many types of flares, which can be divided according to different classification methods. The industry generally has the following classifications:

(1) According to the installation method between the burner and the ground, it can be divided into elevated flares and ground flares.

① Elevated flares refer to flares where the flare head is installed above the ground at a certain height to reduce heat radiation intensity and facilitate diffusion.

② Ground flares can be divided into open ground flares and enclosed ground flares:

- Open ground flares refer to flares where multiple burners are arranged in an array within a ventilated fence and burn in stages.
- Enclosed ground flares refer to flares with a combustion chamber and chimney, and one or more burners are set up in the combustion chamber.

(2) Flares can be divided into smokeless flares and non-smokeless flares according to whether they need to eliminate smoke. The difference between smokeless flares and non-smokeless flares is that the color of the flame during combustion is different. The flame of the non-smokeless flare is long and red, sometimes accompanied by a small amount of black smoke, while the flame of the smokeless flare is short and blue, and even invisible in the light.
(3) High-temperature venting flares and low-temperature venting flares are classified according to the temperature of the venting air.
(4) Single-point combustion flares and multi-point burner flares are classified according to the selection form of the flare burner.
(5) Cold venting flares and hot venting flares are classified according to whether the emitted gas is burned after venting.

In addition, according to whether it can be disassembled, elevated flares can also be divided into detachable flares and non-detachable flares, and there are other classification methods for elevated flares.

9.1.1.1 Elevated Flares

Elevated flares consist of a self-control system, ignition system, steel structure support, and an upright rising pipe. The flame is far from the ground and is ignited and burned remotely at the top. The height of the high-altitude combustion tower can be adjusted, from a minimum of 5 m to a maximum of 200 m. According to the different structural support methods of the elevated flare, the elevated flare is also divided into self-supporting, tensioned support, and tower support, as shown in Fig. 9.2. Since most of the oilfield station fields use elevated flares, this book will focus on elevated flares, please refer to the fourth section of this chapter for details.

9.1.1.2 Ground Flares

The ground flare consists of a ground combustion furnace, ground combustion furnace pillar, ground burner, windproof muffler wall, staged combustion system, and a long-burning lamp automatic ignition device, safety measures, control system, and venting system. The combustion of flare gas is completed within the cylindrical ground combustion furnace. The combustion process is enclosed, invisible to the outside world, without light pollution, and low heat radiation. A certain number of specially structured ground burners are installed in the cylindrical ground combustion furnace. The ground burner adopts a plum blossom-shaped porous structure, which can divide a large amount of flare gas into many small strands, which is beneficial for its mixing with air, increasing the contact area with air, and achieving smokeless combustion. The mixing of air and flare gas mainly relies on the pressure of the flare gas itself and the specially designed burner. The flare head of the ground

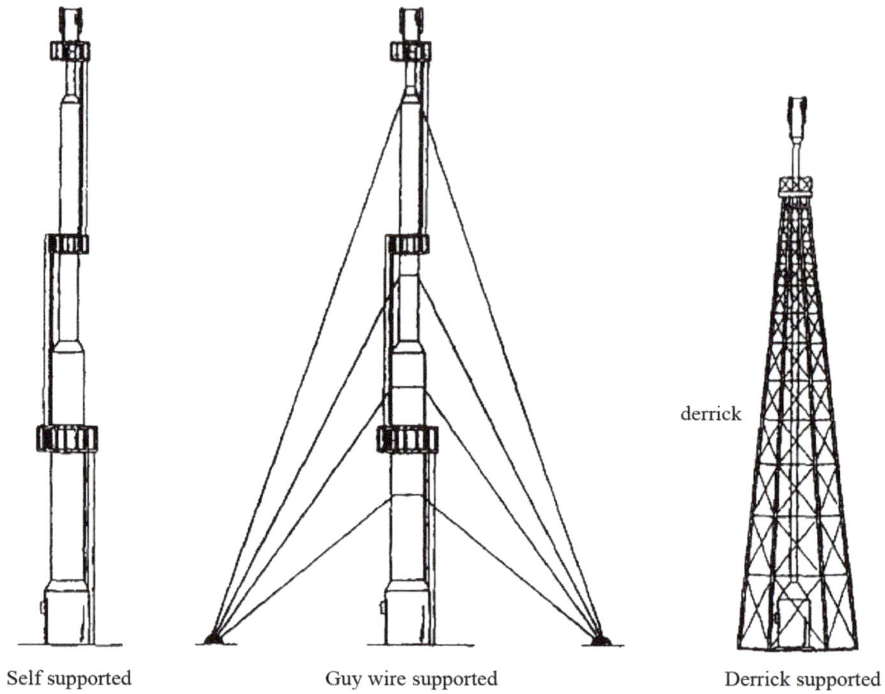

Fig. 9.2 Elevated flare

flare is divided into ordinary combustion type and steam-assisted combustion type (Fig. 9.3).

1. Ground Flare Design Principles

Ground flares can be used to treat flammable gases that are mildly toxic and non-toxic, and should not be used to treat moderately toxic flammable gases. Ground flares must not be used to treat extremely or highly toxic flammable gases. Ground flares should be used to treat flammable gases emitted during start-up and normal production, and should not be used to treat flammable gases emitted in emergencies.

Fig. 9.3 Ground flare

The operating pressure of each staged pipeline should be determined according to the maximum allowable discharge back pressure value of the total discharge pipe before each staged pipeline. The bursting pressure of the bypass of the staged control valve should not be higher than the maximum allowable discharge back pressure of the total discharge pipe. When the bypass of the staged control valve uses a bursting needle valve, the maximum operating pressure should be 90% of the maximum allowable discharge back pressure of the total discharge pipe; when the bypass of the staged control valve uses a bursting piece, the maximum operating pressure should be 75% of the maximum allowable discharge back pressure of the total discharge pipe. The operating pressure of each stage should be within the optimal operating elasticity range of the burner, avoiding jumps between stages; the maximum allowable discharge back pressure value and the number of stages of the total discharge pipe before each staged pipeline should be determined through economic comparison based on factors such as the investment of the total discharge pipe, the staged system, and the consumption of public engineering media.

The control valves on each staged pipeline should be equipped with a bursting needle valve or bursting piece bypass, and the bursting pressure of the bursting needle valve or bursting piece should not be higher than the maximum allowable back pressure before each staged pipeline. When the maximum allowable back pressure value before each staged pipeline is relatively low, a bursting needle valve should be selected for the bypass. The sum of the cross-sectional areas of each staged pipeline should not be less than the cross-sectional area of the total discharge pipe. The nominal diameter of the bursting needle valve or bursting piece bypass can be one level smaller than the staged pipeline, but it should ensure that the pressure before each staged pipeline is less than or equal to the maximum allowable back pressure.

The pressure switch valves on each staged pipeline should use metal hard-sealed butterfly valves, and their opening time should not be more than 1 s. The leakage level of the pressure switch valve and the bursting needle valve on the bypass of each staged pipeline should not be lower than ANSI level V. In addition to the step-by-step opening function, the control system should also have a cross-level opening function. Except for the first two stages of the emission system, each burner is equipped with a long-burning lamp, the number of long-burning lamps in other stages should not be less than two, and the long-burning lamp should be kept burning; each burner of the steam-assisted combustion type burner needs to be equipped with a long-burning lamp.

The flare should take sufficient smoke elimination measures, and the smoke emission should comply with relevant environmental protection requirements. A nitrogen purge system should be installed after each staged pressure switch valve of the ground flare. The continuous burning system should have a continuous nitrogen purge system to prevent backfire. For the low-pressure level emission system, a steam-assisted combustion burner should be used, and the steam should be adjusted according to the emission volume and relative molecular mass of the flare gas. Condensate closed discharge facilities should be set up before and after each staged pressure switch valve.

The allowable heat radiation intensity of the ground flare to the surrounding area is the same as the requirements of the elevated flare.

2. Enclosed Ground Flares

The processing capacity of a single set of enclosed ground flares should not be more than 100 t/h.

The height of the exhaust pipe should meet the following requirements:

(1) The diffusion of flue gas should meet environmental protection requirements.
(2) It should not be less than three times the height of the burner flame.

The heat flux density in the combustion chamber should be controlled at 275–335 kW/m^3 (Standard state). The design should use burners that are resistant to coking, blockage, and high temperature deformation. Oxygen deficiency in the center of the combustion chamber should be avoided. The arrangement of the burners should ensure balanced pressure to prevent flame blowback and flame flashover. The inside of the combustion chamber should be lined with fire-resistant protection, and the outside temperature of the combustion chamber should not exceed 60 °C.

3. Open Ground Flare

The height of the heat radiation metal fence should be higher than the top of each burner flame by 2 m. Low-pressure burners should be arranged in the middle of the heat radiation metal fence, and high-pressure burners should be arranged on both sides. The graded pipelines inside the heat radiation metal fence should take heat radiation protection measures, and a fire observation window and maintenance door should be set on the side near the grading control valve. The installation distance of the burners on the same level pipeline should ensure relay ignition, and the distance between different level pipelines should meet the requirements of smokeless combustion. The distance between the burner arranged by the radiation metal fence and the metal fence should ensure that the flame cannot directly burn the metal fence. The burner and support pipe should be made of high-temperature resistant metal materials.

9.1.2 Composition of the Flare

The flare system includes the following parts (may include all or only some):

(1) Flare, including:

   ① Flare head or burner; The main function of the flare head is to burn the flare emission gas from the venting system as clean as possible according to the design requirements, without raining fire, emitting black smoke, or extinguishing. Depending on the composition of the treated emission gas and the type of flare used, the flare head is mainly divided into the following types:

- Elevated flare: steam smoke elimination type, heat diffusion type, acidic gas heat storage type.
- Ground flare: ordinary combustion type, steam-assisted combustion type.

② Flare body.
③ Tower (elevated flare) or wall (ground flare), flare tower support system.
④ Pilot light, pilot light igniter, and pipeline.

(2) Ignition system.
(3) Flame monitoring (for example: monitoring and pilot light igniter).
(4) Backfire prevention system.
(5) Purging system.
(6) Smoke elimination facilities, including:

① Air-assisted combustion.
② Steam-assisted combustion.
③ Water spraying.
④ Gas-assisted ignition (for low calorific value flare fluid).
⑤ Flare extinguishing.

(7) Oxygen analyzer.
(8) Flow, temperature measurement, and alarm.
(9) Flare gas recovery system.
(10) Liquid seal tank.

9.1.3 Selection of the Flare

The selection of the flare type should consider the following factors:

(1) The nature, frequency, and quantity of the discharge.
(2) Available space.
(3) Impact on surrounding vegetation and residents.
(4) Environmental requirements for smoke, pollution, noise, radiation, and light emission.

9.1.3.1 Site Size and Layout

The flare is a facility with open flames. In engineering practice, the height of the elevated flare is often determined based on the fire spacing determined by domestic and foreign standards, HSE requirements, and safety analysis, as well as the allowable radiation intensity for people or equipment. The fire spacing of the ground flare should be calculated based on the allowable radiant heat intensity for people or equipment, and should not be less than the fire spacing determined by domestic and foreign standards, HSE requirements, and safety analysis. Therefore, for open ground flares, because their burners are close to the ground, they do not have an advantage in terms of land area compared to elevated flares. However, enclosed ground flares can effectively block radiant heat with their combustion chambers and chimneys. While ensuring the safety of people or equipment, their land area is much smaller than that of elevated flares or open ground flares, making them suitable for situations with limited construction sites.

9.1.3.2 Characteristics of the Discharge Medium

The characteristics of the discharge medium include composition, discharge volume, and discharge pressure, etc.

Ground flares can be used to handle mildly toxic and non-toxic flammable gases, but are not suitable for handling moderately toxic flammable gases, and must not be used to handle extremely or highly toxic flammable gases. To ensure stable combustion and handling flexibility, ground flares have a staged combustion system, and the back pressure of the total discharge pipe in front of each stage must meet the pressure difference requirements between the combustion systems. The processing capacity of a single set of enclosed ground flares should not exceed 100 t/h.

Elevated flares do not have these limitations. For acidic gases containing high levels of hydrogen sulfide emitted from desulfurization towers, acidic water vapor extraction devices, etc., in gas treatment plants, acidic gas flares should be set up for separate treatment. The combustion rate of hydrogen sulfide needs to be greater than 99% to reduce environmental pollution. To ensure complete combustion of hydrogen sulfide and consider the ground concentration of sulfur dioxide, such acidic flare gas needs to be burned and discharged through the elevated flare system: as the flare height increases, the ground concentration of sulfur dioxide decreases.

9.1.3.3 Economy

Because ground flares consist of multiple burners, multi-stage control valves, bypasses, control systems, safety interlock systems, combustion chambers (for enclosed ground flares) or radiation shielding metal fences (for open ground flares), the initial investment and operation and maintenance costs are generally higher than those of elevated flares with the same processing capacity.

9.1.3.4 Safety

Ground flares can be used to handle flammable gases emitted during start-up, shutdown, and normal production, but are not suitable for handling flammable gases emitted during emergencies.

9.1.3.5 Public Relations

If elevated flares are set up near residential areas or navigable waterways, the light, heat radiation, and noise emitted by the burning of the elevated flares may affect the lives of residents or the correct navigation of ships.

9.1.3.6 Geographic Environment

For factories located in low-lying areas, using elevated flares is more conducive to the diffusion of combustion exhaust gases.

9.2 Flare System Design Guidelines and Process

9.2.1 Design Guidelines

The endpoint of the flare venting system is the release point of hydrocarbon gases in the environment. The height and position of the flare or vent pipe need to consider multiple factors:

(1) The acceptable radiation heat value for equipment and personnel when the gas is vented and burned.
(2) The lower explosion limit of the cold diffusion of the vented hydrocarbon gas that has not been burned.
(3) The ground concentration when the vented fluid contains the following components:

   ① Nitrogen oxides.
   ② Hydrogen sulfide (unburned) or sulfur dioxide (combustion product).
   ③ Carbon monoxide.
   ④ Unignited aromatic compounds (BTEX).

(4) The pilot light in the factory is a potential ignition point, and the lower explosion limit curve needs to be evaluated based on the venting volume within the station, to determine the placement of the flare.
(5) Noise restrictions (considering on-site and third-party contact). Although customers and stations have different specific standards, and the nature of the vented fluid and the impurities it contains are also different, the design of the flare should comply with the API Std 521 Pressure-relieving and depressuring systems standard. These specific standards should be established at the beginning of the design and annotated in the design basis of the flare system.

9.2.2 Design Process

With the increasing scale of oil and gas fields and the increasing complexity of production, the huge emissions have become a major challenge for the flare system, which puts forward new requirements for the construction of flares. In the flare device, compared with the elevated flare, the ground flare has different degrees of advantages in combustion, noise control, operation and maintenance, but the ground flare is currently affected by the emission volume, generally for small volume flares, and used for relatively single gas emissions.

From an economic and energy-saving perspective, establishing a shared elevated flare has become a trend. In addition, according to production needs, such as further expanding the production scale, it will definitely increase the venting volume of the production device, greatly increasing the risk of flare operation, and putting forward higher requirements for the reasonable configuration of the flare system. There are many available flare designs to choose from, and screening should be done in the early stages of the project. Regarding the design of the combustion pit, unless specifically required, it should not be considered as an available type of flare, because its combustion efficiency is poor and risk management is more challenging. Figure 9.4 shows the design process of the flare system, and Table 9.1 lists the advantages and disadvantages of various flares [1].

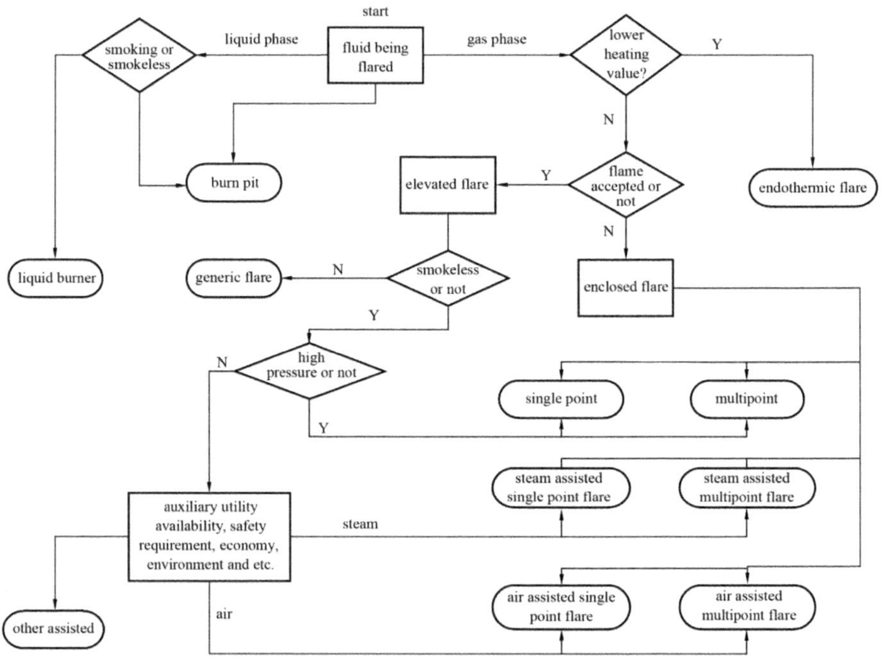

Fig. 9.4 Flare design process

Table 9.1 Analysis of the advantages and disadvantages of various torches

Option	Advantages	Disadvantages
Elevated torch	• Low investment • Simple construction • Best diffusion effect • Large emission, suitable for emergency venting	• High visibility • High flow rate → high noise
Enclosed ground torch	• Flame is invisible • Low noise • Small footprint	• Highest investment • Visible high enclosure wall
Open ground torch	• Medium investment • Can reduce visibility • By using more low-flow, low-speed burners, and radiation shields, noise can be reduced	• Complex process, instrumentation system and control • Larger footprint • Higher protective fence
Pit torch	• Lowest cost • Simple construction	• Only horizontal diffusion • Medium footprint • Very loud noise

9.3 Elements of Torch System Design

9.3.1 Considerations for Torch System Design

The design of the torch system should pay attention to the following points:

(1) Under normal operating conditions, continuous venting should be minimized.
(2) Safety of nearby personnel should be considered under all circumstances, including the full or partial start-up, purging, operation and emergency venting, shutdown, inspection and maintenance of the system.
(3) Protection of devices and equipment near the torch system, including the surface of the torch system itself, should be considered.
(4) Prevent the torch system from being damaged by external events (such as fires).
(5) The torch system should be inherently safe, especially in the following aspects:

① Flammable or explosive mixtures/blockages or flow restrictions.
② Chemical reactions/toxic components.
③ Corrosion, erosion and hydrogen embrittlement/equipment damage.
④ Torch flame stability/ignition safety.
⑤ Safety of the ignition process/switching of different torches.

(6) Attention should be paid to the flow rate, composition, molecular weight, temperature, frequency and duration of any process medium discharged to the torch system, as well as factors affecting flow, such as allowable back pressure and solid deposition. Special attention should be paid to the flow rate when the torch system is depressurized, especially the flow rate when all pressure relief

valves open simultaneously due to fire or utility system failure (if all pressure relief valves are designed to fail open).

(7) The materials used in the torch system should be suitable for operation at the system's lowest temperature and be able to withstand any sudden cooling caused by depressurization.
(8) The service life of the torch system.
(9) The requirements for maintenance and inspection of the torch system and the impact of maintenance and inspection on the entire plant (requirements for backup torch systems).
(10) On-site meteorological and other environmental conditions.
(11) National and local regulations, especially those related to smokeless combustion, torch visibility, pollution and noise restrictions.
(12) Systems for treating condensate hydrocarbons and acidic water from separation tanks and sealed tanks.
(13) The space and layout of the torch system on the master plan.
(14) Systems with liquid/vapor vaporization and heating.
(15) If inert gas needs to be discharged to the torch system during operation or maintenance, consider that the inert gas may cause the pilot light to go out. At this time, fuel gas needs to be supplemented to ensure that the calorific value of the torch gas is at least 11200 kJ/m^3 (300Btu/ft^3, standard state). It should be noted that low-flow inert gas does not need to be supplemented with additional fuel gas, such as purge gas.
(16) The flow rate of the torch head gas should meet the requirements for flame stability, noise and diffusion.

According to experience, the flow rate of the torch head gas is allowed to be above Mach 0.2. However, if it exceeds Mach 0.2, it needs to be confirmed with the torch manufacturer. For emergency torch venting, the upper limit of the flow rate is generally Mach 0.5. If this flow rate is exceeded, the flame may become unstable and rise, which will increase the risk of flame extinction. For sonic torches, a Mach number above 1.0 is allowed, provided that the flame is stable.

9.3.2 Hazards to be Considered in Torch System Design

The design of the torch system should consider the following possible hazards:

(1) Air may enter the torch system through the following paths, forming a flammable and explosive mixture:

   ① The downward flow of gas caused by buoyancy effect, loss of purge gas, molecular seal failure.
   ② If the steam in the torch system condenses or cools (the whole field stops production), air will be sucked into the torch system from the torch head or other venting and purging ports. Because the entire torch system has a large

heat absorption capacity, when the torch gas cools, the volume shrinkage will be large and very rapid.

③ If it rains, the cooling of pipes and tanks originally exposed to the sun by rainwater may also cause air to be sucked into the torch system.

④ Due to the buoyancy of lighter gases, the gas pressure in the lower position of the pipeline may be lower than atmospheric pressure, which will also cause air to enter the torch system through any openings, such as venting ports, purging ports, flanges, etc.

⑤ The vacuum system connected to the torch may also cause the torch system to suck in air, so a high-integrity isolation cut-off mechanism is needed to prevent such accidents.

⑥ Air in the process may also enter the torch system. The cause of such accidents may be control failure of the oxidation device, or air entering the purge gas.

(2) Blockage/flow obstruction:

① Ice formation in liquid seals, condensate in torch pipelines or molecular seals, steam condensation, freezing of torch tips when steam flow is low in winter, low ambient temperature, low temperature emissions or self-cooling are all potential causes of steam pipeline blockage or obstruction.

② Polymer products, hydrates, wax, corrosion products.

③ Solids, catalysts, polymers, etc. brought in from the production system.

④ Liquid retention caused by poor liquid discharge design or liquid level control failure.

⑤ Valve mis-closure or failure to close.

(3) Toxic components: It is advisable to set up separate torch pipelines for gases with a volume fraction of H_2S exceeding 10% or containing other highly toxic substances, and it is best to combine them with the main torch gas flow near the torch head, which can reduce the possibility of H_2S corrosion in the main torch pipeline. The wastewater produced by torch seals, drain pipes, etc. should also be carefully considered.

(4) Chemical reactions in the torch system, self-igniting scale (pyrophoric scale), acetylides, peroxides, etc.

(5) Mechanical damage, hydraulic shock of liquid blockage, propulsion of solid ice, hydrates, shock, low temperature brittleness caused by sudden cooling, external fire damage, backfire of torch head, high temperature gas discharge into torch system.

(6) Liquid accumulation in the torch system: Liquid discharge to the torch system should be avoided, if it cannot be prevented, the potential hazards of liquid entering the torch system should be carefully considered. During emergency venting, a large amount of liquid discharged into the torch may cause slug flow in horizontal pipelines, liquid entering the separator, and may also cause mechanical damage.

9.4 Elevated Torch Design

The elevated torch consists of a self-control system, ignition system, steel structure support and an upright riser. The flame is far from the ground and is ignited remotely at the top. The height of the elevated combustion tower can be adjusted, from a minimum of 5 m to a maximum of 200 m.

9.4.1 Radiant Heat

The size of the safety zone of the torch facility depends on the allowable heat radiation intensity. The principle of determining the maximum emission of torch gas considers a certain safety factor, and the maximum emission duration usually does not exceed 30 min. The highest heat radiation intensity of the sun in a day is 0.79–1.04 kW/m^2, and is greatly affected by the weather; during the start-up and shutdown of the device, due to unstable operation or the inability of downstream devices to start synchronously, a large amount of flammable gas will be continuously emitted to the flare for several days. Therefore, whether the heat radiation of the sun is superimposed on the heat radiation produced by the flare should be treated differently under different conditions.

The following are the setting requirements for radiation intensity:

(1) When determining the safety zone of the flare facility according to the maximum emission load, the heat radiation intensity is allowed not to consider the solar heat radiation intensity.
(2) When calculating the safety zone of the flare facility according to the emission load during the start-up and shutdown of the device, the allowable heat radiation intensity under this condition should consider the solar heat radiation intensity.
(3) In areas where the public, such as residential areas outside the factory, public welfare facilities, villages, etc., are active, the allowable heat radiation intensity should be less than or equal to 1.58 kW/m^2.
(4) The allowable heat radiation intensity in the densely populated areas of neighboring similar enterprises and oil depots, and the administrative management areas within petrochemical enterprises should be less than or equal to 2.33 kW/m^2.
(5) The allowable heat radiation intensity in the sparsely populated areas of neighboring similar enterprises and oil depots, and outside the factory trees and other vegetation should be less than or equal to 3.00 kW/m^2.
(6) The allowable heat radiation intensity of each production device inside the petrochemical factory should be less than or equal to 3.20 kW/m^2.
(7) The allowable heat radiation intensity at the top platform of the flare tower during maintenance should not be greater than 4.73 kW/m^2.
(8) The allowable heat radiation intensity in the layout area of the flare facility's liquid separator, water seal tank, pump, etc. should be less than or equal to

9.00 kW/m^2. When the heat radiation intensity in this area is greater than 6.31 kW/m^2, there should be a safe place for operators or maintenance personnel to hide.

The allowable heat radiation intensity is a function of exposure duration, and it should include factors such as human reaction time and flexibility. In API Std 521 Pressure-relieving and depressuring systems, it is recommended to consider the total exposure time of operators or maintenance personnel as 8–10 s. When the heat radiation intensity is greater than 6.31 kW/m^2, the operators or maintenance personnel in this area do not have enough time to escape, so a safe hiding place should be set up on the spot. The safe hiding place can be buildings such as the cabinet room within a 60 m range nearby, or a specially set up shed to shield radiation heat.

9.4.2 Design Requirements for Flare Head and Auxiliary Facilities

9.4.2.1 General Requirements

The main function of the flare head is to burn the flare emission gas from the relief system as clean as possible according to the design requirements, without rain of fire, black smoke, or extinguishing. According to the composition of the treated emission gas and different application occasions, the flare head of the elevated flare is mainly divided into: ordinary pipe type, steam smoke elimination type, air combustion type, acidic gas heat storage type, and sonic flare head.

1. Basic Requirements for Flare Head Design

The basic requirements for flare head design are as follows:

(1) Can safely burn off flare gas of various conditions (referring to different flow rates, different parameters, and different compositions).
(2) Completely burn the flare gas, and the pollution of the combustion products to the surrounding environment complies with relevant regulations.
(3) On the premise of ensuring the complete combustion of the flare gas, the energy consumption (steam, electricity, or water) is low.
(4) The structure is simple, easy to manufacture, the material selection is appropriate, the service life is long, the weight is light, and it is easy to install and maintain.
(5) The noise and light pollution produced during combustion are small. The design of the flare head should consider reducing the risk of the flame rising into the air under the condition of high-speed discharge or gusts combined with low-speed discharge. The flare head should meet the requirements of smokeless combustion during normal operation and start-up and shutdown of the device. The flare head is equipped with a pilot light. After the pilot light is ignited by the igniter, it will continue to burn. When the emission gas reaches the flare head,

it is immediately ignited by the pilot light. The flare head is installed at the top of the flare tower, which can reduce the heat radiation to the environment and the range of toxic diffusion.

2. Setting Requirements for Flare Head

The setting requirements for the flare head are as follows:

(1) A flame baffle is set at the top of the torch head to improve the stability of the flame during high-speed emission. Its flow restriction area is usually 2% to 10%. The torch head equipped with this flame stabilization device allows the outlet pressure drop to reach 14 kPa. When it exceeds 14 kPa, the stability of the flame is difficult to guarantee.
(2) The upper 3 m part of the torch head (including internal parts) should be made of ANSI 310SS or equivalent material, and the part below 3 m should be made of 304 or equivalent material. At the same time, the design temperature of the upper part of the torch head should not be lower than 1200 °C.
(3) The allowable Mach number of the exhaust gas at the outlet of the torch head depends on the allowable pressure drop of the system, environmental noise standards, flame stability, and combustion characteristics of the gas. For torches with low system emission pressure and strict environmental noise requirements, the Mach number should be controlled below 0.5 during short-term accident emissions, and continuous or frequent emissions during normal factory production should be maintained at a Mach number of 0.2. For systems with high emission pressure and less strict environmental noise requirements, it is suitable to use sonic speed. When the gas velocity at the outlet of the torch head is too low, the flame is greatly affected by the wind, and the flame may fall several meters along the torch head in the low-pressure area downwind, causing the torch head to overheat and corrode. Data published in torch research literature shows that the Mach number for stable combustion of the torch is 0.2–0.5.
(4) For acid gas torches, the main concern is the burnout rate of toxic and harmful substances in the gas. The main acid gas in petrochemical enterprises is hydrogen sulfide gas. At present, the design of acid gas torches generally adopts the method of low speed and maintaining an appropriate combustion temperature. It can also use a high-speed torch head with a Mach number above 0.5 to fully mix the acid gas with air to meet the burnout rate requirements of hydrogen sulfide.
(5) A certain distance should be maintained from the outlet of the torch head to the top platform of the steel tower to avoid damage to the top platform of the torch tower under the action of wind during low emission conditions. The temperature is high when hydrocarbon compounds burn, and the temperature is relatively low when acid gas, pure hydrogen gas, and other low calorific value gases burn.

9.4.2.2 Auxiliary Facility Design Requirements

The design requirements for the auxiliary facilities of the torch body are as follows:

(1) The auxiliary design of the steel tower should meet the following requirements:

① Ladder platforms should be set up in sections. When using straight ladders, each section should be 5–10 m high.
② The steel tower should be equipped with aviation obstruction lights in accordance with relevant regulations.
③ The top platform should have an area and passage for torch head maintenance, and facilities for hoisting the torch head should be provided.

(2) The installation of process thermodynamic pipelines laid on the steel tower or torch body should meet the following requirements:

① Steam pipelines, insulated heating pipelines, ignition tubes, and fuel gas pipelines should be designed with thermal compensation measures and corresponding fixed brackets.
② Process thermodynamic pipelines laid on the steel tower or torch body should not have liquid accumulation points.
③ At least one fixed bracket should be set for normal temperature pipelines.
④ Ignition tubes and fuel gas pipelines at the bottom of the torch should be connected to horizontal pipelines with tees, and flanges and flange covers should be set at the end of the vertical pipelines.

(3) The distance from the outlet of the torch head used for burning hydrocarbon compounds to the top of the steel tower should not be less than 7 m, and the distance from the outlet of the torch head used for burning acid gas, pure hydrogen gas, and other low calorific value gases to the top of the steel tower should not be less than 5 m.
(4) The bottom of the torch body should have space for accumulating rainwater, condensate, and rust, and should be equipped with a manhole, sewage outlet, condensate outlet, and level gauge.

9.4.2.3 Torch Calculation

1. Torch Diameter

API Std 521 Pressure-relieving and depressuring systems recommend using formula (9.1) to calculate the torch diameter (inner diameter) based on the determined Mach number:

$$Ma = 3.23 \times 10^{-5} \times \left(\frac{q_{\mathrm{m}}}{p_2 d^2}\right) \cdot \left(\frac{ZT}{kM}\right)^{0.5} \tag{9.1}$$

where

Ma—Mach number;

q_m—mass flow rate, kg/h;

p_2—Flow pressure of the torch head (absolute pressure), kPa;

d—Torch diameter (inner diameter), m;

Z—Compression factor;

T—Flow temperature, K;

M—Average relative molecular mass of the steam;

k—Specific heat ratio of the gas.

The torch diameter calculated according to the specifications is the effective cross-sectional area of the torch head outlet. The actual area of the torch head outlet should also include the equivalent area of other internal components, which is considered by the torch head supplier. The diameter of the torch body should be finally determined by the allowable pressure drop. When emission pipes of different pressures are connected to the same torch body, the operating conditions of simultaneous emissions from different pressure systems should be calculated to ensure that the emissions from the lower pressure system are not hindered.

2. Torch Height

The principles for determining the torch height are as follows:

(1) The maximum allowable heat radiation intensity of the receiving points in different ground areas must meet domestic and international specification requirements.
(2) Calculate the noise generated on the ground during torch emission based on the height, and calculate the torch height according to domestic and international environmental noise limits.
(3) Calculate the ground concentration of flammable and toxic gases based on the torch height, and calculate the torch height according to domestic and international atmospheric pollutant emission standards.

(1) Minimum Distance from the Center of the Torch to the Exposed Target

API Std 521 Pressure-relieving and depressuring systems recommend using the equation derived by Hajek and Ludwig to determine the minimum distance from the torch to the target that must limit radiant heat, see formula (9.2):

$$D = \sqrt{\frac{\tau F Q}{4\pi K}} \tag{9.2}$$

where

D—The minimum distance from the center of the torch to the considered target, m;

τ—Heat intensity conduction coefficient;

F—Heat radiation coefficient;
Q—Released heat, kW;
K—Allowed radiant heat, kW/m^2.

(2) Determine the Center of the Flame

The main methods to determine the center of the flame are as follows: API 521 Simple method, Brzustowski's and Sommer's method, and G. R. Kent (GPSA) method, etc.

(1) API 521 Simple method

The API 521 Simple method has been used for many years since it was published in API Std 521 Pressure-relieving and depressuring systems in 1969. This method is based on a logarithmic coordinate graph of flame length and gas low calorific value made from several sets of actual measurement data, and an approximate relationship graph of lateral wind at the torch body outlet and flame deformation. The position of the flame center point is obtained through the graphical method, and the height of the torch is calculated based on the geometric relationship between the flame center point and the receiving point. This method requires the use of two empirical graphs, both of which are based on actual observation data of various hydrogen and hydrocarbon compounds. In addition, compared with the actual flame length of the torch, the data calculated by this method has a larger deviation, especially at larger emission volumes, the calculation results are not very reasonable. Therefore, this method is not suitable for larger emission volumes, or emission gases containing alcohols and fats. This method assumes 1, that is, it is assumed that the radiant heat generated by the torch does not propagate in the atmosphere, so the API 521 Simple method is conservative. For the specific calculation method, refer to Appendix C2.2 of API Std 521 Pressure-relieving and depressuring systems.

(2) Brzustowski's and Sommer's method

Brzustowski's and Sommer's method is a single-point method to determine the center of the flame based on jet diffusion theory. This method is based on the flame length calculation method of flammable gas jet mixing in lateral wind and explosion lower limit in air. The center point of the flame is defined by the center point of the center curve of the flame shape under the action of lateral wind. This method requires the use of an empirical graph to determine the coordinates of the torch center point. According to the product of the explosion lower limit of the emission gas and the air thrust and the jet thrust at the torch outlet, the result is fitted by the curve. Therefore, this calculation method requires a clear composition of the emitted torch gas. In addition, the empirical formula for calculating is obtained by cross-plotting the absorbance calculated from the Hottel chart, which is applicable to most flare gases except H_2 and H_2S, because H_2 and H_2S hardly or do not emit light radiation when burning. For the specific calculation method, refer to Appendix C 2.3 of API Std 521 Pressure-relieving and depressuring systems.

(3) G. R. Kent (GPSA) method

The G. R. Kent (GPSA) method conducts experiments on flame length in still air through small diameter pipes, concluding that the flame length is fixed starting from Mach number ≥ 0.2, and is approximately 118 times the diameter of the gas outlet. This method assumes that the heat of the flame is evenly released along the length of the flame, so the center of the torch flame in windless conditions is approximately at 1/3 of the flame length, obtained through integration. This method requires the use of Ivanov's jet trajectory equation to determine the center of the flame, assuming the flame as a rigid body, and the calculation results are slightly conservative compared to Brzustowski's and Sommer's methods. In addition, this method is included in the standard SH 3009—2013 "Design Specification for Combustible Gas Emission Systems in Petrochemical Industry". When calculating the height of the torch using this method, the proportion of radiant heat weakened after passing through the atmosphere is not considered, so the calculation results are slightly conservative. For specific calculation methods, refer to Chap. 5 of GPSA.

(4) Other methods

There are other methods that can be used to calculate the height of the torch. Complex models that consider wind speed, torch outlet gas speed, flame shape, and flame segment analysis are more suitable for special situations, especially for large emission systems. The FLARESI software also proposes the Point Source method, which is a multi-point extension of the API 521 Simple method. In this method, the flame is assumed to be completely transparent, so the radiation from one point does not interfere with or obscure another point. The flame is divided into a series of smaller point source elements, and the contributions of these point source elements add up to the total radiation of the flame. In practice, this method usually gives more practical values than the API 521 Simple method. However, it tends to over-predict the thermal radiation in the near field.

(3) Radiation Coefficient of the Torch

Based on the above methods to calculate the center of the flame, and then through the maximum allowable thermal radiation intensity of different areas on the ground, the final height of the torch is obtained. In addition, factors affecting the thermal radiation of the torch on the ground also include the thermal radiation coefficient F. Since not all the heat released by the torch flame can be transmitted to the ground by radiation, the composition of the gas, the type of flame, the fuel/air mixing state, the type of torch head, etc., will all affect the value of the thermal radiation coefficient. Therefore, the selection of the thermal radiation coefficient is also one of the factors that need to be considered when calculating the torch height. When the torch radiation is calculated by the torch manufacturer, the manufacturer must provide the radiation coefficient for calculating the torch radiation. Experience shows that some manufacturers give too low radiation coefficients, so the value of the radiation coefficient must be carefully checked. The radiation coefficient usually used by manufacturers does not consider the impact of droplet carryover, because they believe that the torch knockout drum can perform sufficient gas–liquid separation. Therefore, the size of the droplets in the torch knockout drum and the amount of liquid carryover should be clearly stated for the torch manufacturer to calculate. The following are the reference values of the thermal radiation coefficient F given in Chapter 13 of the Total GS EP ECP 103 Process Design Guidelines:

(1) Tube torch:

- Natural gas with a molar mass of 18 g/mol: 0.21;
- Natural gas with a molar mass of 21 g/mol: 0.23;
- Ethane: 0.25;
- Propane: 0.30.

(2) Sonic torch: When the mass carryover rate of droplets does not exceed 5% (mass fraction), the radiation coefficient *F* should be 0.15 (conventional conditions). The higher the droplet carryover rate, the larger the radiation coefficient. If there is no droplet carryover, the minimum radiation coefficient *F* should not be less than 0.13.

Radiation should consider both full and reduced quantities. The radiation coefficient F increases with the decrease in flow rate, which is provided by the torch head supplier. The radiation coefficient can be obtained from Fig. 9.5.

Since torch burning takes place in an open environment, pure mathematical analytical calculations are almost impossible. There are many methods for calculating the height of torches at home and abroad, all of which use geometric simplification models. However, none of the methods mentioned above can accurately calculate the height of the torch, and each method has its applicable range. In the process of specific projects, you can choose according to the different applicable conditions of each method, then calculate the height of the torch through the ground concentration of the gas emitted by the torch and the noise generated by the torch on the ground, and finally determine the height of the elevated torch.

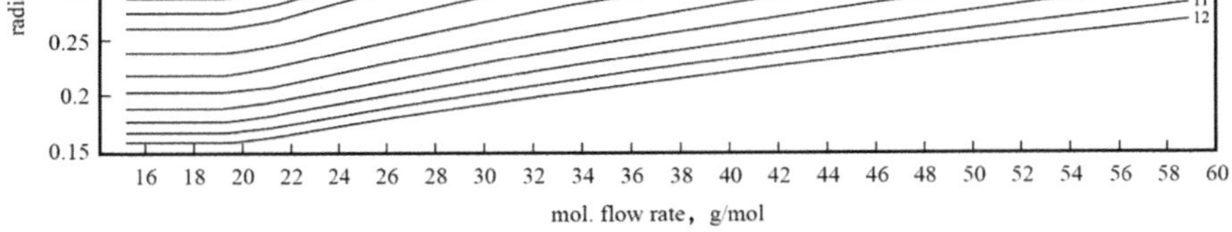

Fig. 9.5 Radiation coefficient corresponding to different flow rates/molar masses

The determination of torch height is related to many factors such as emission medium characteristics, maximum emission and heat release, atmospheric environment, allowable ground heat radiation intensity, heat radiation radius, allowable atmospheric pollutant ground concentration, torch self-structure design, etc. For a specific process, the wind speed at the torch outlet, the allowable ground heat radiation intensity and the heat radiation radius are the main influencing factors. In addition, the determination of torch height should also take into account investment, safety, environmental protection, operation, maintenance and local planning.

9.4.2.4 Software Simulation

With the increasing complexity of oilfield surface engineering, the manual superposition method of radiation heat calculation can no longer meet the needs of oilfield torch design. In order to meet the safe, reliable and strict environmental protection requirements, the torch model is designed using special tool software, which can comprehensively consider the effects of wind direction, wind speed, humidity, atmospheric pressure, atmospheric stability, radiation intensity, SO_2 ground concentration and other factors, and quickly and conveniently complete the calculation of torch height and diameter.

FLARESIM software is a set of torch design tool software developed by SOFTBITS in the United States. By inputting the composition of combustion gas, volume flow rate, environmental related parameters, initial values of diameter and height of torch riser, initial values of outer diameter of torch head, type of torch head, *F* factor, maximum allowable radiation value and other basic parameters, the software can automatically calculate the key output parameters such as pressure drop, flame length, radiation intensity, noise, SO_2, CO_2, H_2S diffusion concentration, etc., and can reflect them intuitively in the form of charts. Given the increasing emphasis on environmental protection, torches are often required to burn smokelessly, and FLARESIM software can also calculate the amount of combustion air. If the above results do not meet the requirements of relevant specifications and laws, the diameter and height parameters of the torch can be modified and recalculated until the optimal value is obtained.

PHAST software is a process hazard consequence analysis tool launched by the Norwegian Classification Society, which is the most widely used quantitative risk analysis software in China. PHAST can be used for consequence assessment of material leakage, atmospheric diffusion of gas, jet fire and pool fire. The API Std 521 model in the "Jet Fire" model of this software is suitable for the calculation and analysis of heat radiation after the torch is ignited in the station yard.

9.5 Ignition Equipment Design

Ignition equipment is an indispensable device for safely and reliably igniting torch gas and ensuring the safe combustion of torch gas. If the ignition equipment cannot ignite the discharged torch gas in time, the torch gas will gather in the atmosphere and at some place on the ground (the heavier components in the torch gas may fall at some downwind place), which will be one of the factors causing fire hazard.

9.5.1 Ignition Means

Ignition means can be divided into high-altitude automatic ignition, high-altitude manual ignition and ground manual ignition.

9.5.1.1 High-Altitude Automatic Ignition

High-altitude automatic ignition is the primary method of torch ignition. It uses the emission signal detector installed on the torch main pipe as the ignition signal source, the flame telemetry detector installed on the ground to detect the flame of the torch, and the thermocouple on the torch head to detect the temperature of the torch head, thus forming a closed-loop control system. When the torch is always in the emission state, but the pilot light or torch automatically extinguishes for some reason, the system will automatically re-ignite the torch. At the same time, the system also provides semi-automatic and hard manual ignition operation modes for field debugging or special circumstances. The automatic ignition of the torch requires timely and reliable, low noise, complete combustion, and meets environmental protection requirements.

9.5.1.2 High-Altitude Manual Ignition

The ignition method is the same as high-altitude automatic ignition. When the torch discharges gas, manually open the valve on the fuel gas pipeline, spray the ignition gas into the high-altitude igniter, and the flame sprayed from the top of the high-altitude igniter ignites the pilot light or torch.

9.5.1.3 Ground Manual Ignition

The ground internal combustion igniter is an alternative means of torch ignition. The igniter uses instrument air and fuel gas, which enter the mixing chamber through their respective pipelines and flow restriction orifices (or pressure reducing valves).

When the concentration of the two reaches the chemical concentration range, they are ignited by the electric spark generated by the high-voltage discharge of the high-energy semiconductor nozzle and detonate. The combustion material is rushed by the shock wave into the flame propagation tube to the pilot light, igniting the pilot light.

9.5.2 *Setting Requirements*

(1) Elevated torches should be equipped with high-altitude electric igniters and ground internal combustion igniters.
(2) Igniters should be equipped with an uninterruptible power supply.
(3) The number of high-altitude electric igniters should be the same as the number of pilot lights; each torch head should be equipped with one ground internal combustion igniter, and its ignition tube should be set up separately from the igniter to each pilot light.
(4) The number of torch pilot lights should meet the following requirements:

① When the diameter of the torch head is less than or equal to 0.5 m, no less than 2 pilot lights should be used.
② When the diameter of the torch head is greater than 0.5 m and less than or equal to 1.0 m, no less than 3 pilot lights should be used.
③ When the diameter of the torch head is greater than 1.0 m, no less than 4 pilot lights should be used.

(5) Pilot lights should be equipped with temperature detection instruments.

9.5.3 *Design Requirements*

A fixed ignition system should be provided for permanently used torches or combustion pits. This system must be reliable, even if the backup high-altitude automatic ignition is prioritized over other devices, unless it is determined that they can provide good performance. In any case, backup equipment should allow operators to ignite the treatment system at a safe distance.

(1) The torch head or burner should be equipped with an ignition device (pilot light) capable of igniting the torch gas under relevant flow conditions and environmental conditions.
(2) The pilot light should be ignited by a reliable ignition system that can operate under all relevant environmental conditions.
(3) The automatic ignition system is the main method of ignition, but it is still recommended to set up an independent manual backup system. The torch gun cannot be used as an independent manual backup system.

9.5.4 *Fuel Gas Supply*

(1) The supply of pilot light fuel gas should be highly reliable, and automatic backup gas supply can be used to improve overall reliability.
(2) The supplier should confirm the range of relative molecular mass and calorific value of the pilot light fuel gas to ensure that the torch can be normally ignited without adjusting the air and torch gas flow.
(3) The gas volume of each igniter should be sufficient to prevent the igniter from extinguishing under the specified maximum wind speed. If necessary, a reliable backup fuel source should be provided for black start (such as other parts of the device or propane gas cylinder group). If possible, the pilot light fuel gas should be directly taken from the factory fuel gas main pipe. The gas is best sourced from natural gas pipelines or liquefied petroleum gas evaporators, etc. The pilot light fuel gas pipeline should be installed at the top of the pipeline bracket, and two parallel filters should be set up, or a double filter should be used, and a valve should be set up to allow cleaning/replacement of the filter.
(4) In order to check the pressure drop through the filter, a differential pressure gauge should be installed on the filter, or a bypass filter should be installed for maintenance during torch operation; the mesh size of the filter element should be about 0.5 mm (0.020 in). The pipeline and fittings downstream of the filter should be 321 or 347 stainless steel to avoid blockage by corrosion products, and there should be an automatic pressure reducing valve downstream of the filter.
(5) The fuel gas consumption of a single pilot light should not exceed 4 m^3/h (standard state).
(6) A pressure regulating valve should be installed on the main pipe of the pilot light fuel gas supply pipeline, the pressure of the fuel gas source should be greater than or equal to 0.35 MPa, and the pressure after the pressure regulating valve should be stable at 0.2 MPa. The fuel gas supply pipeline for each pilot light should be connected separately from the bottom of the torch to the fuel gas inlet of the pilot light.
(7) If an internal combustion ignition device (Flame front generator) is used, check valves should be set up separately on the air pipeline and the pilot light fuel gas pipeline. This can prevent fuel gas or air from flowing back into the fuel gas system when the flame propagation device is blocked.

9.6 Liquid Seal Tank Design

9.6.1 Function of Liquid Seal

The torch gas sealing system includes water (liquid) seal and gas seal, both of which are set up to prevent the backflow of exhaust gas and the explosion and combustion accident caused by air flowing back into the torch system. The liquid seal can prevent the entry of air into the torch pipe network due to thermal shrinkage, and can be used to transfer gas (for example, for torch gas recovery system), it is usually used in combination with continuous gas blowing.

Under special circumstances, the flare system may experience negative pressure conditions. Water seal tanks are typically used as a means to prevent backfire due to negative pressure, in order to avoid extensive damage to the flare system. When the emission of combustible gas is nearing its end, the amount of water in the water seal tank must meet the demand for effective sealing water to prevent air from flowing back into the flare system from the flare head, causing an explosion. In addition to normal operation water replenishment, the maintenance of the water volume in the water seal tank can also be achieved by instantly adding a large amount of water when the flare discharge pressure is close to atmospheric pressure, or by using an appropriate water seal tank size to prevent a large amount of water from being lost from the tank, in order to meet the water volume requirements in the water seal tank. Although both methods can meet the water volume requirements in the water seal tank, the latter is an inherently safe measure that does not require the aid of an instrument control system. Therefore, considering the importance of the water seal tank, inherent safety should be the first choice in design.

The following situations may not require a water seal:

(1) The back pressure allowed by the exhaust equipment is very low, so that the depth of the inlet insertion tube is less than 100 mm.
(2) The temperature of the exhaust gas is very low, so that it may cause the water seal to freeze (without a heating station).

There should be a certain gas phase space in the water seal groove to prevent water entrainment. The speed of water replenishment in the water seal should be appropriate and not too fast. The water discharged from the overflow port of the water seal groove should be recycled. The water seal groove in cold regions should adopt anti-freezing measures and prevent hydrocarbon compounds from covering the liquid surface. When the water seal groove and the flare foundation are set together, the water seal groove should be as close as possible to the flare chimney. According to the structure, water seal tanks can be divided into horizontal and vertical types.

In situations where the flare gas discharge volume is very large, liquid sealing may not be able to prevent backfire. Some accidents have shown that flames can propagate back through continuous bubbles. In this case, continuous purging or other methods should also be used to ensure that there is no air in the flare manifold.

9.6.2 Types of Liquid Seals

(1) If the ambient temperature or depressurization flow temperature is not lower than 0 °C (32°F), water seals are usually used. To prevent freezing in cold weather, anti-freezing measures need to be taken, such as equipping with an automatic heating device (which can be electrical or steam coil).
(2) For colder fluids, ethylene glycol or other suitable media should be used for liquid sealing. Whether to use pure antifreeze liquid or use it with water should be determined based on the expected gas temperature.
(3) The selected liquid seal fluid should be compatible with all discharged gases.

9.6.3 Principles of Liquid Seal Tank Design

The water seal tank is set between the flare separator and the flare body, and is set near the root of the flare. When there are two or more flares operating simultaneously in the same emission system, each flare should be equipped with a water seal tank, and the water seal heights of different flare water seal tanks should be set in layers. Two mutually backup flares should be equipped with a common water seal tank, but a safe purge required for the switching operation of the two flares should be set. In some cases, some special matters may affect the size of the seal tank, one of which is a large amount of hot air flowing into the vent pipe manifold.

The vacuum generated by these hot gases cooling will suck in enough sealing liquid into the manifold to break the seal, therefore, air will enter the flare system. To avoid this situation, the inlet pipeline should be made into a vacuum leg shape. At the inlet on the seal tank, the vertical height of the vacuum leg can be determined based on the expected maximum vacuum degree. At the maximum vacuum degree, the amount of liquid in the inlet pipeline should be obtained from the seal tank. To meet this requirement, the size of the seal tank may need to be increased. When selecting the size of the seal tank, the maximum back pressure allowed at the outlet of the vent pipe manifold should be determined first.

According to the requirement of ISO 23251 Petroleum, petrochemical and natural gas industries—Pressure-relieving and depressuring systems or API RP 52 Land drilling practices for protection of the environment, the effective water seal volume in the water seal tank should be at least enough to fill a 3 m vertical pipe at the entrance of the water seal tank when negative pressure occurs in the combustible gas discharge network. The minimum immersion depth of the entrance to the torch manifold should be 10 cm (4 in). The gas inlet of the water seal tank should adopt an effective gas distribution structure to prevent torch pulsating combustion caused by fluctuations in the seal water. The release of gas flow pulses, manifested as flame jumping and loud noise, can make the flame very noticeable and make it difficult to achieve smokeless combustion of the flame using steam. The conventional design is to open long slots or V-shaped notches at the end of the water seal pipe, so that the flow area increases with the increase of gas flow, which can minimize the pressure

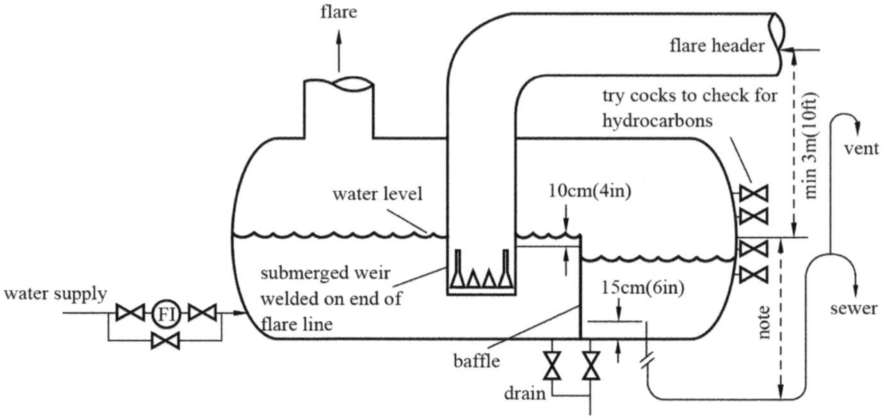

Fig. 9.6 Effective water seal volume in water seal tank. *Note* The minimum pressure of the sewage pipe seal should be designed to be 175% of the maximum operating pressure of the seal tank

fluctuations in the seal tank. Another design to control pulses is to control the liquid sway during the release process by setting up a splash guard. The function of the splash guard is to suppress any pressure fluctuations caused by the liquid flow in the tank without causing noticeable pulsation. When the bottom of the water seal tank gas inlet adopts a toothed end face, the distance from the bottom of the inlet pipe to the bottom of the water seal tank should be greater than or equal to 0.25 times the inner diameter of the gas inlet (Fig. 9.6).

It is not advisable to use a liquid barrier to divide the space in a horizontal water seal tank to remove condensate accumulated on the water surface. If this structure is used, it should be ensured that the effective water seal volume in the water seal tank, after subtracting the maximum volume used for liquid removal divided by the liquid barrier, meets the water seal design requirements.

The water seal tank should be equipped with a U-shaped overflow pipe (no cut-off valve should be installed), the water seal height of the overflow pipe should be greater than or equal to 1.75 times the maximum operating pressure (gauge pressure) of the gas phase space in the water seal tank, and the minimum diameter of the overflow pipe should be DN50mm. The inner surface of the pipe at its highest point should be at the same horizontal height as the required water seal liquid level.

A DN25mm vacuum breaking pipe should be installed above the high point of the U-shaped overflow pipe, and its height should be greater than or equal to 300 mm. No cut-off valve should be installed on the vacuum breaking pipe. The overflow outlet of the U-shaped overflow pipe should be sealed and connected to an oil or sulfur-containing sewage system, and a sight glass should be installed on the overflow pipe. The overflow water replenishment of the water seal tank should be restricted by using a flow limiting orifice plate, and the flow rate should not exceed 50% of the self-flow capacity of the U-shaped overflow pipe. The design pressure of the water seal tank should not be less than 0.7 MPa (gauge pressure), and negative pressure conditions are not considered. When the average temperature in the coldest month is below 5 °C, the water seal tank should take anti-freezing measures. When the temperature of the combustible gas discharge is greater than 100 °C, the water seal tank should be equipped with a low liquid level alarm and automatic water replenishment measures to maintain the water seal volume. The water seal tank should also be equipped with liquid level, temperature, pressure instruments and high liquid level alarm. The water seal tank should be a horizontal tank, and the ratio of its length to diameter should be between 2.5 and 6.0.

The radial cross-sectional area of the gas flow in the horizontal water seal tank should be greater than or equal to 3 times the cross-sectional area of the inlet pipe. The height of the gas phase space in the vertical water seal tank should be greater than or equal to the inner diameter of the water seal tank, and should not be less than 1 m.

The type of gas inlet and outlet channels of the water seal tank should be one of the following:

(1) In a horizontal tank, the gas enters from one end of the vertical upper part of the tank axis and is discharged from the other end.
(2) In a horizontal tank, the gas enters from the middle of the vertical upper part of the tank axis and is discharged from both ends.
(3) In a vertical tank, the gas enters radially from the tank body and is discharged from the top of the vertical axis of the tank body.

The water seal tank should be equipped with equipment to maintain the liquid level of the sealant, and the design of its sealing system should meet the following conditions:

(1) Prevent the accumulation of hydrocarbons.
(2) Prevent the displacement of the sealing liquid.
(3) Maintain the sealing liquid level within the working pressure range.
(4) Consider continuous replacement of the sealing liquid to prevent the accumulation of H_2S and CO_2.

If water is used for sealing, the treatment of excess water should consider the toxic and harmful substances in the flare gas (such as H_2S) potential pollution. If a water circulation system is used, consideration should be given to allowing for fluid replenishment and checking the sealant fluid stock. If antifreeze is used, a water circulation system should be considered. If the requirements for fluid replenishment are not high (i.e., static fluid sealing), an antifreeze system can also be used.

For the calculation method of water seal tank, please refer to API Std 521 Pressure-relieving and depressuring systems and SH 3009 "Design Specification for Combustible Gas Emission Systems in Petrochemical Industry".

9.7 Purging and Sealing of Flare Systems

9.7.1 Purging System

The flare system should have a continuous purging system, and the purging gas can be a reliable source of inert gas (such as nitrogen), fuel gas or natural gas. If feasible, it is best to use inert gas (if a gas with a low relative molecular mass is used as the purging gas, a larger volume of gas is required to achieve the same effect of preventing backfire and air inflow). If inert gas is chosen, attention should be paid to the possibility that inert gas may lead to the emission of unburned gas into the atmosphere. If self-made nitrogen from the plant is used, the possibility of oxygen entering the nitrogen system should be evaluated.

9.7.1.1 Source of Purging Gas

The purging gas should have a reliable source:

(1) The purging gas should have an automatic backup system.
(2) A failure of a gas source or an injection point in the purging system should not cause danger to the flare system.

9.7.1.2 Minimum Purging Rate of Flare Tip

The minimum purging rate of the flare tip should be the higher of the following two:

(1) To prevent wind from blowing air into the flare tip.
(2) To prevent backfire inside the flare tip.

The purging rate required to prevent wind from blowing air into the flare system can be calculated according to the Husa formula recommended in API Std 521 Pressure-relieving and depressuring systems. The calculation of the gas purging rate is based on the following condition: the concentration of oxygen at 8 m (26 ft) from the top of the flare tip should be less than half of the lower flammable limit, which is to prevent flame backfire. If inert gas is used as the purging gas, the speed of backfire will be significantly reduced. This can significantly reduce the risk of destructive explosions. If inert gas is used for purging, there is a maximum purging rate that can prevent backfire. This purging rate may be significantly lower than the purging rate calculated using the Husa formula. This maximum purging rate refers to the purging rate that can ensure that the flame ejection speed is higher than the flame backfire speed.

9.7.1.3 Maximum Oxygen Content

If the relative molecular mass of the purging gas or flare gas is low (for example, it contains a high concentration of hydrogen), the maximum oxygen content should not exceed the following values:

(1) For flare gas with a relative molecular mass greater than 6 but not greater than 8, the maximum volume fraction of oxygen is 5%.
(2) For flare gas with a relative molecular mass greater than 4 but not greater than 6, the maximum volume fraction of oxygen is 4%.
(3) For flare gas with a relative molecular mass not greater than 4, the maximum volume fraction of oxygen is 3%.

9.7.1.4 Prevention of Internal Combustion

For flare tips without fire-resistant lining, the flare supplier should specify the minimum purging amount required to prevent internal combustion of the flare tip. This flow rate should also consider the design of different flare tips or molecular seals. If the flare tip has a fire-resistant lining, the calculation of the purging rate only needs to consider the purging rate required to prevent wind from blowing air into the flare.

When the entire flare system is put back into operation, a large initial purging rate may be required to blow air out of the flare system. To detect whether the flare gas can maintain a safe oxygen level, the flare can be equipped with an oxygen monitoring system; however, the reliability of these systems should be considered.

(1) For flammable purging gases that are heavier than air, the minimum purging rate can theoretically be achieved at very low flow rates. This may lead to internal combustion in the flare, increasing the temperature of the flare and shortening its lifespan, or causing the flame to extinguish.
(2) The flare manufacturer should write the minimum purging rate that can prevent internal combustion in the flare tip on the flare data sheet, and confirm whether it is possible to prevent internal combustion in the flare tip using nitrogen according to the Husa calculation method. Otherwise, alternative solutions should be considered, such as cooling the flare tip, or upgrading the metal material of the flare tip.

When considering whether to cool the flare tip, upgrade the metal material of the flare tip, or increase the purging rate, an economic comparison should be made, and the lifespan of the flare and local laws and regulations should be considered.

9.7.1.5 Prevention of Vacuum Formation

Emergency purge gas should be provided to the torch system to prevent the formation of a vacuum. Possible causes of vacuum formation include steam condensation, system cooling gas contraction due to factors such as rainfall, etc.

(1) The emergency purge gas to prevent vacuum formation does not need to be continuous, but it needs to be able to automatically control based on factors such as pressure and temperature.
(2) An alternative method is to use liquid sealing. If liquid sealing is used, the formation of a vacuum due to gas cooling upstream of the liquid seal tank does not necessarily pose a danger. The prerequisites are:

① The design of the equipment takes into account the maximum vacuum conditions that may occur.
② An increase in pressure difference will not trigger additional flow.

The end of the torch manifold of each unit should be connected with purge gas to ensure that there is no air in the torch system before introducing hydrocarbons. During normal operation, continuous purge gas can be connected at the end of the torch manifold or near the separator. Generally speaking, continuous purge gas should be connected near the separator, so it is closer to the torch head and can better prevent air from entering.

9.7.2 Torch Sealing

During low-load operation of the torch gas, the mixture of air entering the torch and the incoming torch gas may explode. There are two common types of mechanical seals, usually located at or below the torch head, used to reduce the amount of continuous purge gas required to prevent air from infiltrating into the torch (Fig. 9.7).

9.7.2.1 Molecular Seal

This type of seal uses the relative molecular mass difference between the purge gas and the infiltrating air to form a gravity seal, thereby preventing air from entering the torch tower. The annular baffle arranged in the seal forces the incoming air to make two 180° turns (one up and one down) before it can enter the torch tower. If the purge gas is lighter than air, it will accumulate at the top of the seal to prevent air from infiltrating into the torch system; if the purge gas is heavier than air, it will accumulate at the bottom of the seal to prevent air from infiltrating into the torch system. This type of seal typically reduces the required purge gas velocity through the torch head to 0.3 mm/s (0.01 ft/s). Moreover, for some components of the purge gas, this flow rate will limit the oxygen content to be lower than the oxygen content

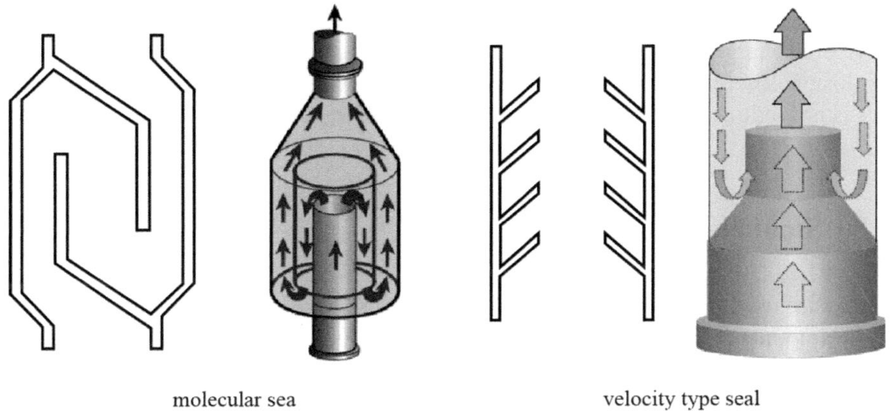

Fig. 9.7 Torch seal

of the device and less than 0.1%. However, these low purge flow rates cannot prevent the torch head lining from burning, which will shorten the life of the torch head. This effect is called "cancer" because the damage to the metal arm of the torch head of the molecular seal is hidden until the flame burns through the torch head or seal, at which point an emergency shutdown and immediate repair are required.

In the molecular seal, when using a purge gas that is lighter than air, the buoyancy of the purge gas forms an area greater than atmospheric pressure at the top of the seal, preventing air from entering the torch body. Heavier purified gas will "flood" the seal. However, due to the possibility of rainwater ingress and condensation, molecular seals need to have a liquid discharge device. Ice, carbon deposits, or refractory materials may block the sealing device, so this type of seal is not recommended.

9.7.2.2 Velocity Seal

This type of seal works on the premise that infiltrating air enters, passes through the torch head close to the inside of the torch tower, the velocity seal is a conical head obstacle, and has one or more baffles, which force the air to leave the inner wall of the torch tower, where it encounters concentrated purge gas flow and is flushed out of the torch head. This type of seal typically reduces the purge gas velocity through the torch head to between 0.5 mm/s (0.02 ft/s) and 1 mm/s (0.04 ft/s), thus keeping the oxygen concentration below the seal requirement of 4–8% (approximately 50% of the limiting oxygen concentration required to produce a combustible mixture).

If the outflow speed of the torch gas always exceeds the backfire speed, backfire from the torch flame to the torch body will not occur. The outflow speed can be increased by using a perforated plate with one or more holes.

Although called seals, these two types of seals cannot completely prevent backflow, only reduce it. They are both installed directly below the torch head. However, when the torch gas condenses or cools at a rate that exceeds the rate of purge and device discharge, it can no longer prevent air from entering, and there is a risk of explosion. Therefore, in design, even if a sealing device is used, the amount of purge gas is not reduced.

9.8 Smokeless Torch

Flares are used to control the processing of excess gases flowing continuously and large amounts of gas that suddenly increase in emergencies. For the airflow that is expected to occur during normal operations each day, flares are usually required to burn smokelessly. Some environmentally sensitive areas require 100% smokeless operation or even require a fully enclosed flare. Various techniques can be used for smokeless operation, most of which are based on the premise that smoke is produced by rich fuel combustion and can be eliminated by promoting the uniform distribution of air throughout the flame. In addition to the requirements for smokeless operation, strict combustion regulations (government and local) are constantly evolving, and in most areas, especially regarding low noise, limiting smoke emissions, continuous ignition monitoring, limiting flare tip exit velocity and (or) the minimum heat content of flare gas, detailed current regulations should be consulted during design. Smokeless operation is usually the primary requirement for designing flare system burners, but sometimes smokeless design is not required according to local regulations. If the color of the flare combustion is not darker than a No. 1 Ringelmann chart (refer to the Ringelmann blackness chart), it should be considered smokeless combustion.

Almost every flare design is to promote a group of flare gases to operate smokelessly under suitable conditions. In order to promote the smooth distribution of air throughout the flame (thus preventing the formation of smoke), energy is needed to generate turbulence and mix combustible air into the flare gas when igniting the flare gas. This energy may exist in the gas, manifested in the form of pressure, or can be applied to the system, that is, when the gas is discharged from the flare tip, another medium is injected into the gas, such as injecting high-pressure steam, compressed air or blowing low-pressure blower air into the gas. To create conditions conducive to smokeless combustion, the range of flare design is very complex, including from a simple open pipe section with an ignition source to a segmented flare system with a complex control system. The following briefly outlines the types of smokeless systems most commonly seen in oil and gas field stations.

9.8.1 Steam Smoke Elimination

The most common type of smokeless flare tip is a flare tip that controls smoke generation with steam (Fig. 9.8). Steam can be injected through a single pipe nozzle installed in the center of the flare, through a series of steam or air injection ports in the flare, or through a pipe manifold installed around the flare tip, or for special applications, a combination of the three methods can be used. Injecting steam into the flame section can generate turbulence and (or) attract air into the flame section with the help of steam injection, thus improving the distribution of air. The combination of air and steam leads to water–gas movement interaction, making the reaction of flare gas faster, thus eliminating the rich combustion conditions that cause smoke generation. The patented flare tip designs used come from various manufacturers, who provide unique steam injection methods and different steam efficiencies. The amount of steam required is mainly a function of gas composition, flow rate, steam pressure, and flare tip design. The general range of steam required is 0.11–0.45 kg of steam per 0.45 kg of flare gas. ISO 23251 Petroleum, petrochemical and natural gas industries—Pressure-relieving and depressuring systems or API Std 521 Pressure-relieving and depressuring systems provide preliminary steam injection design rates that can promote smokeless combustion. In specific projects, the flare supplier should be consulted for the steam requirements of their specific flare tip design.

In cold weather, internal steam nozzles may cause condensate to enter the flare manifold and collection pipe, causing them to freeze. In some cases, this can lead to complete blockage of the flare or flare manifold. If using steam, the following points should be noted:

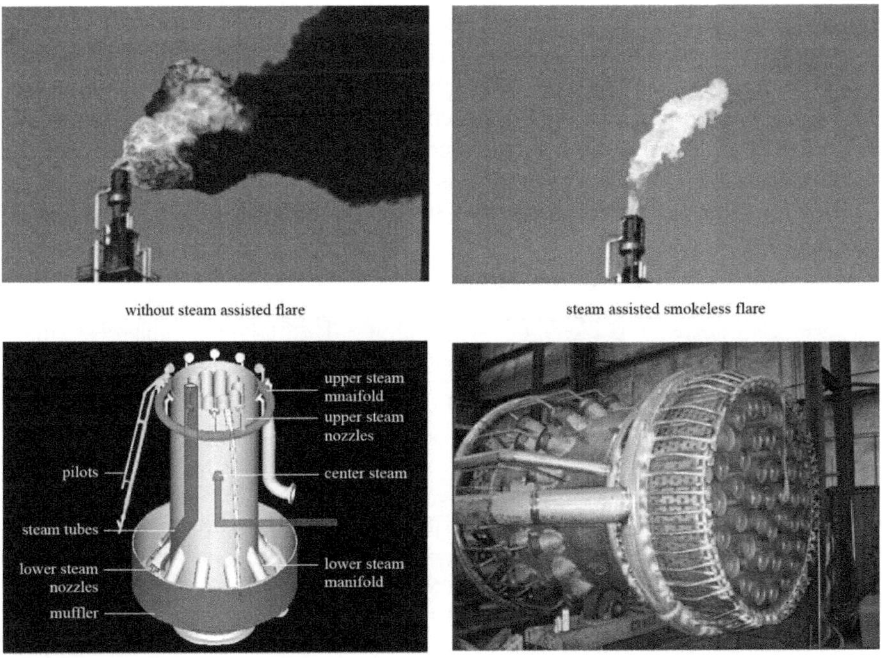

Steam smoke elimination flare tip structure

Steam smoke elimination flare tip

Fig. 9.8 Steam smoke elimination flare tip

(1) Steam pipes should use insulation materials to ensure that the steam provided to the flare tip is dry steam.
(2) Condensation of steam at the flare tip should be avoided, which can cause the pilot light to go out or mechanical damage.
(3) Steam traps should be installed at the low points of the steam pipeline, and they should have anti-freeze design.
(4) Steam flow can be automatically or manually controlled based on the flow of flare gas or the visible characteristics of the flame.
(5) To cool the top piping system, a bypass should be set up outside the steam control valve to maintain the minimum steam flow specified by the supplier.
(6) Expansion joints should be installed in the steam riser as required.

9.8.2 Air Smoke Elimination

If the site resources are not suitable for assisting in smokeless operation, low-pressure air systems are usually the first to be evaluated. Selection of methods. When the gas is ignited, the system generates turbulence in the flame section by injecting low-pressure air provided by a blower through the torch head, thereby promoting the smooth distribution of air throughout the flame. Typically, air flows coaxially with the torch gas to the torch head and mixes there under a pressure of 50.8–152.4 mm water column (WC pressure). This system has a higher initial investment because it requires a double-layer torch tube and an air blower, as shown in Fig. 9.9. However, the operating cost of this system is much lower than that of the steam-assisted system design (only a blower needs power). The amount of air provided by the blower for smokeless operation is generally 10–30% of the equivalent air required for saturated hydrocarbons and 30–40% of the equivalent air required for unsaturated hydrocarbons.

Fig. 9.9 Air smoke torch

9.8.3 Sonic Torch

The sonic torch uses the pressure energy inherent in the torch gas itself (typically 35–135 kPa at the torch head) to eliminate rich combustion conditions and achieve smokeless operation. It varies with the manufacturer and design characteristics. This method injects the torch gas into the atmosphere under high pressure, generates turbulence in the flame section, and promotes the smooth distribution of air throughout the flame. Because no external usable gas is needed, these systems are generally beneficial for handling large amounts of vent gas, whether for the economy of smokeless operation or for controlling flame conditions. The discharge of a single torch head is relatively small, and the design of a larger discharge system may require multiple torch heads to be connected together with a manifold. Under adjustment conditions, maintaining sufficient torch head pressure is key, and it is often necessary to appropriately control the number of torch heads used based on the relationship of gas flow, using a staged system. The staged torch system can be installed on the ground or elevated, but for larger systems, ground design may be needed because many torch heads are needed and these torch heads must be evenly distributed to allow air to enter the system. In specific projects, consult the torch supplier for their specific torch head design (Fig. 9.10).

The above are all about the torch equipment for handling exothermic torch gas, that is, the gas has a high enough heat value (usually for non-assisted torch heat value greater than standard state 200Btu/ft^3, for assisted torch heat value greater than 300Btu/ft^3) to maintain their own combustion, without adding any auxiliary fuel. Endothermic gases may also be treated in thermal incineration systems, in some cases, the preferred method is to use special torch designs—these torches use auxiliary fuel gas to ignite the torch gas. For small gas flows, a simple method to increase the heat value of the torch gas is to add fuel gas to the torch manifold to increase the net heat value of the mixture to meet the requirements; in other cases, it may be necessary to add fuel gas injection manifolds around the torch head (similar to steam manifolds) or at the outlet end of the torch head where the gas must flow

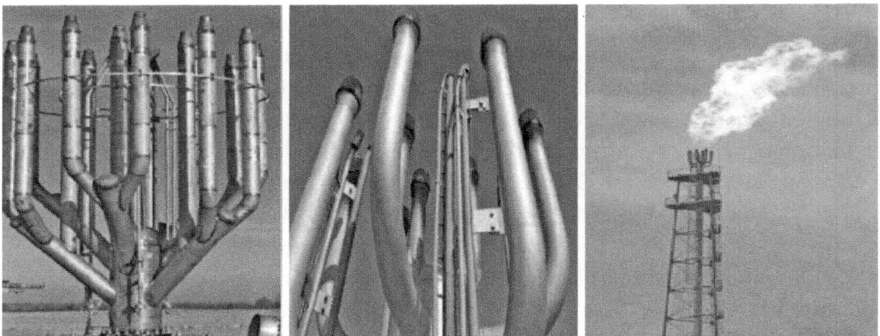

Fig. 9.10 Sonic flare

to ignite the torch gas. Common fuel gases are diluted hydrogen and high heat value gases containing small amounts of H_2S.

The simplest elevated torch is often referred to as a utility flare or pipe flare, consisting of a pipe equipped with a flame stabilizer and a pilot light. Smokeless elevated torch heads usually use steam to achieve smokeless operation, and some will use air to reduce the formation of smoke. The speed limit of these torches is usually Mach 0.5. High-pressure torches do not require steam or utility air to promote smokeless combustion, but use the energy of the torch gas itself. High-pressure torches can operate at Mach 1.0 speed. However, high-pressure torches need to consider back pressure to meet the requirements of pressure relief devices.

A closed ground torch refers to any non-elevated torch enclosed in a shell or fence (blocked by a solid wall to prevent heat radiation). It can also be called a multi-point torch. The advantage of a ground torch is that it hides the flame, monitors combustion emissions, and reduces noise. If space permits, the discharge of a ground torch can be made very large.

(1) Ground multi-point torches can be ignited or shut down in stages within a large flow range (high turndown ratio). Care must be taken when specifying the minimum opening of each stage to avoid backfire in the torch pipe when the torch load drops below the minimum value. Due to the small nozzles, multi-point torches can usually burn "smokeless", they can entrain air into the burning gas without using steam.
(2) A significant disadvantage of a ground torch is that if the torch ignition system fails, the torch gas will be released on the ground. Dispersion and consequence analysis should be conducted to assess the impact of unburned torch gas release in the event of a ground torch point system failure.
(3) The outlet design of ground torch gas needs to ensure sufficient diffusion of combustion products.

According to the U.S. Environmental Protection Agency (EPA) document AP 42 Volume 1 and Chapter 13.5, the minimum calorific value of flare gas should be 11250 kJ/m^3 (300Btu/ft^3) to ensure efficient combustion of the flare. If the calorific value of the flare gas is less than 9300 kJ/m^3 (250Btu/ft^3), additional fuel gas is needed for complete combustion. Domestic flare designs usually set acidic, high-pressure, and low-pressure flares according to the characteristics and pressure of the medium. The combustible gas emitted by each unit of each device in the entire plant area must be processed through these types of flares. According to relevant regulations, the safety distance between the flare and adjacent facilities in the plant area must be greater than 120 m, and the layout of the flare should also consider heat radiation. According to the requirements of SH 3009 "Design Specification for Combustible Gas Emission Systems in Petrochemical Industry", the allowable heat radiation between the flare and the production equipment inside the chemical enterprise should be less than 3.2 kW/m^2. The increase in the total amount of emission gas added during the expansion and reconstruction of the enterprise will lead to an increase in the intensity of radiation heat, so the impact of heat radiation values on the layout of the flare cannot be underestimated.

Because the combustion of combustible gas in the flare leads to energy waste, people hope to achieve full recovery and reuse of flare gas by optimizing the setting of the flare gas recovery system, fundamentally eliminating the safety hazards of the flare. However, in actual production, the changes in the emission volume and the changes in the composition of the flare gas are very large, so it is difficult to implement the plan to fully recover and reuse waste gas only by optimizing the flare setting. The flare system is still an integral part of the process design.

9.9 Flare Noise and Backup

9.9.1 Flare Noise

The ground noise (at positions accessible to personnel) generated by the flare head during combustion should meet the following requirements, and the flare supplier is required to provide noise information of the flare under the maximum emergency venting flow and maximum smokeless combustion flow:

(1) Less than or equal to 90 dB during normal operating conditions (including start-up and shutdown).
(2) Less than or equal to 115 dB during the maximum emission conditions of the whole plant emergency accident.

The main source of noise from smokeless flares is the noise of steam injection. Therefore, generally speaking, the lower the proportion of steam to flare gas, the quieter the combustion.

9.9.2 Flare Backup

The backup of the flare should consider maintenance, inspection, and failure. If the flare only serves one unit, it can be maintained and inspected during the normal shutdown maintenance period. It is generally believed that the risk of flare failure is not great. If the failure of the flare only leads to the shutdown of one unit, it is acceptable. If two or more flares are available, the flare pipeline can be configured to allow one of the flares to stop using. In this case, it can be used as a backup in the form of atmospheric emissions.

If a backup is actually needed, that is, there is an extra backup flare, it can replace the main flare during maintenance, inspection, or failure. The most economical method is to use a common bracket to support multiple flare heads, each of which can be taken down and maintained separately, while the rest of the flare heads continue to work. This type of flare can replace risers, nozzles, and flare auxiliary equipment during maintenance.

9.10 Safety Design of Flare System

9.10.1 Safety Analysis of Flare System

The flare is an important safety facility for oil and petrochemical enterprises, and its normal operation is crucial to the production and safety of the enterprise. The safety of the flare system mainly includes two aspects: one is the safety of the flare system itself, including whether the flare processing capacity meets the needs of the factory's combustible gas and harmful gas safe emission, the burnout rate of harmful gases, noise, smoke elimination effect, the reliability of ignition facilities, anti-backfire facilities, whether the system material selection meets the requirements, etc.; the second is the operation safety of the flare system, including the timely cleaning of condensate, the monitoring of water seal level, the maintenance and regular inspection of ignition facilities, the control of smoke elimination steam, the regular maintenance and inspection of the oil gas recovery interlock control system, the monitoring of the burning condition of the pilot light, etc. The effective combination of inherent safety in design and effective production operation safety can ensure the safe operation of the flare system. The safety analysis of the flare system is shown in Table 9.2.

Based on the above analysis, to ensure the safe operation of the torch system, it is necessary to design safety controls for torch ignition and combustion, backfire prevention, and prevention of liquid rain.

9.10.2 Monitoring of Torch Ignition and Combustion

9.10.2.1 Ensuring the Success Rate of High-Altitude Igniter Ignition

There are many reasons for the unsuccessful ignition of high-altitude igniters. The main risks in the process design and operation of high-altitude ignition systems are the instability of the pilot light fuel gas and the carbon deposition of the high-altitude igniter. To solve the problem of unstable pilot light fuel gas, the pressure gauge signal of the fuel gas pipeline after the filter should be remotely transmitted to the control room, and a low-pressure alarm should be added to detect and take measures in time when the pressure of the fuel gas pipeline is low, such as replacing the filter. It is also recommended to add backup fuel gas, such as natural gas or liquefied gas, to ensure a reliable and stable source of pilot light fuel gas. To solve the problem of carbon deposition in high-altitude igniters, high-altitude igniters with anti-carbon deposition functions can be used to ensure the success rate of ground deflagration ignition.

The ground deflagration ignition system is an emergency response system when the high-altitude ignition system fails, requiring manual operation under the flare stack. This operation is quite dangerous, especially during the release of acidic gas,

Table 9.2 Safety analysis of flare system

Serial number	Phenomenon	Reason	Consequence	Protective measures
1.1	No flame in the cylinder	(1) No flame in the pilot light or unsuccessful ignition (2) The emission gas contains non-combustible gas (Purge nitrogen)	(1) Emission of hydrocarbon flare gas accidents, flash explosions may occur when exposed to open flame (2) Frequent ignition leads to damage to the ignition equipment	(1) Configure a reliable automatic ignition system and regularly test it (2) Regularly clean the camera and equip it with a telescope (3) Coordinate the purging and displacement process of the scheduling device to control the emission of non-combustible gases
1.2	Large flame in the cylinder	Large venting volume	(1) High heat radiation (2) Incomplete combustion, the flare produces black smoke (3) High noise (4) Light pollution (5) Damage to the flare tip	(1) Set up smoke elimination steam, ejector steam, and central steam system (2) Apply silicone oil 10 m below the flare tip to protect the flare tip (3) Install a steam silencer on the flare tip
1.3	Small flame in the cylinder	(1) Small amount of gas emission and high external wind speed (2) Smoldering or internal combustion in the cylinder	(1) Smoldering damages the flare tip (2) Internal combustion in the cylinder can easily lead to backfire	(1) Raise the flame with central steam and ejector steam to prevent smoldering (2) The flare tip is made of 310SS material

(continued)

Table 9.2 (continued)

Serial number	Phenomenon	Reason	Consequence	Protective measures
2	Large smoke	(1) Large amount of device venting (2) Insufficient smoke elimination steam volume (3) Combustion gas composition is heavy (4) Poor separation effect of the liquid tank, liquid hydrocarbons are emitted to the flare tip	(1) Pollute the environment (2) Fire rain may occur (3) May burn out the flare tip or other accessories	(1) Adjust the steam volume in time to ensure steam balance (2) Regularly drain the liquid tank to ensure good separation effect two-stage separation can be adopted (the first stage at the device outlet, the second stage before entering the flare)
3.1	Large steam flow	(1) Valve is mistakenly opened wide (2) Instrument error or damage	(1) When the venting volume is small, the flare may be blown out (2) High noise (3) Causes unnecessary steam loss	(1) Strictly follow the operating procedures, monitor the flame smoke situation in time, and adjust the steam volume in time (2) The flare tip is equipped with a steam silencer
3.2	Small or no steam flow	(1) The valve is mistakenly closed or reduced (2) No steam or low steam pressure (Pipeline leakage, etc.) (3) The steam pipeline itself is not capable enough (4) Steam pipeline condensation	Large smoke, black smoke, poor cooling effect, causing damage to the flare tip and polluting the environment	(1) Set a bypass for the regulating valve (2) Insulate the steam pipeline and drain condensate in time (3) Strictly implement anti-freezing measures in winter

(continued)

Table 9.2 (continued)

Serial number	Phenomenon	Reason	Consequence	Protective measures
4.1	High-altitude igniter fails to ignite	(1) Power outage (2) High-voltage ignition electrode aging (external corrosion) or carbon deposition (long-term combustion or repeated ignition), unable to generate electric sparks (3) High-voltage igniter ignition electrode wiring fault (loose wiring or mis-grounding, etc.) (4) Improper installation of high-voltage igniter electrode (angle and distance) (5) The high-voltage end of the high-voltage generator cannot reach the rated voltage or the low-voltage end voltage is low (6) No fuel gas, or insufficient fuel gas concentration, or too high or too low pressure, or a large amount of liquid (7) Ignition signal error or unable to effectively transmit	Failure to ignite the pilot light may cause accidental emissions	(1) Equipped with UPS power supply (2) High-voltage generator adds a sealed protective box and desiccant inside the box (replaced regularly) (3) Install the electrode correctly as required (angle and distance) (4) Strengthen the maintenance check of the solenoid valve (5) Fuel gas solenoid valve fails to open (6) Regularly blow the high-altitude ignition fuel pipe (once a week), regularly drain the fuel gas separator, and regularly test ignition (once a week) (7) Insulation of fuel gas pipeline and winter heating (8) Regular inspection (fuel gas pressure) and calibration of instruments (9) Drain condensation during each inspection of the instrument air

(continued)

Table 9.2 (continued)

Serial number	Phenomenon	Reason	Consequence	Protective measures
4.2	Delayed ignition of high-altitude igniter	(1) Slow fuel gas supply (valve blockage) (2) Small amount of liquid in fuel gas	May cause deflagration	(1) Regularly drain condensation from fuel gas (2) Regular maintenance of the valve
5	Ground deflagration ignition system fails to ignite	(1) No purified air (2) Power outage (3) Fuel gas carries liquid, is blocked or has low pressure (4) The ratio of purified air to fuel gas is inappropriate (5) Spark plug failure or aging (6) Liquid accumulation in the open flame tube	Unable to ignite the pilot light, causing torch gas emissions under accident conditions	(1) Equipped with UPS power supply (2) Fuel gas solenoid valve fails to open (3) Purified air has a manual pressure regulator (4) Regularly drain the fuel gas separator, blow the deflagration pipe with nitrogen (once a week), regularly test the ground deflagration ignition system (5) Regular inspection (fuel gas pressure) and calibration of instruments

(continued)

Table 9.2 (continued)

Serial number	Phenomenon	Reason	Consequence	Protective measures
6	Failure to ignite the pilot light	(1) Failure of high-altitude igniter to ignite, see this Table 4.1 (2) Failure of ground deflagration ignition, see this Table 5 (3) The relative position of the high-altitude igniter and the pilot light is inappropriate (4) The pilot light fuel gas pipeline is blocked or there is no fuel gas (5) Harsh environment causes the pilot light to go out (6) The pilot light thermocouple is burnt out, unable to detect the temperature signal, and cannot trigger the ignition signal	Unable to ignite the torch, which may cause accidental emissions	(1) See this 4.1 "Failure of high-altitude igniter to ignite" and 5 "Failure of ground deflagration ignition system to ignite" (2) Check the installation quality of the high-altitude igniter (3) Regularly blow the fuel gas pipeline (4) Use a pilot light that can resist certain wind speed and rainfall (5) Use armored thermocouples, or other reliable pilot light flame monitoring systems
7	Small nitrogen flow	(1) Nitrogen pipeline blockage or valve failure (2) No nitrogen or insufficient nitrogen flow (3) Nitrogen flow limiting orifice plate is too small	(1) Backfire may occur during venting (2) Under normal circumstances, an explosive gas environment may form inside the torch body	(1) Install a pressure gauge on the nitrogen pipeline to monitor the nitrogen pressure in time (2) After venting, promptly blow the torch body with nitrogen

which can easily cause poisoning injuries to personnel near the acidic gas flare due to gravity diffusion. Therefore, the acidic gas flare cannot rely on the ground deflagration ignition system. It is recommended that the acidic gas flare adopt the method of pilot light ignition, or change from a manual ignition system to an automatic deflagration ignition system.

9.10.2.2 Monitoring the Combustion Status of the Pilot Light and Flare

To ensure the safe operation of the flare system, a thermocouple can be set up on the flare head to alarm when the temperature reaches the lower limit. The on-site igniter can be equipped with an indicator light for the combustion status of the pilot light, and the switch status signal of the pilot light can be led out from the igniter to the DCS system in the control room. A television monitor can be set up in the control room to observe the combustion situation and smoke elimination effect of the flare in real time. The combustion degree of the flare can be judged from the color and height of the flame, and the change in flame length can also reflect the change in flare gas flow, thereby reflecting the operation of the related equipment.

9.10.2.3 Control of Steam Flow

When the smoke elimination steam and ejector steam are insufficient, it will lead to incomplete combustion of the flare gas, producing a large amount of smoke. To solve this problem, the method of timely adjusting the amount of smoke elimination steam and ejector steam can be adopted. By injecting steam to produce a heat absorption effect, the temperature of the flame combustion zone is reduced, the oxidation time of the hydrocarbon medium is extended, and its relative molecular mass is reduced.

An appropriate amount of steam can promote the combustion reaction, thereby achieving smokeless combustion; while an excessive amount of steam not only causes waste and significantly increases noise, but also causes flame pulsation, making the combustion unstable or even extinguishing. Therefore, excessive steam should be avoided or prevented as much as possible. At the same time, when the flare is ignited, a continuous supply of a certain amount of steam should be maintained at the flare head, which can cool and protect the flare head. Even when there is no emission gas, the supply of protective steam is not allowed to stop. When the emission volume is large, the control valve should be adjusted in time to increase the steam volume.

Determining the steam flow based on the flare gas discharge volume can solve the problem of insufficient combustion when the flare gas discharge volume is large, and can also solve the problem of the flame being easily blown out by steam when the flare gas volume is small. Therefore, it is recommended to add a control of the ratio of flare gas discharge volume to steam volume according to the relative molecular mass of the flare gas, and automatically adjust the steam volume.

9.10.2.4 Adjustment of Fuel Gas and Air Flow

Fuel gas is used for ignition and lighting the pilot light, and the role of air is to ignite. When the flare device is put into operation, the pilot light must be ignited first. By adjusting the ratio of air to fuel gas flow, a spark is generated by the igniter to quickly and reliably ignite the pilot light. The role of the pilot light is to ignite the flare gas discharged from the flare tube in time. During the normal operation of the production device, a small part of the gas may be discharged to balance production. Although the start and stop of the production device are predictable, accidents in the production device are unpredictable. In order to maintain the normal operation of the production device and quickly eliminate accidents, the extinguishing of the pilot light is very critical. Therefore, a fixed value adjustment system for fuel gas flow should be set up, and a pressure detection instrument should also be installed on the fuel gas pipeline. When the pressure is lower than the set value, an alarm will be triggered, so that the operator can monitor the operation in the control room and take appropriate measures to ensure the normal operation of the flare device.

9.10.3 Measures to Prevent Backfire

The flare system is an important facility to ensure normal production in the factory and in case of accidents. Taking measures to prevent backfire can prevent the flare system itself from backfiring and exploding, ensuring the safe production of the factory. Reliable methods to prevent backfire should be adopted in the design of the flare system, and the following methods can be used individually or in combination:

(1) Gas purging.
(2) Liquid sealing (water seal).
(3) Velocity seal.

The main purpose of the above methods is to prevent air from entering the flare tube. It should be noted that gas sealing (i.e., velocity sealing) itself is not sufficient to prevent backfire. If the torch gas condenses, a large amount of air is likely to be sucked in; if the torch manifold is suddenly cooled by heavy rain, gas will also be sucked in. The water seal tank is an important facility and means to prevent torch backfire explosions, which can cause damage to the flammable gas discharge pipe network and its connected equipment. The closer the location of the water seal tank to the base of the torch or discharge tower, the smaller the range of damage caused by backfire explosions. The purge rate used to prevent air from being sucked into the torch system in the above situations may be several orders of magnitude higher than the purge rate used to prevent air from being blown into the torch tube by the wind. If a torch gas recovery device is used, a gas purge and liquid seal should be installed.

If air (or oxygen) enters the torch system and forms a flammable gas mixture within the system, this mixture will be ignited by the pilot light of the torch head. If the flame backfire speed exceeds the flame ejection speed, an explosion will occur.

If the flame ejection speed is very close to the flame backfire speed, a fairly stable combustion may occur inside the torch tube, which could lead to overheating and mechanical damage (this kind of internal combustion can continue for a long enough time to overheat the torch tube body, provided there is enough air entering; this will happen if there is not enough purge gas to prevent air from entering the torch tube body). The following situations lead to the formation of a flammable mixture:

(1) The vacuum system is connected to the torch.
(2) Lighter than air gases, especially hydrogen, enter the torch system.
(3) Condensation or rapid cooling occurs within the torch system (this can be reduced or even prevented by heating or insulating the torch pipeline, but this may be costly and unreliable).
(4) The pressure relief valve is removed.
(5) Two or more torches share a manifold, and there is no liquid seal in between.
(6) Air or oxygen is used in the process equipment connected to the torch system.

If the flame ejection speed is greater than the backfire speed, the flame will not return to the torch tube (even if the torch gas has pre-mixed with air before reaching the torch head). To achieve this, purge gas or inert gas can be used. Purge gas can increase the speed of the torch ejection, and inert gas can reduce the speed of the flame return. The position where the inert gas is introduced should be as close to the torch head as possible, and a full mix with the torch gas should be achieved. However, it is necessary to ensure that the calorific value of the mixed gas is higher than the required calorific value, otherwise the pilot light and flame may be extinguished. The use of inert gas should also consider whether there is a reliable source of inert gas and its cost-effectiveness. Speed seal and inert gas introduction can be combined.

The advantage of using inert gas is to prevent backfire, but the disadvantage is that when the torch gas flow rate is low, the introduction of inert gas will reduce the combustibility of the torch gas. If unburned toxic and odorous gases are discharged into the atmosphere, it may cause environmental pollution and harm to personnel.

9.10.3.1 Gas Purging

Under torch conditions, any oxygen-free gas that will not reach the dew point can be used as a purge gas, such as nitrogen, natural gas, and methane-rich fuel gas are ideal purge gases. If the relative molecular mass of the purge gas is less than 28, then the volume of the purge gas needs to be increased. In addition, steam is not recommended as a purge gas, because when steam condenses, its volume will shrink, which will draw air into the torch system, and the condensed water from the steam will remain in the torch system, posing a risk of freezing, blocking part of the system, and accelerating material corrosion.

For torches with high-speed combustion or wide explosion limit characteristics (such as hydrogen, acetylene, and ethylene oxide, etc.), acidic gas torches, and toxic medium torches, using fuel gas as a purge gas can improve their combustion characteristics and further enhance the safety of torch operation. Layered supply of backfire prevention purge gas can reduce the consumption of fuel gas.

The speed sealer should be installed near the inlet flange at the lower half of the torch head to avoid being in a high-temperature area for a long time. Damage can also be avoided by preventing air from entering too deeply below the torch head, while also facilitating maintenance. When the torch gas flow decreases to a certain value, due to the torch gas heating and then cooling, and due to the temperature difference between day and night, the recombination fraction in the torch gas will undergo condensation, possibly creating a vacuum, causing air to flow back into the cylinder from the top of the cylinder. If the torch gas carries oxygen, under certain conditions, it can cause the gas in the torch system to reach the explosive limit range. If it encounters a burning long-burning lamp or other sufficient energy source at this time, an explosion or backfire will occur. Therefore, a certain flow of purge gas should be maintained at the torch head outlet. A pressure regulating valve, orifice plate, and pressure detection instrument should be set up on the purge gas pipeline. If the pressure is lower than the set value, an alarm will be triggered to ensure the normal operation of the torch device.

9.10.3.2 Water Seal

The design of the water seal tank should ensure that it is not damaged in the event of a backfire explosion. Theoretically, the pressure produced by the explosion of hydrocarbon gas in a closed space is 7–8 times the gas pressure (absolute pressure). The explosion of the torch usually occurs at the end of the discharge. At this time, the pressure in the water seal tank is close to atmospheric pressure. Considering the large gap between the allowable stress of equipment design and the strength limit of metal, the design pressure of the water seal tank should not be lower than 0.7 MPa (gauge pressure).

With the enlargement of oil and gas field stations, the volume of gas discharged from the torch is increasing, and the size of the corresponding water seal tank is also becoming larger. Using a vertical tank may affect the height of the system pipe gallery; in addition, when the volume of discharged gas is large, the stability of the water seal liquid level in the vertical tank is far less than that in the horizontal tank, which can easily cause the overflow water volume to be too large.

Under certain conditions, there are two types of negative pressure conditions in the combustible gas discharge pipe network: one negative pressure condition is when the high-temperature gas discharge stops encountering rainfall, the temperature of the gas in the pipeline drops sharply, causing the entire pipe network to appear negative pressure. If the seal water volume is insufficient, it will cause air to enter the pipe network system from the torch head; another negative pressure condition is when the discharge gas with a density less than the air density is slowly flowing or not flowing under the effect of atmospheric pressure elevation difference, the water seal tank to any point at the torch outlet is in different negative pressure states. If the water seal water volume is insufficient and the system pipe network maintains positive pressure measures fail at this time, the entire combustible gas discharge system will appear negative pressure. This negative pressure is self-balancing and will not cause air to enter the pipe network system from the torch head, but it can cause air to enter the system from the vent pipe or holes formed by corrosion on the equipment. For combustible gases containing a large amount of hydrogen, acetylene, ethylene oxide, etc., which have an abnormally fast combustion rate, once oxygen enters the system pipe network to form explosive gas, and backfire flash explosion occurs after the torch water seal tank, the water seal cannot block the flame from propagating to the system in front of the water seal tank.

When two or more torches operate simultaneously in the same venting system, there will be a pressure difference between different torches. When the torch gas discharge volume is small, it may cause mutual suction between the torches, leading to air entering the torch tube and causing an explosion accident (it is very easy to occur when the hydrogen content is high). Therefore, water seal tanks must be used between torches to prevent gas backflow in the torch tube. Setting the water seal height in layers is beneficial to reduce the torch head smoldering problem during small gas volume conditions. Two mutually backup torches only exist in a short time simultaneous use condition during switching. The backup torch is isolated from the torch in use with a valve and blind plate after switching, and it is economical and reasonable to share a water seal tank. However, to avoid accidents during the connection of the two torches during the switching period, sufficient safety purge gas should be provided during the switching operation of the two torches.

The start and stop of the condensate pump are controlled by the liquid level in the separator tank. When the liquid level is high, the pump automatically starts, and when the liquid level is low, the pump automatically stops. If the pump's automatic start and stop fail, an alarm will be triggered when the liquid level reaches the high limit and low limit, so that the operator can take appropriate measures in time to prevent accidents. The liquid level of the water seal tank is maintained by the liquid flow. The liquid level and temperature of the water seal tank can also be monitored in the control room. In cold weather conditions or when there is a possibility of discharging low-temperature gas, in order to prevent the water seal from freezing, heating steam can be automatically controlled to enter based on temperature parameters, or other heating measures can be taken.

9.10.4 Preventing Torch Rain

Torch rain is caused by the combustion of liquid in the torch gas. This situation is very likely to cause accidents, especially when the torch is set up in the device area. The fundamental method to prevent torch rain is to strictly control the discharge of the device. Combustible liquids must be allowed to enter the torch system only after the evaporator, and the discharge of heavy hydrocarbon liquids into the torch system is strictly prohibited. When designing the separator tank, it should be ensured that there is enough volume, and the inlet filter of the condensate pump should be checked regularly to prevent debris and polymers from blocking the pump inlet, and the liquid level of the separator tank should be checked regularly.

9.10.5 Other Safety Matters

9.10.5.1 Control of Navigation Lights

The anti-aircraft signs and light protection of the torch should be implemented according to relevant regulations. The activation of the navigation lights should be automatically controlled, and its operation signal should be sent to the control room.

9.10.5.2 Other

The torch should avoid being placed in wind-trapping areas to facilitate the diffusion of emissions. The heat radiation, light radiation, noise, and pollutant concentration produced by the torch should not exceed the values stipulated by relevant standards. Elevated torches should be equipped with navigation lights as required. External torches and their ancillary equipment should be enclosed with wire mesh or walls.

9.11 Torch Gas Recovery System and Environmental Protection Measures

9.11.1 Recovery System

The hydrocarbons and other combustible gases burned in the torch are considerable. In order to alleviate energy tension, reduce product costs, and reduce environmental pollution, countries around the world have begun to modify the torch system and recycle and reuse torch gas. This not only improves economic benefits but also eliminates smoke, noise, and exhaust pollution caused by torch combustion, extending the life of the torch head. Many devices in our country have recycled and reused torch gas, achieving significant economic benefits. In most cases, the recovered torch gas is processed and transported to the plant's fuel gas system. The torch gas recovery system consists of two parts: the water seal system and the torch gas compressor unit.

9.11.2 Water Seal System

The torch is usually set outside the process unit boundary. The torch gas from the torch gas manifold enters the torch liquid separator before burning in the torch, further separating larger droplets carried in the torch gas, and then enters the water seal tank. The water seal tank is generally set before the torch, serving as a safety facility to prevent backfire in the torch system, and also as a torch gas recovery and pressure control device to prevent the compressor from pumping empty. The water seal tank is usually one of the equipment components of the torch system, and its control is considered by the torch system as a whole.

Under normal circumstances, when the torch gas pressure is low, the torch gas is sealed by the water seal, and the torch gas enters the torch gas recovery device for recycling. In the event of abnormal or major accident emergency discharge, the torch gas breaks through the water seal and is directed to the torch, and the gas is discharged from the torch tube to the torch head for combustion and venting.

9.11.3 Compressor Unit

In the past, gas cabinets were used to recycle and reuse torch gas, but gas cabinet operation was unstable, investment was high, and the footprint was large. In recent years, compressors have been used to directly suck and compress combustible gases (i.e., torch gas) and send them to the fuel gas system. This process is simple, takes up less space, is easy to operate, and has better benefits. The torch gas recovery compressor unit consists of a compressor and related auxiliary equipment and piping systems.

9.11.3.1 Typical Design of Torch Gas Compressor

A typical torch gas recovery system is shown in Fig. 9.11. This typical system consists of one or more reciprocating compressors, with the compressor inlet directly connected to the torch manifold. The compressed gas is usually transported to a treatment system suitable for gas components, and then to the fuel gas or treatment system.

9.11.3.2 Recovery System Load

Torch recovery systems rarely calculate using emergency torch loads. Usually, economic factors determine the processing capacity provided by normal torch flow, and excess gas is burned by the torch. Torch loads vary greatly over time. If it is an existing device, before choosing a compressor to recover torch gas, the flow and

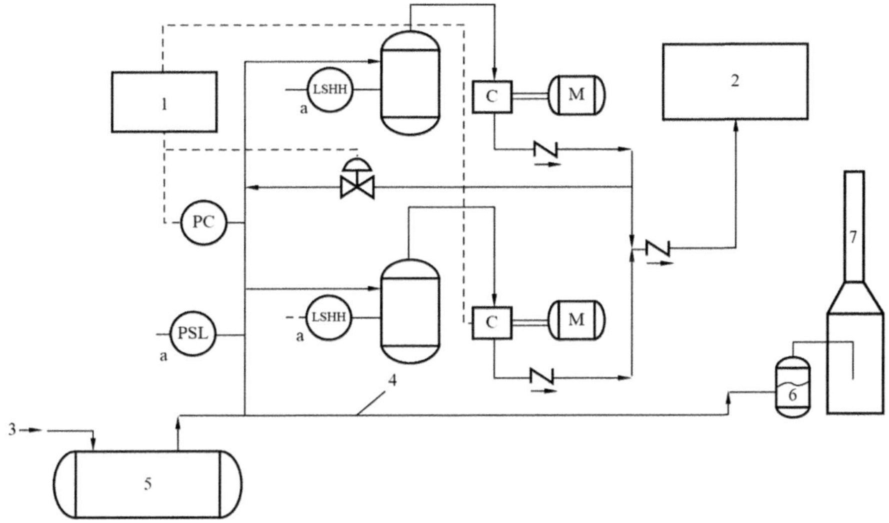

Fig. 9.11 Typical torch gas recovery system. 1—Compressor load control; 2—Torch gas treatment system; 3—Torch liquid separator from process unit; 4—Torch gas manifold; 5—Torch liquid separator; 6—Water seal; 7—Torch; a—Compressor shutdown

composition of the torch gas should be measured for a long time, then the average value is calculated, and the compressor is selected based on the average value. If the compressor is chosen too small, the torch gas cannot be fully recovered; if it is chosen too large, some gas will often be circulated, consuming more electricity. For new devices, the compressor can only be selected based on the experience of similar devices. The compression ratio of general hydrocarbons does not change much, but changes in density and adiabatic index *K* will affect the power of the compressor and the temperature at the compressor outlet.

9.11.3.3 Recovery System Design Requirements

The torch gas of the process unit generally comes from the discharge of unbalanced materials, the discharge of leaked materials, the discharge of safety valves, and the discharge of emergencies. These discharged materials are all flammable and explosive media, so special attention should be paid during processing. The torch gas recovery system is considered to add a torch gas recovery device on the basis that the torch system can safely discharge, and it should be able to recover torch gas and ensure the safety of the torch system. The following measures are taken to ensure safety in torch gas recovery.

(1) The typical torch gas recovery system should be located downstream of the main torch manifold connected to all device manifolds and the manifold pressure A position that generally does not change with the load.

(2) The torch gas may carry oxygen, and when the oxygen content reaches a certain value, it may form an explosive mixture. To prevent explosions and ensure safety, a continuous oxygen content analyzer should be installed on the compressor inlet pipeline. When the oxygen content exceeds a certain value, an alarm is triggered, and the compressor is interlocked to stop. At the same time, a temporary sampling port should be installed to periodically analyze the oxygen content in the torch gas to calibrate the accuracy of the oxygen content analyzer.
(3) A water seal tank should be set up on the torch manifold before the torch. First, it serves as a measure to prevent the torch from backfiring; second, it serves as a pressure control device for the torch gas recovery system to prevent the compressor from being evacuated. At the same time, the nitrogen replenishment point of the torch head gas seal (molecular seal or fluid seal) should be set or moved to the torch gas manifold after the water seal tank, which not only ensures the normal use of the torch top gas seal but also prevents the recovery of torch gas containing a large amount of nitrogen.
(4) The torch gas recovery system must be designed as a bypass from the torch manifold. The main torch gas flow cannot pass through any compressor separation or inlet pipeline. The connection pipeline to the torch gas recovery system should be led out from the top of the torch pipeline to reduce the possibility of liquid entry, as the possibility of carrying a large amount of liquid in the torch system is usually high. A liquid separation container should be provided for the compressor, and the compressor should be automatically shut down when the high inlet separator liquid level is reached. The compressor also needs other mechanical protection systems, which can shut down the compressor or reasonably unload the compressor.
(5) The torch gas recovery system must maintain positive pressure and measures must be taken to prevent air from entering the torch gas recovery system from the torch. All compressors should be equipped with high-reliability, low-inlet pressure shutdown controllers, and auxiliary instruments should also be considered for installation on the manifold between the torch and the compressor inlet to detect backflow and automatically shut down the torch gas recovery system. To prevent the compressor from being evacuated, a low-pressure alarm interlock and compressor inlet and outlet pressure regulation facilities should be set up on the compressor's inlet pipeline. To ensure the safety of the fuel gas network, when the compressor outlet pressure reaches a certain value, the compressor inlet valve closes, the compressor internally recirculates, and when the outlet pressure exceeds a certain value, the compressor interlocks to stop.
(6) All on-site instruments and electrical facilities in the torch gas recovery facility should be explosion-proof, and lightning protection measures should also be considered. A combustible gas detector is also installed on-site to detect combustible gas leaks in a timely manner.
(7) A safety valve is set on the gas–liquid separation tank of the compressor unit. When the pressure exceeds the set value, the safety valve jumps and the fuel gas is discharged to the torch system.

(8) In cold and humid places, to extend the service life of the equipment and facilitate operation and maintenance, the torch gas recovery compressor unit needs a closed plant or rain shelter.

9.11.4 Environmental Measures

The torch gas recovery facility itself has an environmental protection function. The torch gas it processes is pressurized by the compressor, separated from droplets, and then sent into the fuel gas system, reducing the pollution caused by torch combustion to the atmosphere. But the recovery of torch gas also produces oily wastewater [2].

If the torch gas recovery facility is arranged in the ethylene unit or other process unit boundary area or nearby, the oily wastewater produced during the recovery of torch gas also returns to the sewage system of the nearby related process unit. Otherwise, other measures should be taken, such as setting up a waste oil tank, separating oil and water, the wastewater enters the sewage system, and the waste oil is recycled. When the torch gas is discharged to the water seal tank, it may carry some hydrocarbon condensate, so the water seal tank drainage generally drains into the production sewage pipe, which is treated by the sewage treatment facility.

References

1. API Standard 521 Pressure-relieving and depressuring systems, Seventh Edition, June 2020.
2. OSHA29 CFR Part 1910.119 Process safety management of highly hazardous chemicals.

Chapter 10
Cryogenic Analysis

10.1 Cryogenic Effects

In the process of oil and gas production, there are sometimes conditions of lower temperatures. Low temperatures increase the risk of brittle failure of metal materials such as carbon steel, leading to damage to pipelines or other process facilities. Especially when pipelines or process facilities are at a higher stress level, or there are some initial defects, the metal embrittlement caused by low temperatures may cause destructive accidents. In engineering history, accidents caused by low temperatures are very rare, but once they occur, they often cause huge losses and harm [1].

There are many situations in the production process that may lead to low temperatures. For example, during the process of system depressurization, the upstream high-pressure system undergoes "isentropic expansion" cooling, and the temperature of the downstream system will also decrease due to the Joule–Thomson expansion ("isenthalpic expansion") caused by the large pressure difference at both ends of the valve. In addition to system depressurization, other causes of low temperatures may include system pressurization, normal/abnormal operations, venting, emergency releases, and changes in other environmental factors. These causes of low temperatures will be detailed in subsequent chapters. Low-temperature condition analysis of process systems is very important for both new and existing facilities. It should be noted that for existing facilities, the initial design may not have considered the low temperatures caused by changes in subsequent operating conditions. This chapter provides a low-temperature research method for predicting the lowest temperatures of process media and equipment.

The significance of this research is:

(1) To judge whether the materials used in existing facilities can withstand the lowest temperatures in the production process.
(2) To select materials with appropriate temperature ranges for facilities to be built.

K. Fang, *Onshore Oilfield Surface Facilities: Process Safety Engineering Guide*,
https://doi.org/10.1007/978-981-96-7830-3_10

(3) To determine whether solid particles that may harm production and pose safety risks, such as hydrates, will be produced due to a decrease in temperature during the process production.

10.1.1 Low-Temperature Damage

The main cause of low-temperature damage to carbon steel pipes and pressure vessels is brittle fracture of steel. Brittle fracture is a sudden and rapidly developing form of damage. When steel is at a low temperature, its toughness decreases and its ability to deform plastically decreases. Brittle fracture often starts from an initial defect in the material. The stress at this defect point (which may be externally applied stress or residual stress inside the material) cannot be eliminated by plastic deformation, so it quickly develops into a larger crack, eventually causing the material to break as a whole. In general, the low-temperature failure of carbon steel requires the coexistence of the following three conditions:

(1) Initial defect.
(2) High stress that drives the defect to develop into a fracture.
(3) The material lacks toughness due to being at a low temperature and cannot prevent the development of the fracture process.

Defects may be spherical or crack-like physical defects, or crack-like impurities, such as manganese sulfide. Defects in materials cannot be completely eliminated, and we usually assume that defects always exist.

High stress can be caused by the pressure of the medium inside the pipe or equipment, or by the inherent stress in the pipe, For example, self-weight, load applied during pipe connection, residual stress caused by welding, or stress caused by thermal expansion and fluid momentum. Therefore, simply relying on depressurizing the pipeline may not be enough to reduce the applied stress below the critical value.

In addition, although temperature significantly affects the toughness of carbon steel materials, temperature does not significantly affect the toughness of stainless steel and aluminum. If the temperature of a carbon steel pipe or vessel is below the material's minimum temperature, failure may not necessarily occur, but each occurrence of low temperature increases the probability of an accident.

10.1.2 Low-Temperature Accidents

In 1998, an explosion caused by low-temperature embrittlement of carbon steel occurred at a natural gas processing plant in Australia. The cause of the accident was identified as the catastrophic failure of the GP-905 shell-and-tube heat exchanger (Fig. 10.1).

Before the accident, the rich temperature flow in the heat exchanger was cut off, so the heat exchanger was cooled below the metal's minimum design temperature.

Fig. 10.1 Shell-and-tube heat exchanger GP-905 with fracture

When the rich temperature flow was reconnected, the shell of the heat exchanger heated up quickly, and the high stress caused by thermal shock damaged the heat exchanger head. The initial defect in this accident was a welding defect left during equipment manufacturing. The crack started from this defect, quickly spread to the shell wall, and caused a fracture. Subsequently, volatile hydrocarbon fluids were released from the heat exchanger, eventually leading to an explosion and fire.

10.2 Causes of Low Temperature

From a thermodynamic perspective, the main reasons for the system or equipment to produce low temperatures include:

(1) The isentropic expansion process caused by pressure reduction.
(2) The isenthalpic expansion process (Joule–Thomson process) caused by fluid passing through a throttle valve.
(3) Auto-refrigeration (temperature reduction caused by liquid evaporation due to pressure reduction).
(4) Cold fluid entering the system.
(5) Heat flowing into the system from the outside.

10.2.1 Depressurization (Emergency Release or Maintenance Venting)

When the process system is in an emergency condition or undergoing maintenance, equipment and pipelines may experience low temperatures due to pressure reduction. When analyzing the low temperatures caused by pressure reduction in the system, each situation that may cause pressure reduction (including some abnormal operation situations) should be analyzed to find the worst case, i.e., the condition that may produce the lowest temperature. If the operating temperature of the system is above the ambient temperature, it is necessary to consider the situation of isochoric cooling of the system to the ambient temperature before venting.

10.2.2 Pressurization

When the process system starts trial operation, or restarts after depressurization, low temperature conditions caused by pressurization may occur, which appear in the initial stage of system pressure increase. At this time, high-pressure fluid enters the low-pressure system (pressurized system) through a valve or throttle element, creating a large pressure difference on both sides of the valve or throttle element, and isenthalpic expansion occurs due to rapid pressure reduction, which is the main cause of low temperature.

When analyzing the lowest temperature that may occur in the system during pressurization, it should be based on the worst operating conditions of the upstream (high-pressure) system, and isenthalpic flash evaporation should be carried out under this condition. The lowest temperature obtained under this condition is the lowest temperature during the pressurization process of the system.

10.2.3 Normal Abnormal Conditions

When analyzing the lowest temperature of the system, not only should the temperature changes under normal operating conditions be considered, but also the low temperatures that may be caused by some abnormal conditions. The following are some abnormal conditions that may cause low temperatures:

(1) Loss of heat load (e.g., furnace failure).
(2) Excessive cold load (e.g., the flow rate of the process medium passing through the cooler decreases, but the cold load provided by the cooler does not decrease accordingly, resulting in excessive cooling of the process medium).
(3) Control valve or control loop failure.
(4) Cold medium entering the hot system.

(5) Deviation of pressure difference changes from the normal design range during the Joule–Thomson cooling process (reasons for the deviation of pressure difference from the normal design value include: changes in pressure difference before and after the control valve or throttle orifice plate, gas breakthrough caused by liquid level drop, pressure reduction caused by safety valve opening).
(6) Changes in heat load during the chemical reaction process due to changes in operating temperature and pressure of the feed.
(7) System shutdown and startup.
(8) Backflow.
(9) When inert gas enters a volatile liquid system, the partial pressure of the volatile liquid in the gas phase space decreases, resulting in vaporization cooling.

10.2.4 Purging

Although purging is generally considered a normal operating condition, special consideration should still be given to it when analyzing low temperature conditions in the system. This is because the purge point is usually the boundary between high and low pressure, and the existence of pressure difference may lead to low temperature conditions. There are two situations of purging:

(1) Purging during maintenance.
(2) Purging during operation.

During maintenance, if purging is done under pressure, the possibility of low temperature conditions should be considered. For purging under operating conditions, when calculating the possible lowest temperature, the lower of the following two should be taken.

(1) The lowest temperature after the fluid under operating conditions undergoes "isenthalpic flash evaporation" when passing through the purge valve.
(2) The lowest temperature that may occur in the system corresponding to the purge point during depressurization.

10.2.5 Emergency Release

When analyzing the lowest temperature of the system during emergency release, the release rate used should refer to the pressure release study of the relevant system. When calculating the lowest temperature of the outlet pipe of the pressure release valve, it should be based on the lowest temperature that occurs during the isenthalpic flash evaporation process when the fluid reaches the release pressure (including overpressure) and flows through the release valve to the atmosphere.

10.2.6 Environmental Factors

Under extreme weather conditions, the lowest environmental temperature may also become a determining factor for the lowest temperature of the system. Changes in environmental conditions may also affect the cooling load, as well as the temperature of the air consumed by the process system, which will affect the operating temperature. For example:

(1) The temperature or wind speed of the air will affect the cooling load of the cooling equipment (such as air coolers or cooling towers), the process involving air (such as the oxidation regeneration process of catalysts), and the process consuming air (such as the compression process driven by gas turbines or gas engines).
(2) The temperature or flow rate of water will affect the temperature of equipment that needs to dissipate heat with water.
(3) The temperature of the soil can affect the temperature of buried facilities (such as buried pipelines).

10.3 Low Temperature Analysis

10.3.1 Design Process

When performing low temperature calculations, all conditions that may cause low temperatures should be considered. For the condition that may cause the lowest temperature, detailed process calculations should be performed to predict the lowest temperature. This section introduces the general steps to determine the lowest temperature of the entire process system (Fig. 10.2).

(1) Make a table to list all conditions that may produce low temperatures. Different systems should be divided according to the process flow, and the inventory and main pressure and temperature changes of each system should be determined.
(2) Preliminary screening of low temperature conditions that may occur in each system.

    ① First, isenthalpic flash is performed on each system to preliminarily predict the possible lowest temperature.
    ② Isenthalpic flash can basically predict the low temperature that may occur at the high and low pressure interface (for example, at the throttling orifice plate or valve where there is a high and low pressure interface).
    ③ For the condition of depressurization, the calculation method of isentropic flash is more suitable. HYSYS Dynamic Depressuring Utility can be used for depressurization simulation calculation to obtain the lowest temperature.

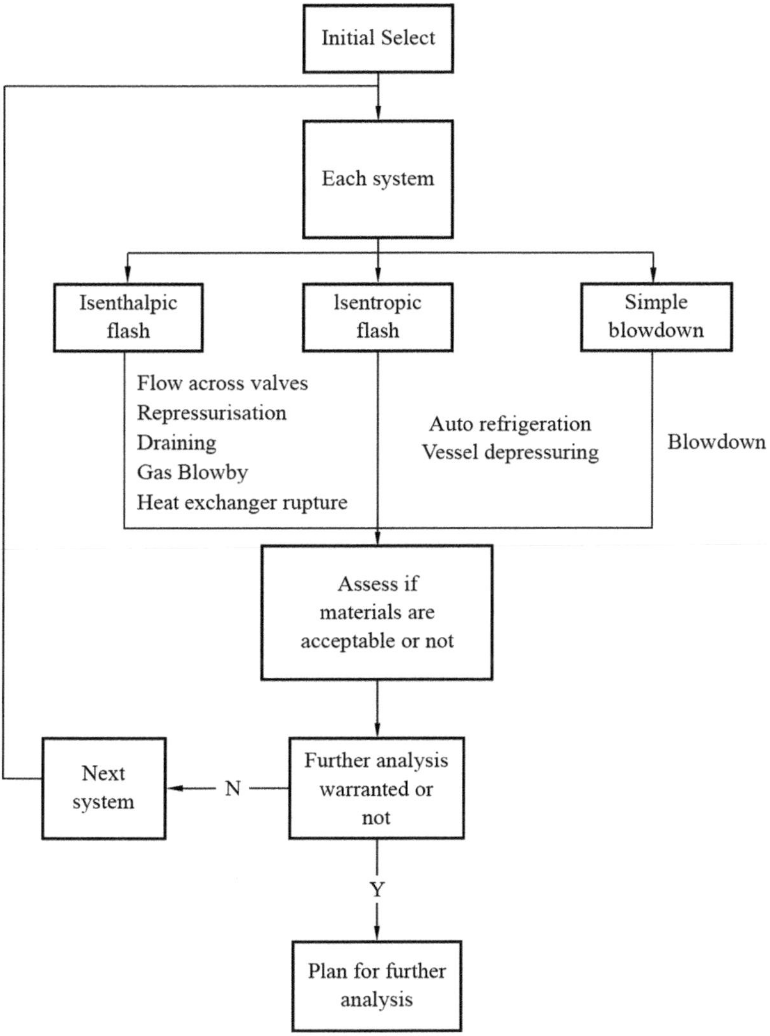

Fig. 10.2 Low temperature analysis steps

If the lowest temperature obtained through the above steps is much higher than the lowest temperature that the metal material can withstand, it is generally considered that no subsequent calculations are needed; if the lowest temperature calculated through the above steps is very close to the lowest temperature of the metal material, more detailed calculations are needed. Whether subsequent calculations are needed requires the engineer to judge based on experience.

(1) According to the data table and steady-state calculation results, recheck the input conditions of the low temperature simulation calculation, such as vessel size, thickness, composition, material, and operating and design conditions, etc.
(2) Perform detailed low temperature calculations for conditions that may exceed the lowest temperature of the material.
(3) If the results obtained from such calculations still have uncertainties, or the temperature obtained is still very close to the limit of the lowest temperature of the material, further in-depth low temperature research is still needed.

Each project should have a design basis document, which stipulates how to conduct low temperature analysis. This basic document should make specific requirements for the low temperature analysis process in combination with specific process flows, geographical locations, relevant laws, and the owner's requirements.

The accuracy required for low temperature calculations depends on the expected use of the calculation results. Feasibility and conceptual design may require lower accuracy than front-end and detailed design. For preliminary studies, engineers can use simple methods to predict low temperatures, and then add a relatively large design margin. In detailed design, more stringent methods should be used, and smaller design margins should be used.

The choice of design margin should be based on similar engineering experience, rather than arbitrarily specifying a design margin. This is because the results of low temperature calculations are very sensitive to initial conditions, the layout of pipelines and equipment systems, the composition of fluids, etc. Engineers can also conduct a sensitivity study of the calculation results to the input conditions to determine the design margin.

10.3.2 Determining the Lowest Design Temperature

The lowest design temperature should be the lower of the following temperatures:

(1) The lowest ambient temperature.
(2) 5 °C below the lowest operating temperature under steady state.
(3) For situations where low temperatures are produced during depressurization or pressurization, a margin should be subtracted from the possible lowest temperature, and this margin should be determined according to different projects.

For the process of depressurization liquid boiling, the boiling point of the liquid can be designated as the lowest temperature, and no margin is needed in this case. This is because for boiling liquid, the lowest temperature is the boiling point of the liquid.

For the situation where a container filled with gas produces low temperature due to depressurization, a margin should be subtracted from the lowest temperature calculated by the software. Generally, a 10 °C margin is sufficiently conservative. But if the temperature obtained after subtracting the 10 °C margin is lower than the lowest design temperature of the metal material, more detailed calculations should be performed.

For the situation where a container containing liquid depressurizes, if the liquid boils during the depressurization process, then the lowest design temperature is the lowest temperature that appears during the depressurization process of the liquid, and no margin needs to be subtracted.

It is generally believed that the temperature of the depressurization nozzle and the container is the temperature of the fluid, unless there is special research to prove that the temperature of the nozzle and the container is different from the temperature of the fluid.

In some cases, if the pressure borne by the pipeline or equipment is low enough, it is permissible for the metal temperature to be lower than the minimum design temperature. This is based on the principle that brittle failure only occurs when the metal stress is high enough. Different mechanical specifications have different regulations for this (for example, specifications for pressure vessels, piping, and pipelines). When using the "low stress" principle for low-temperature design, care should be taken to have sufficient preheating time before the equipment is repressurized.

10.3.3 Pressurization and Depressurization

System depressurization and pressurization are transient processes, which cannot be simulated by steady-state analysis software. This section will discuss some concepts and methods of transient analysis [2].

API Std 521 Pressure-relieving and depressuring systems stipulate (also applicable to the pressurization process): "The depressurization process without heat exchange with the outside world is usually considered an isentropic process, and the minimum temperature of the depressurization process can be obtained by looking up charts or assuming an isentropic efficiency for flash calculation."

In fact, the minimum temperature obtained by the flash calculation method is greatly related to the assumed isentropic efficiency. Unless an accurate isentropic efficiency can be obtained, this calculation method can only be used for the following purposes:

(1) A screening means to roughly predict the minimum temperature.

(2) If the minimum temperature calculated under the worst conditions is not very low, and there is no need to use special materials. Then this calculation method is also acceptable.

In many cases, a more rigorous calculation method should be used. API Std 521 Pressure-relieving and depressuring systems also have the following instructions:

> Appropriate process simulation software can be used for more accurate minimum temperature calculations, especially when considering the heating of the container wall by the surrounding environment (i.e., the minimum design temperature is not the minimum fluid temperature). Especially when the minimum temperature is between the material selection boundaries (for example: between carbon steel and stainless steel), it is recommended to use strict simulation calculation methods (for example: time-related nonlinear thermodynamic analysis methods). Detailed analysis calculations can consider all the factors that affect the cooling effect. This calculation can also generate transient temperature curves of the depressurized volume and the corresponding temperature curves of the fluid in the release system

Figure 10.3 illustrates the difference between the assumed isentropic expansion and the actual process. The isentropic expansion line represents the process of isentropic expansion of the gas assuming that it is not affected by the outside world, and the real process line reflects the actual process. The real process line shows that when the gas cools, the heat transfer from the pipeline/equipment and the environment prevents the gas from reaching the "isentropic temperature".

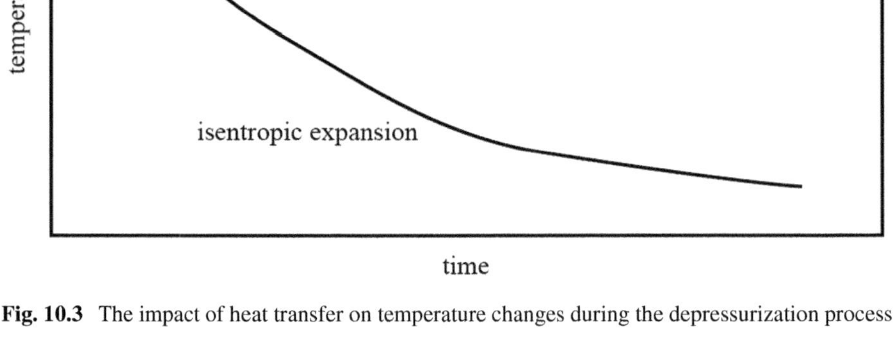

Fig. 10.3 The impact of heat transfer on temperature changes during the depressurization process

10.3.4 Isentropic and Isenthalpic

The isentropic and isenthalpic depressurization processes described in this chapter are adiabatic (i.e., no heat transfer), the difference lies in reversibility. The isentropic process is completely reversible (entropy remains constant), while the isenthalpic process is completely irreversible (enthalpy remains constant).

10.3.4.1 Isentropic Expansion

As stated in API Std 521 Pressure-relieving and depressuring systems, the depressurization process in the absence of heat transfer is close to an isentropic process. In fact, heat transfer always exists, so the actual thermodynamic process cannot be completely isentropic. Regardless, this expansion process is still the main cause of low temperatures. In this section, "isentropic expansion" (in quotes) is used to describe system depressurization expansion, and isentropic expansion (without quotes) is used to describe the strict thermodynamic process.

10.3.4.2 Isenthalpic Expansion

Another cause of low temperatures is the pressure drop when the fluid passes through the relief valve or the throttling orifice plate. This adiabatic, irreversible isenthalpic process is called the Joule–Thomson effect. According to Joule–Thomson's description, the temperature of real gas drops when the pressure decreases (unlike ideal gas, the temperature remains constant). For most gases in oil and gas field processes, this rule applies. However, hydrogen and helium will increase in temperature during the pressure drop process.

In this section, "isenthalpic expansion" (in quotes) is used to describe the expansion process of the fluid through the valve or the throttling orifice plate, and isenthalpic expansion (without quotes) is used to describe the strict thermodynamic process.

The temperature drop upstream of the throttle valve or throttle orifice plate is mainly due to the isentropic expansion process. The temperature drop downstream of the throttle valve or throttle orifice plate is the combined effect of upstream "isentropic expansion" and "isenthalpic expansion" that occurs during the fluid flow through the valve process.

10.3.4.3 Component Effects

The degree of temperature drop caused by isentropic and isenthalpic flash evaporation depends on the composition of the gas. For example, Table 10.1 compares the temperatures of methane and ethane after isentropic and isenthalpic flash evaporation. The cooling effect of isentropic flash evaporation of methane is more pronounced, while

Table 10.1 Comparison of temperature changes in isenthalpic and isentropic flash evaporation

Gas components	Methane	Ethane (°C)
Inlet pressure (gauge pressure)	1500 kPa	
Inlet temperature	50 °C	
Outlet pressure (gauge pressure)	0kPa	
Outlet temperature of isentropic process	−110 °C	−79
Outlet temperature of isenthalpic process	+44 °C	+35

the cooling effect of isenthalpic flash evaporation of ethane is more pronounced. This difference reminds designers that when calculating the minimum temperature, the possible impact of component changes on the calculation results should be considered.

An example is the case of different components flowing through a natural gas pipeline. When calculating the minimum temperature during pressure reduction inside the pipeline, the temperature obtained from lighter components may be lower, and when calculating the minimum temperature of the venting pipeline, the temperature obtained when using heavier components may be lower.

10.3.5 Kinetic Energy Effects

10.3.5.1 Influence of Medium Temperature

Kinetic energy may affect the temperature of the fluid and metal. If the gas flow rate is very high (such as the gas downstream of the pressure relief valve or the flow restriction orifice), a part of the total energy will be converted into kinetic energy, causing a decrease in fluid temperature.

In general, the energy existence Eq. (10.1) shows a balance:

$$H_{\text{in}} + E_{\text{Kin}} + E_{\text{Pin}} = H_{\text{out}} + E_{\text{Kout}} + E_{\text{Pout}} \tag{10.1}$$

where

H—Enthalpy, $H = U + pV$ (U represents internal energy, p represents pressure, V represents volume);

EK—Kinetic energy;

EP—Potential energy;

The subscript "in" "out"—respectively represent in and out.

For pressure relief valves or flow restriction orifices, it is generally considered that the EKin term can be ignored, EPin = EPout, so Eq. (10.1) can be simplified to Eq. (10.2).

$$H_{\text{in}} = H_{\text{out}} + E_{\text{Kout}} \tag{10.2}$$

The above equation indicates that higher flow rates may cause a significant drop in gas temperature (the temperature drop may exceed 30 °C). Therefore, it is necessary to consider the impact of kinetic energy on fluid temperature.

10.3.5.2 Influence of Metal Temperature

Considering the impact of kinetic energy on gas temperature will help engineers more conservatively predict the minimum temperature of pipe walls or equipment shells. Generally, the minimum temperature of the gas will occur at the connection between the valve or flow restriction orifice and the downstream pipeline. This is because the gas has the highest flow rate when passing through the valve or orifice. Increasing the size of the tailpipe can alleviate the low temperature caused by this local high flow rate.

For the pressure reduction process, the pressure of the pressure source will decrease over time, and the temperature of the pressure source will also decrease, but the mass flow rate and flow rate of the exhaust tailpipe will also decrease with the temperature. Therefore, as shown in Figs. 10.4 and 10.5, kinetic energy does not have a significant impact on the temperature of the metal. Therefore, the minimum gas temperature can be used to estimate the temperature of the metal, and no margin needs to be subtracted.

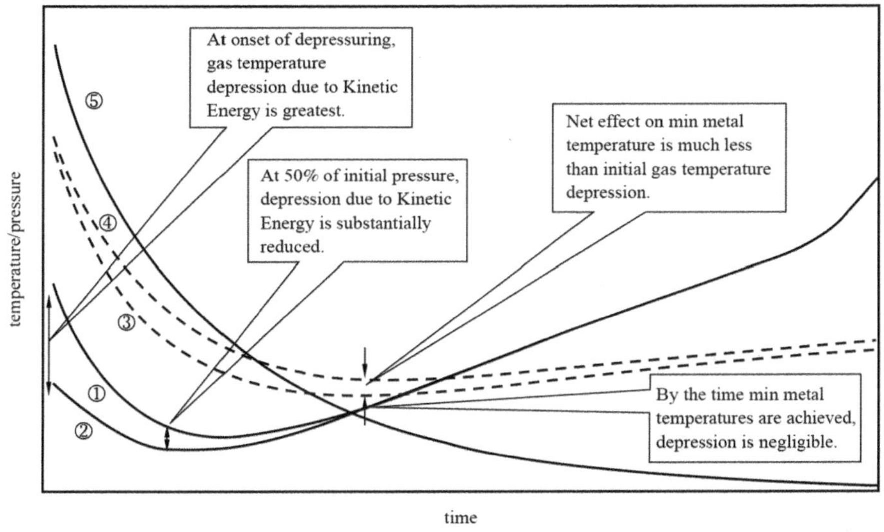

Fig. 10.4 Kinetic energy effect of pressure reduction process—large size tailpipe. ① Gas temperature (including kinetic energy); ② Gas temperature (excluding kinetic energy); ③ Metal wall temperature (including kinetic energy); ④ Metal wall temperature (excluding kinetic energy); ⑤ Pressure

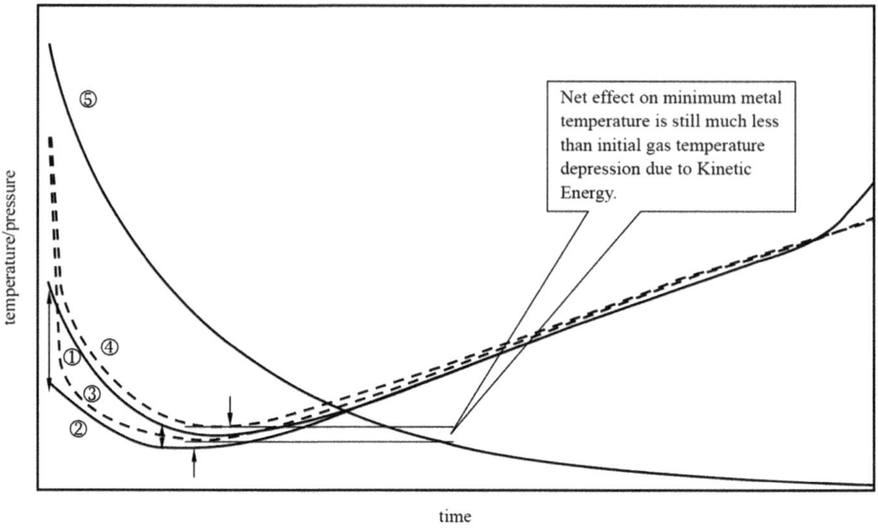

Fig. 10.5 Kinetic energy effect of pressure reduction process—small size tailpipe. ① Gas temperature (including kinetic energy); ② Gas temperature (excluding kinetic energy); ③ Metal wall temperature (including kinetic energy); ④ Metal wall temperature (excluding kinetic energy); ⑤ Pressure

The pressurization process is different from the pressure reduction process. During the pressurization process, the pressure and temperature of the pressure source are constant. If the system being pressurized is a large system, the gas flow rate and downstream pressure will remain constant for a long time. In this case, the temperature of the pipes and equipment of the pressurized system will gradually approach the temperature of the gas. In this case, it is appropriate to subtract a margin of 5 °C from the calculated minimum gas temperature. If the temperature calculated above Approaches the boundary temperature for material selection, then a more detailed analysis as shown in Figs. 10.4, 10.5 and 10.6 should be performed.

10.3.5.3 The Impact of Flow Restriction Orifice Plates and Valves

Flow restriction orifice plates are usually made of low-temperature resistant stainless steel, so the orifice plates themselves do not have a problem with low-temperature brittleness. In addition, the coldest gas temperature near the orifice plate occurs at the hole contraction downstream of the flow restriction orifice plate, which is a certain distance from the orifice plate.

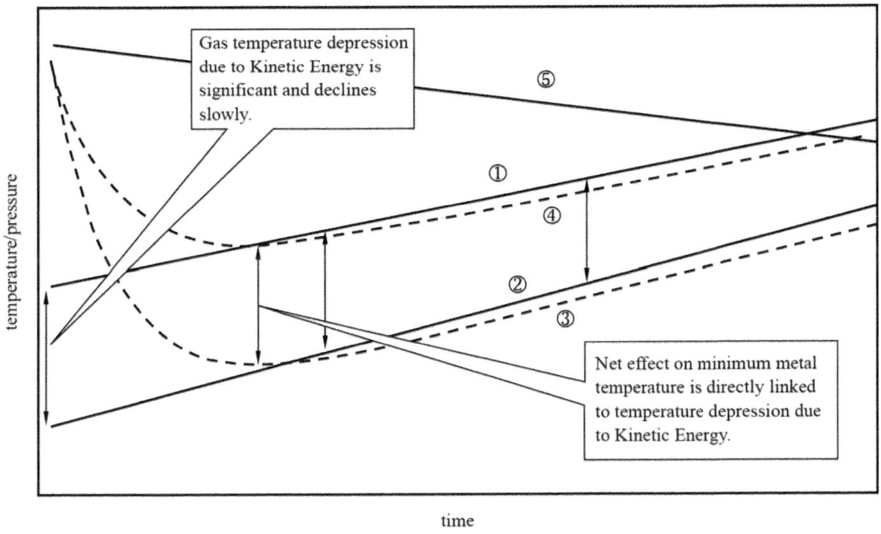

Fig. 10.6 Kinetic energy effect of pressurization process. ① Gas temperature (including kinetic energy); ② Gas temperature (excluding kinetic energy); ③ Metal wall temperature (including kinetic energy); ④ Metal wall temperature (excluding kinetic energy); ⑤ Flow rate

Valves contain both throttling elements and sealing materials, so the ability of the valve to withstand low temperatures needs to be considered. In terms of low-temperature resistance, the valve body material should be at least not lower than the material selected for the downstream pipeline. However, if the kinetic energy effect inside the valve is greater than the kinetic energy effect inside the pipeline, the gas temperature inside the valve body may be lower than the gas temperature in the downstream pipeline.

10.3.5.4 Kinetic Energy Simulation

HYSYS does not include kinetic energy in the energy balance, so if the gas flow rate is very high, the prediction of the gas temperature downstream of the valve/orifice using HYSYS may be overly optimistic. The Aspen Hydraulics subprocess can be used for rigorous hydraulic analysis, but it is not specifically designed for flare system modeling, so it can be a bit clumsy to use.

Aspen Flare System Analyzer (formerly known as Flare Net) can be used to simulate flare systems and includes the impact of kinetic energy on temperature. Its calculation results can be used to check the temperature downstream of the throttle orifice plate or other locations obtained in the HYSYS Dynamics simulation, and use this information to manually modify the original HYSYS Dynamics temperature results.

10.3.5.5 Simulation Software

Kinds of software can be used to simulate the process of equipment and pipeline depressurization and pressurization. Table 10.2 lists some simulation software:

The choice of simulation software should consider the specific situation of the system being simulated and the requirements for the accuracy of the simulation. Generally speaking, customers will have a preference, and it should be determined as early as possible with the customer which software to choose, and this should be explained in the design basis documents.

HYSYS Dynamic is quite successful in simulating the depressurization of gas phase systems. However, some information indicates that when condensate is present, the system depressurization simulation of HYSYS Dynamic is not very accurate. HYSYS Dynamic Depressuring Utility provides a simple and easy-to-use interface, so it is favored by many engineers.

When using this module, it should be noted that: this model is quite accurate for the depressurization simulation of a single vessel, but it is not very suitable for other systems. When there are many pipelines (for example: inlet and outlet manifolds, pipelines), or there are many vessels and the pipelines connecting them, this analysis tool should not be used for simulation, for the following reasons:

Table 10.2 Software for transient simulation

Purpose	Software
Preliminary screening	HYSYS dynamic depressuring utility
	PRO II STEADY STATE
Advanced release simulation	HYSYS dynamics
	Aspen blowdown
	Unisim blowdown tool
	Unisim dynamics
	BLOWDOWN-Royal Institute of Technology, London
	gFLARE-Process Systems Enterprise
Fire survivability modeling	Vessfire-Petrell
	Saffire-genesis
Pipeline transient simulation	OLGA 2000 (mainly used for lines)

(1) When a system contains many vessels and pipelines, the software calculation will convert the entire system's volume into an average vessel. This assumption cannot accurately predict the lowest temperature that may occur in the real system. API Std 521 Pressure-relieving and depressuring systems state that the temperature in different locations of a system is different.
(2) The heat transfer coefficient used in the calculation of the vessel unit is the natural convection correlation, which is not suitable for the calculation of pipelines. The heat transfer correlation required for pipeline calculation is very different from the correlation used for vessels.
(3) The calculation of the separator unit uses the vertical dimension of the vessel as the input condition for calculating the internal heat transfer coefficient. This dimension is related to the direction of the vessel (for example, for a horizontal vessel, the vertical dimension is the diameter, and for a vertical vessel, the vertical dimension is the height). But when there are many vessels and pipelines, it is impossible to clearly define the direction and size.

The Blowdown software developed by the Royal Institute of Technology in London can simulate the entire plant. Although HYSYS Dynamic has uncertainty in predicting the lowest temperature, the HYSYS Dynamic Depressuring Utility program is still an effective tool for estimating depressurization time, because the gas flow rate through the orifice is not very sensitive to the gas temperature.

10.3.6 Liquid Treatment

10.3.6.1 Hydrocarbon Liquids

The system's liquid will vaporize during the depressurization process, affecting the thermodynamic processes of the gas and liquid phases. Similarly, the condensation of droplets during depressurization will also affect the thermodynamic processes of the gas and liquid phases. For containers containing liquid, a more conservative approach in simulation calculations is to assume that the liquid in the container is at a low liquid level. The lower the liquid level, the lower the fluid temperature during depressurization, but the higher the temperature of the metal wall. Since the minimum design temperature of the container is generally based on the minimum fluid temperature, selecting a low liquid level for calculation is a more conservative method. It should be noted that although the low liquid level is not the normal operating liquid level, during maintenance depressurization, the liquid level in the container is usually reduced first. Therefore, the low liquid level can serve as an appropriate starting point.

Additionally, API Std 521 Pressure-relieving and depressurizing systems also mentions that "During the depressurization process of high-pressure pure gas phase fluids (especially dense gases, such as supercritical fluids), special attention needs to be paid to the possible condensation of the liquid phase. These liquids may accumulate at low points (such as the connection between the bottom of the container and the discharge pipeline, etc.). The wall temperature of the metal may be higher than the temperature of the droplets, causing the droplets to boil and local low temperatures. If this situation occurs, a more detailed analysis may be needed to determine the minimum temperature."

When gas is depressurized through a pressure relief valve and orifice plate, it may also produce condensing droplets, which are carried in the gas phase to the flare system, affecting the outlet temperature of the pressure relief valve.

10.3.6.2 Water

The presence of water will affect the temperature obtained through simulation software calculations. Some past work on systems containing gas phase/light hydrocarbon components has shown that, in the absence of free water and saturated water, the fluid temperature obtained by HYSYS simulation will be higher. Other work on coalbed methane (without liquid phase light hydrocarbons) has yielded exactly the opposite results. In the absence of free water and saturated water, the temperature obtained by HYSYS simulation will be lower.

HYSYS simulation can obtain a sub-zero depressurization endpoint temperature, and it shows that water is still in the liquid state. Ice may exist at the interface of hydrocarbons/water, hydrocarbons/container. This will insulate the heat transfer between hydrocarbons and water, thus making the temperature of hydrocarbons lower than the overall temperature.

Due to the latent heat in the icing process and its impact on the heat transfer coefficient between the fluid and the metal wall, predicting the temperature of hydrocarbons will be difficult. If engineers believe that water will have a significant impact on the temperature of the depressurization process, it is necessary to hire a professional institution for strict calculations.

10.3.7 Initial Conditions

In the process of calculating the minimum temperature of system depressurization, it is necessary to consider how to stipulate the initial conditions of the system. Usually, the initial conditions that can obtain the lowest depressurization temperature need to be selected. This is usually the condition with the highest initial pressure and the lowest initial temperature. For refrigeration equipment (such as refrigeration systems, expansion machines, etc.), using the operating conditions of the system as the initial conditions can obtain the lowest discharge temperature. For systems operating at temperatures higher than the ambient temperature, it is necessary to consider the situation of depressurization after the system is shut down and cooled to the ambient temperature at constant volume; it is also necessary to consider that due to leakage problems, even if the system has cooled to the ambient temperature, the pressure still maintains the original operating pressure. Therefore, depressurization calculations can start with the maximum shutdown pressure, or consider using the design pressure as the starting pressure.

When considering the low temperature generated during the depressurization process, it is necessary to consider the depressurization caused by pressurization, which will make the system start temperature lower than the normal temperature. When considering the low temperature caused by the pressurization process, it is necessary to consider the situation of immediate pressurization after depressurization, which will also make the system start temperature lower than the normal temperature.

10.3.8 Depressurization Time

The duration of the pressurization and depressurization process will also affect the temperature of the fluid and metal, but this effect is difficult to predict. In fact, changes in duration may cause the temperature at some locations in the same system to decrease, while the temperature at some locations increases. Given the complexity of the thermodynamic and heat transfer interactions between different phases of the fluid and the environment, container, and pipeline, the impact of the duration of the pressurization and depressurization process on the minimum temperature must be predicted based on strict calculation results. The following example can be used to illustrate this concept. In this example, the initial conditions of the system are the ambient temperature and a fixed initial pressure. The endpoint temperatures corresponding to different depressurization times are shown in Fig. 10.7.

This example is divided by theoretical points *A* and *E*, the definitions of points *A* and *E* are as follows:

(1) *A*: Instantaneous Pressure Drop: Instantaneous expansion pressure drop will result in the lowest fluid temperature. However, there is no time for heat transfer between the metal and the fluid, so the metal temperature remains at the initial temperature T_0.
(2) *E*: Infinitely Slow Pressure Drop: The system pressure drop is very slow, and the heat transfer from the environment can keep the gas and metal temperature at the ambient temperature T_0.
(3) Points *B*, *C*, and *D* reflect the actual pressure drop time (the actual pressure drop is neither instantaneous nor infinitely long), resulting in the lowest metal temperature below the initial/ambient temperature T_0.

When evaluating system pressure drop, it is recommended to perform a sensitivity analysis on the duration of the pressure drop to determine whether the design is at point *C* (most conservative), or at point *B* or *D* (at point *B* or *D*, changes in the duration of the pressure drop may result in a lower minimum temperature).

Also note that the relationship in Fig. 10.7 also applies to different points in the same system. For example, for a given pressure drop time, a container in a system

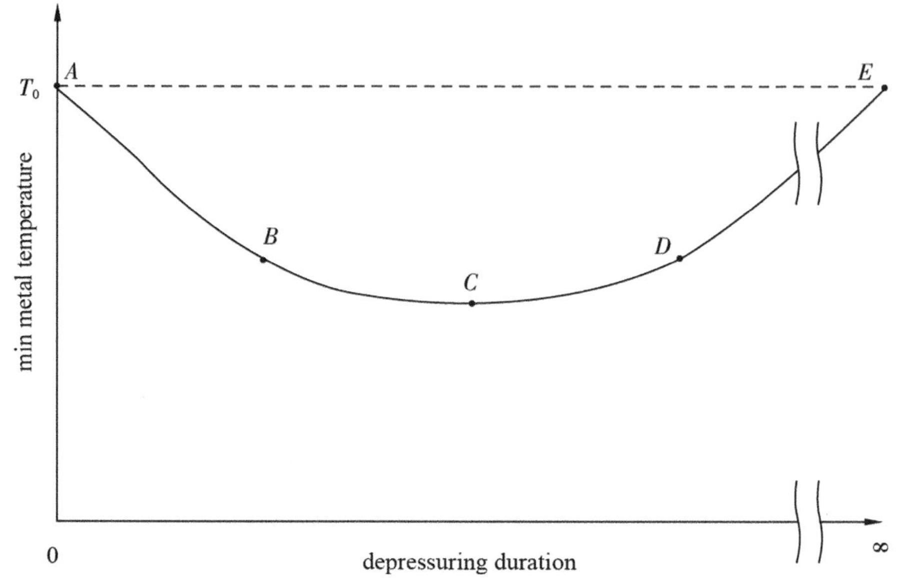

Fig. 10.7 Pipeline of depressurization duration and minimum metal temperature

may be at point *B*, but the pipeline of the flare system may be at point *D*, in this case, increasing the pressure drop time may lower the temperature of the container and raise the temperature of the flare system pipeline.

Understanding the sensitivity of temperature to the duration of pressure drop helps to understand the choice of margin in the design. If the duration changes, it should be understood whether it will affect the minimum temperature.

10.3.9 Distillation Tower

When performing low-temperature analysis of the distillation tower, the initial release temperature, fluid composition, and volume of the pressure relief system need to be considered. Past projects have shown that it is appropriate to use the composition of the feed at ambient temperature. This can make the calculation results more conservative. Because some light components will flow down from the tower tray, the final release components include both the heavy components at the bottom of the tower and the light components at the top of the tower. For the reboiler, the components of the reboiler under normal operating conditions can be cooled to ambient temperature as the initial condition.

The inventory of liquid remaining on the tower tray, compared to the inventory of liquid at the low level of the distillation tower, is not negligible. But because the lower the liquid level, the lower the minimum temperature obtained, ignoring the inventory of liquid on the tower tray can result in more conservative calculation results. However, for distillation towers operating at low temperatures such as demethanizer, the inventory of liquid on the tower tray should be considered.

10.4 Emergency Pressure Relief Piping

10.4.1 Inlet Piping

The piping immediately upstream of the relief valve or throttle orifice is usually small in diameter and short in length. During the pressure drop or boost process, due to the high gas velocity and lack of heat loss in the pipe, the metal temperature of this section of pipe will be very close to the gas temperature. The lowest metal temperature should usually be specified as the lowest gas temperature, which is not overly conservative and can simplify modeling.

For the pressure drop process, engineers should consider the setting of the pressure relief piping. If the pressure is relieved from a high point on the pipeline from the vessel, the pipeline may be overcooled. Consider relieving pressure from other locations (such as the nozzle of the vessel or a more low-temperature resistant pipe, or some large diameter or thick-walled pipes, where the temperature change is not significant during pressure drop). For the pressurization process, since the pressure of the upstream pressure source does not drop, there is no low-temperature risk for the pipe, so the location of the pipe take-off is not important.

10.4.2 Outlet Piping (Tail Pipe)

The piping immediately downstream of the relief valve or throttle orifice experiences the lowest temperature, and the temperature of the downstream flare system branch and manifold gradually increases. This is because the gas flowing in the flare system absorbs heat from the pipe wall along the way and heats up, and as the pipe diameter increases, the flow rate decreases, the kinetic energy effect weakens, and the temperature rises. To minimize the requirement for low-temperature materials, the reducer should be placed upstream of the orifice plate, as shown in Fig. 10.8. This can reduce the impact of the kinetic energy of high-speed gas on temperature.

The tailpipe usually needs to choose low-temperature resistant materials (such as stainless steel), but the torch manifold can use carbon steel materials. In this way, when the vent tailpipe is merged into the torch manifold, a carbon steel/stainless steel connection interface will appear. This interface may be at a low temperature. During the piping process, the temperature here should be avoided from dropping below the lowest temperature of carbon steel.

Fig. 10.8 Reduce the use of stainless steel pipes by changing the position of the flow restriction orifice plate

One solution is to use a reducer at the stainless steel vent tailpipe to enlarge the pipe diameter and then connect it with the flange, as shown in Fig. 10.8. This can avoid too low temperature by increasing the thermal inertia of the pipe, reduce the kinetic energy effect, and reduce the speed of the cold fluid impacting the pipe wall opposite the flange, alleviating the low temperature of the torch manifold pipe wall caused by this. The material used for the manifold tee can also be changed to stainless steel to ensure that the impact of cold air will not cause any problems, and more advanced analysis can be carried out to ensure that there will be no problems at this connection. For example, the torch manifold can keep the vent gas at an appropriate temperature by transferring heat to the vent gas, without any low-temperature resistance treatment [3].

10.4.3 Wall Thickness

The design should consider whether the corrosion allowance is included in the simulation, that is, whether the pipeline is in a "new" or "corroded" state. Lower wall thickness usually results in lower metal temperatures. There are exceptions, in the pressure relief pipeline, a thicker wall thickness can increase the impact of kinetic energy on temperature.

ASME B31.3 "Process Piping" and AS 1210 Pressure Vessels stipulate the applicable temperature range of pipeline and vessel materials (Fig. 10.9). It should be noted that the commonly mentioned minimum temperature of carbon steel material −29 °C is only for thin-walled pipes and vessel walls.

As shown in Fig. 10.9, curve B applies to carbon steel materials (ASTM A106 Grade B) commonly used in general engineering. This material can only be used at −29 °C when the wall thickness is less than 12mm; similarly, for carbon steel materials used in vessels, it can only be used at −29 °C when the wall thickness is less than 6mm. Generally speaking, the specifications of pipeline materials usually include carbon steel, low-temperature carbon steel, and stainless steel. The temperature application range of these three materials is shown in Table 10.3.

In the design, special attention should be paid to two material grade boundary temperatures: −29 and −45 °C. In engineering standards, the influence of wall thickness on the lowest temperature that the material can withstand is usually considered.

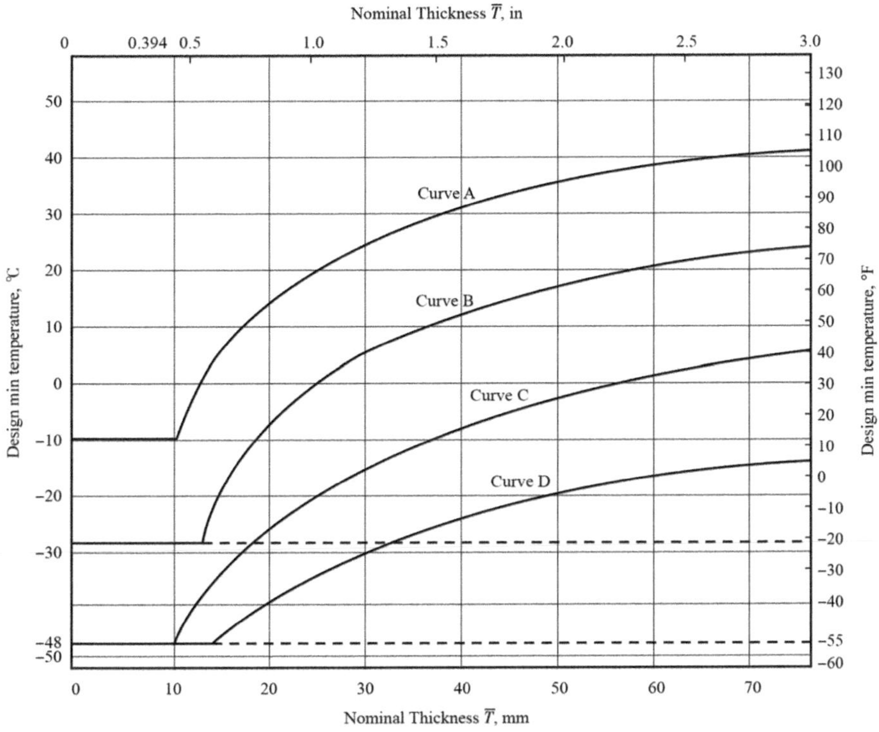

Fig. 10.9 Applicable temperature range of pipes and vessels (according to ASME B31.3)

Table 10.3 Minimum design temperature of different materials

Material	Temperature
Carbon steel	Minimum temperature −29 °C (−20°F)
Low temperature carbon steel	Minimum temperature −45 °C (−50°F)
Stainless steel	Minimum temperature −100 °C (−150°F)

For pipelines or vessels that need to use a larger wall thickness due to a larger diameter or higher pressure, higher-grade materials (such as ASTM A333) are usually required, or lower-grade materials (such as ASTM A106B) are required and a Charpy hardness test is required to determine whether the material is suitable for −29 °C.

References

1. API Standard 521 Pressure-relieving and depressuring systems, Seventh Edition, June 2020.
2. SY/T 10043 Guide to Pressure Relief and Depressurization Systems.
3. OSHA 29 CFR Part 1910.119 Process safety management of highly hazardous chemicals.

GPSR Compliance

The European Union's (EU) General Product Safety Regulation (GPSR) is a set of rules that requires consumer products to be safe and our obligations to ensure this.

If you have any concerns about our products, you can contact us on ProductSafety@springernature.com

In case Publisher is established outside the EU, the EU authorized representative is:

Springer Nature Customer Service Center GmbH
Europaplatz 3
69115 Heidelberg, Germany

Batch number: 10704305

Printed by Printforce, the Netherlands